1 MONTH OF
FREE
READING

at
www.ForgottenBooks.com

By purchasing this book you are
eligible for one month membership to
ForgottenBooks.com, giving you
unlimited access to our entire
collection of over 1,000,000 titles via
our web site and mobile apps.

To claim your free month visit:
www.forgottenbooks.com/free88837

ISBN 978-0-332-14078-0
PIBN 10088837

TECHNOLOGY QUARTERLY

AND

PROCEEDINGS OF THE SOCIETY OF ARTS.

VOLUME XI.

BOSTON:
MASSACHUSETTS INSTITUTE OF TECHNOLOGY.
1898.

THE MASSACHUSETTS INSTITUTE OF TECHNOLOGY held its first meeting on April 8, 1862. By the act of incorporation, which was accepted at this meeting, the SOCIETY OF ARTS was created as a part of the Institute coördinate with the School of Industrial Science.

The objects of the Society are to awaken and maintain an active interest in the sciences and their practical applications, and to aid generally in their advancement in connection with the arts, agriculture, manufactures, and commerce. Regular meetings are held semi-monthly from October to May.

The Society discontinued the publication of the *Abstracts of Proceedings* in 1891, and since then has published its proceedings and the principal papers read at its meetings in the *Technology Quarterly*. The present volume contains the proceedings from October, 1897, to May, 1898, inclusive.

The *Quarterly* contains, also, the results of scientific investigations carried on at the Institute, and other papers of interest to its graduates and friends.

Neither the Massachusetts Institute of Technology nor the SOCIETY OF ARTS assumes any responsibility for the opinions or statements in the papers.

CONTENTS.

MASSACHUSETTS INSTITUTE OF TECHNOLOGY

Society of Arts.

OFFICERS, 1898–1899.

President of the Institute.

JAMES M. CRAFTS.

Secretary of the Society of Arts.

ARTHUR T. HOPKINS.

Executive Committee.

GEORGE W. BLODGETT, *Chairman.*

EDMUND H. HEWINS.　　DESMOND FITZGERALD.

FRANK W. HODGDON.　　CHARLES T. MAIN.

THE PRESIDENT.　　THE SECRETARY.

Board of Publication.

WILLIAM T. SEDGWICK, *Chairman.*

CHARLES R. CROSS.　　A. LAWRENCE ROTCH.

DWIGHT PORTER.　　ROBERT P. BIGELOW.

Editor of the Technology Quarterly.

ROBERT P. BIGELOW.

BY-LAWS.

THE objects of the Society are to awaken and maintain an active interest in the practical sciences, and to aid generally in their advancement and development in connection with arts, agriculture, manufactures, and commerce.

The Society invites all who have any valuable knowledge of this kind which they are willing to contribute to attend its meetings and become members. Persons having valuable inventions or discoveries which they wish to explain will find a suitable occasion in the Society meetings, subject to regulations hereafter provided ; and while the Society will never indorse, by vote or diploma or other official recognition, any invention, discovery, theory, or machine, it will give every facility to those who wish to discuss the principles and intentions of their own machines or inventions, and will endeavor at its meetings, or through properly constituted committees, to show how far any communications made to it are likely to prove of real service to the community.

SECTION I. — ADMINISTRATION.

The immediate management and control of the affairs of the Society of Arts shall be exercised by an Executive Committee, consisting of the President of the Institute and the Secretary of the Society (who shall be members *ex officiis*), and five other members, who shall be elected by the Society of Arts at each annual meeting, to continue in office until other persons have been chosen in their place.

SECT. II. — DUTIES OF THE EXECUTIVE COMMITTEE.

The Executive Committee shall elect its chairman, prescribe his duties, and, with the concurrence of the Treasurer of the Institute, fix his compensation when the interests of the Society require that he should be paid for his services ; they may invite any person to preside at any ordinary meeting who is well versed in the subjects to be discussed ; they shall appoint the days and times of meeting, when not fixed by the Society, and determine the subjects to be considered at the meetings and the mode of conducting the discussions ; they may, with the concurrence of the President of the Institute, make such arrangements for reporting and publishing the proceedings of the Society as they may deem best suited to advance its interests ; they may receive moneys in behalf of the Society in aid of its objects, by subscription, donation, or bequest ; they shall make a report of their doings to the Society at its annual meeting and at such other times as a report may be called for by a majority of the members present at any meeting ; they shall also make a report of their doings to the President of the Institute prior to the annual meeting, and at such other times as the Corporation may require it. Three members shall constitute a quorum for the transaction of business.

SECT. III. — DUTIES OF THE PRESIDENT AND SECRETARY.

1. It shall be the duty of the President of the Institute to preside at the annual and the special meetings of the Society, and also at its ordinary meetings when the Executive Committee does not invite a special chairman to preside.

2. It shall be the duty of the Secretary of the Society to give notice of and attend all meetings of the Society and of the Executive Committee ; to keep a record of the business and orders of each meeting, and read the same at the next meeting ; to keep a list of the members of the Society, and notify them of their election and of their appointment on committees ; and generally to devote his best efforts, under the direction of the Executive Committee, to forwarding the business and advancing the interests of the Society. He shall also record the names of the Executive Committee attending each meeting.

SECT. IV. — FUNDS OF THE SOCIETY.

All the fees and assessments of members, and all moneys received by subscription, donation, or otherwise, in aid of the Society, shall be paid into the treasury of the Corporation, to be held and used for the objects of the Society under the direction of the Executive Commit·

tee, and shall be subject to the order of its Chairman, countersigned by the President of the Corporation.

SECT. V. — MEETINGS OF THE SOCIETY.

1. The annual meeting of the Society shall be held at the Institute on the second Thursday in May. The ordinary meetings shall be held semi-monthly, or whenever deemed expedient by the Society or by the Executive Committee, excepting in the months of June, July, August, and September.

2. If from any cause the annual meeting shall not have been duly notified or held as above required, the same shall be notified and held at such time as the Executive Committee may direct.

3. A special meeting of the Society may at any time be called by the Secretary on a written request of ten members. Twelve members of the Society shall constitute a quorum for the transaction of business.

SECT. VI. — MEMBERS AND THEIR ELECTION.

1. Members of the Society of Arts shall be of three kinds — Associate, Corresponding, and Honorary Members.

2. Candidates for Associate Membership shall be recommended by not less than two members, whose signatures shall be affixed to a written or printed form to that effect. Each nomination shall be referred to the Executive Committee, and when reported favorably upon by them, and read by the Secretary, may be acted upon at the same meeting; the election shall be conducted by ballot, and affirmative votes to the number of three fourths of the votes cast shall be necessary for an election.

3. Corresponding and Honorary Members may be elected in the same way, on nomination by the Executive Committee.

4. Associate Members shall pay an admission fee of three dollars before being entitled to the privileges of membership, and an annual assessment of three dollars on the first of October of each year, this sum to include subscription to the *Technology Quarterly and Proceedings of the Society of Arts.*

An Associate Member who shall have paid at any one time the sum of fifty dollars, or annual assessments for twenty years, shall become a member for life, and be thereafter exempted from annual assessments.

A member neglecting to pay his annual assessment for six months after being notified that the same is due shall be regarded as having withdrawn his membership, unless otherwise decided by the Executive Committee, which shall be authorized, for cause shown, to remit the assessments for any one year; and which shall moreover be empowered to exempt particular members from assessments whenever their claims and the interests of the Society make it proper to do so.

Sect. VII. — Election of the Executive Committee and of the Secretary.

1. At an ordinary meeting of the Society, preceding the annual meeting, a nominating committee of *five* shall be chosen, whose duty it shall be to nominate candidates for the Executive Committee, to post a list of the names selected in the office of the Secretary, and to furnish printed copies thereof to the members at or before the time of election.

2. At a meeting at which an election is to take place the presiding officer shall appoint a committee to collect and count the votes and report the names and the number of votes for each candidate, whereupon he shall announce the same to the meeting.

3. A majority of the votes cast shall be necessary to an election.

4. In the first organization under these By-Laws, the Executive Committee may be elected at an ordinary or special meeting.

5. Vacancies in the committee occurring during the year may be filled by the Society at an ordinary meeting.

6. The Secretary shall be elected by the Society, on nomination by the Executive Committee, at each annual meeting of the Society, or, in case of a vacancy during the session, at such other time as the Executive Committee may appoint; and he shall be reëligible in the same way at the pleasure of the Society.

7. The compensation of the Secretary shall be fixed from year to year by the Executive Committee with the concurrence of the Treasurer of the Institute.

Sect. VIII. — Committees of Arts.

1. The Members of the Society of Arts may be enrolled in divisions, under the following heads, according to the taste or preference of

the individual; each division to constitute a committee upon the subjects to which it appertains :

(1) On Mineral Materials, Mining, and the Manufacture of Iron, Copper, and other Metals.

(2) On Organic Materials — their culture and preparation.

(3) On Tools and Implements.

(4) On Machinery and Motive Powers.

(5) On Textile Manufactures.

(6) On Manufactures of Wood, Leather, Paper, India Rubber, and Gutta Percha.

(7) On Pottery, Glass, Jewelry, and works in the Precious Metals.

(8) On Chemical Products and Processes.

(9) On Household Economy ; including Warming, Illumination, Water-Supply, Drainage, Ventilation, and the Preparation and Preservation of Food.

(10) On Engineering, Architecture, and Ship-building.

(11) On Commerce, Marine Navigation, and Inland Transportation.

(12) On Agriculture and Rural Affairs.

(13) On the Graphic and Fine Arts.

(14) On Ordnance, Firearms, and Military Equipments.

(15) On Physical Apparatus.

2. Any member may belong to more than one of the above-named Committees of Arts, but shall not at the same time be eligible as chairman in more than one.

3. It shall be competent for each Committee of Arts, of ten or more members entitled to vote, to organize ; to elect annually in October, or whenever a vacancy shall occur, a chairman ; to appoint its own meetings ; and to frame its own By-Laws, provided the same do not conflict with the regulations of the Society of Arts.

SECT. IX. — AMENDMENT AND REPEAL.

1. These By-Laws may be amended or repealed, or other provisions added, by a vote of three fourths of the members present at any regular meeting of the Society ; provided that such changes shall have been recommended and approved in accordance with the By-Laws of the Corporation (see extract from By-Laws of Corporation as

printed below) and presented in writing at a preceding meeting of the Society.

2. These By-Laws shall take effect immediately after their approval by the Corporation and adoption by the Society, and all previous By-Laws are hereby repealed.

As amended December 9, 1897.

EXTRACT FROM THE BY-LAWS OF THE CORPORATION.

SECT. VI. — There shall be a Committee on the Society of Arts consisting of five members, appointed at the annual meeting of the Corporation, to hold office for one year, who shall have the general charge and supervision of the organization and proceedings of the Society, subject to the approval of the Corporation. It shall be their duty, in connection with a committee chosen by the Society, to frame By-Laws for the government of the Society, which shall take effect when adopted by the Society and approved by the Corporation.

February, 1894.

LIST OF MEMBERS.

DECEMBER, 1898.

Members are requested to inform the Secretary of any change of address.

LIFE MEMBERS.

Addicks, J. Edward . . . 903 Harrison Building, Philadelphia, Pa.
Atkinson, Edward . . . . . . 31 Milk Street, Boston, Mass.

Beal, James H. . . . . . . 104 Beacon Street, Boston, Mass.
Bowditch, William I. . . . . . 28 State Street, Boston, Mass.
Breed, Francis W. . . . . . 111 Summer Street, Boston, Mass.
Bullard, W. S. . . . . . . 3 Commonwealth Avenue, Boston, Mass.

Cummings, John . . . . Cummingsville, Woburn, Mass.

Dalton, Charles H. . . 33 Commonwealth Avenue, Boston, Mass.
Dewson, F. A. . . . . 53 State Street, Boston, Mass.

Eastman, Ambrose . . . . . . 53 State Street, Boston, Mass.
Endicott, William, Jr. . . . . . 32 Beacon Street, Boston, Mass.

Foster, John . . . . . . 25 Marlborough Street, Boston, Mass.

Gaffield, Thomas . . . . . . 54 Allen Street, Boston, Mass.
Griffin, Eugene . . . . 323 State Street, Albany, N. Y.
Guild, Henry . . . . . . 433 Washington Street, Boston, Mass.

Haven, Franklin . . . . 97 Mount Vernon Street, Boston, Mass.
Henck, J. B. . . . . Montecito, Santa Barbara Co., Cal.
Hewins, Edmund H. . . . . . 625 Tremont Street, Boston, Mass.

Johnson, Samuel . . . . 7 Commonwealth Avenue, Boston, Mass.

Little, James L. . . . . . . Goddard Avenue, Brookline, Mass.
Little, John M. . . . . Hotel Pelham, Boston, Mass.
Lowe, N. M. . . . . 88 Court Street, Boston, Mass.

Mack, Thomas . . . . 269 Commonwealth Avenue, Boston, Mass.
Matthews, Nathan . . . . . 145 Beacon Street, Boston, Mass.
May, F. W. G. . . . . 127 State Street, Boston, Mass.
May, John J. . . . . . . Post-Office Box 2348, Boston, Mass.
McPherson, W. J. . . . . 10 Clarendon Street, Boston, Mass.

Ordway, John M. . . . . New Orleans, La.

Pickering, E. C. . Harvard College Observatory, Cambridge, Mass.
Pickering, H. W. . . . . . . 249 Beacon Street, Boston, Mass.
Prang, Louis . . . . 16 Centre Street, Roxbury, Mass.
Pratt, Miss . . . . Watertown, Mass.

Ròss, Waldo O. . . . . . . 1 Chestnut Street, Boston, Mass.
Runkle, John D. . . Mass. Institute of Technology, Boston, Mass.

Sawyer, Edward . . . . Newton, Mass.
Sawyer, Timothy T. . . . . 319 Dartmouth Street, Boston, Mass.
Sayles, Henry . . . . 42 Beacon Street, Boston, Mass.
Sears, Philip H. . . . . 85 Mount Vernon Street, Boston, Mass.
Sherwin, Thomas . . . . . Revere Street, Jamaica Plain, Mass.
Shurtleff, A. M. . . . . . 9 West Cedar Street, Boston, Mass.
Sinclair, A. D. . . . . 35 Newbury Street, Boston, Mass.
Stevens, B. F. . . . . 91 Pinckney Street, Boston, Mass.
Sullivan, Richard . . . . . 35 Brimmer Street, Boston, Mass.

Tufts, John W. . . . . . . 27 Concord Square, Boston, Mass.

Vose, George L. . . . . Paris, Me.

Ware, William R. . . . . Columbia College, New York, N. Y.
Watson, William . . . . 107 Marlborough Street, Boston, Mass.

Wentworth, Arioch . . . . . 332 Beacon Street, Boston, Mass.
Whitaker, Channing . . . . Lowell, Mass.
Wing, Charles H. . . . Ledger, Mitchell County, North Carolina.

ASSOCIATE MEMBERS.

Adams, Henry S. . . . . Arlington, Mass.
Alden, Charles H., Jr. . . . 1024 Tremont Building, Boston, Mass.
Alden, George A. . . . . . 87 Summer Street, Boston, Mass.
Alden, John . . . . Lawrence, Mass.
Allen, C. Frank . . Mass. Institute of Technology, Boston, Mass.
Allen, Samuel E. . . . . . . 67 Chauncy Street, Boston, Mass.
Allen, Walter S. . . . . 34 South 6th Street, New Bedford, Mass.
Allen, William Henry . 291 Commonwealth Avenue, Boston, Mass.
Andrews, Clement W. . . . . John Crerar Library, Chicago, Ill.
Appleton, Charles B. . . 207 Aspinwall Avenue, Brookline, Mass.
Atwood, Frank W. . . . . 98 Commercial Street, Boston, Mass.

Baker, C. F. . . . . 439 Albany Street, Boston, Mass.
Baker, J. B. . . . . 602 Centre Street, Newton, Mass.
Bardwell, F. L. . . Mass. Institute of Technology, Boston, Mass.
Barnes, Herbert H. . . . . . . Hotel Brunswick, Boston, Mass.
Barstow, George E. . . . . . . 27 Union Street, Lynn, Mass.
Bartlett, Dana P. . . Mass. Institute of Technology, Boston, Mass.
Bartlett, Spaulding . . . . Webster, Mass.
Barton, George H. . Mass. Institute of Technology, Boston, Mass.
Barton, Howard R. . . . . Englewood, N. J.
Bassett, William H. . . . . New Bedford, Mass.
Batcheller, Robert . . 55 Commonwealth Avenue, Boston, Mass.
Bigelow, Charles H. . . . . Salem, Mass.
Bigelow, Otis . . . . . . Avenel, Montgomery Co., Maryland.
Bigelow, Robert P. . Mass. Institute of Technology, Boston, Mass.
Binney, Amos . . . Room 416, 53 State Street, Boston, Mass.
Birkholz, Hans . . . . . Care E. P. Allis Co., Milwaukee, Wis.
Bixby, George L. . . . . Foxboro, Mass.
Blackmer, Adelaide Sherman . . 31 Devon Street, Roxbury, Mass.
Blodgett, George W. . . Boston and Albany Railroad, Boston, Mass.
Blood, Grosvenor T. . . . . . . 125 Milk Street, Boston, Mass.

Blood, John Balch   . . .   22-A Equitable Building, Boston, Mass.
Brackett, Dexter . . . . 3 Mount Vernon Street, Boston, Mass.
Bradlee, Arthur T.   . . . . Chestnut Hill, Mass.
Braley, Samuel T.   . . . . Rutland, Vt.
Brophy, William . . .   17 Egleston Street, Jamaica Plain, Mass.
Bryden, George W. . .   N. E. Structural Co., East Everett, Mass.
Burns, Peter S.   . .   Mass. Institute of Technology, Boston, Mass.
Burton, A. E. . . .   Mass. Institute of Technology, Boston, Mass.

Caldwell, Eliot L. . . . . 3 Head Place, Boston, Mass.
Cameron, J. A.   . . . . Forge Village, Mass.
Carson, Howard A. . . . .   20 Beacon Street, Boston, Mass.
Carter, Henry H. . . . . . . Hotel Ludlow, Boston, Mass.
Carty, J. J. . . . . . . 18 Cortlandt Street, New York, N. Y.
Chandler, F. W. . .   Mass. Institute of Technology, Boston, Mass.
Chandler, S. C. . . . . Cambridge, Mass.
Chase, Charles H. . . . . Stoneham, Mass.
Chase, F. D. . . . . Versailles, Pa.
Clark, John M.   . . . . . 47 Court Street, Boston, Mass.
Clifford, H. E. . . .   Mass. Institute of Technology, Boston, Mass.
Cody, Lewis P.   . . 9 South Division Street, Grand Rapids, Mich.
Coffin, C. A.   . . . . . General Electric Co., Boston, Mass.
Coffin, F. S.   . . . . . 152 Congress Street, Boston, Mass.
Collins, R. B. . . . . 20 Weld Hill Street, Forest Hills, Mass.
Crafts, James M. . .   Mass. Institute of Technology, Boston, Mass.
Craig, J. Holly, ... 69 Broad Street, Boston, Mass.
Crosby, William O. .   Mass. Institute of Technology, Boston, Mass.
Crosby, William W. . . . Lowell Textile School, Lowell, Mass.
Cross, C. R.   . .   Mass. Institute of Technology, Boston, Mass.
Curtis, Henry P. . . . . Parker House, Boston, Mass.
Cutter, Louis F. . . . . Winchester, Mass.

Dana, Gorham . . . . . 41 Allerton Street, Brookline, Mass.
Dates, Henry B. .   Clarkson School of Technology, Potsdam, N. Y.
Davis, Frank E . . Care Washburn & Moen Co, Worcester, Mass.
Day, Nathan B. . . . . 280 Newbury Street, Boston, Mass.
DeLancey, Darragh . . . . Rochester, N. Y.
Densmore, Edward D. . . . 7 Exchange Place, Boston, Mass.
DeWolf, John O. . . 519 Massachusetts Avenue, Cambridge, Mass.

Doolittle, Orrin S. . . . . . . 445 Oley Street, Reading, Pa.
Draper, William L. . . Dodge Mf'g. Co., 137 Purchase Street,
Boston, Mass.
Drown, Thomas M. . . . Lehigh University, South Bethlehem, Pa.
Dudley, P. H. . . . . 80 Pine Street, New York, N. Y.
Dunn, Edward H. . . . . . . 30 South Street, Boston, Mass.
Dutton, Edgar F. . . . . Newton Centre, Mass.
Dwelley, Edwin F. . . . . . 25 Baltimore Street, Lynn, Mass.

Ellis, John . . . . Lonsdale, R. I.
Ely, Sumner B. . . . . 11 West 88th Street, New York, N. Y.
Eustis, W. E. C. . . . . . . 55 Kilby Street, Boston, Mass.
Eustis, W. Tracy . . . . . 950 Beacon Street, Brookline, Mass.
Evans, Robert D. . . . . . . 114 Bedford Street, Boston, Mass.

Fales, Frank L. . . . . 1 Willow Street, Boston, Mass.
Farnham, Isaiah H. . . . . . . 125 Milk Street, Boston, Mass.
Faunce, Linus . . . Mass. Institute of Technology, Boston, Mass.
Fish, Charles H. . . . . Dover, N. H.
Fiske, J. P. B. . . . . . . 164 Devonshire Street, Boston, Mass.
FitzGerald, Desmond . . 3 Mount Vernon Street, Boston, Mass.
FitzGerald, Francis A. J. . Carborundum Co., Niagara Falls, N. Y.
Flinn, Richard J. . . . . West Roxbury, Mass.
Forbes, Eli . . . . Clinton, Mass.
Forbes, Fred B. . . . . . . 92 Orchard Street, Somerville, Mass.
Forbes, Howard C. . . . . . . 31 State Street, Boston, Mass.
Foss, E. N. . . . . . . 8 Everett Street, Jamaica Plain, Mass.
Freeland, J. H. . . . . Hotel Brunswick, Boston, Mass.
Freeman, John R. . . . . . 4 Market Square, Providence, R. I.
French, E. V. . . . . 31 Milk Street, Boston, Mass.
Fuller, George W. . . . 2106 Grand Avenue, Walnut Hills,
Cincinnati, Ohio.
Fuller, William B. . . . . . 57 Lumber District, Albany, N. Y.

Garfield, A. S. . . . . 5 Villa Michon, Paris, France.
Gaylord, W. K. . . . . Pasadena, Cal.
Gilbert, F. A. . . . . 74 Ames Building, Boston, Mass.
Gill, A. H. . . . . Mass. Institute of Technology, Boston, Mass.
Gilley, Frank M. . . . . . . 100 Clark Avenue, Chelsea, Mass.

Goodell, George H. . . . . Susquehanna, Pa.
Goodspeed, Joseph H. . 382 Commonwealth Avenue, Boston, Mass.
Goodwin, H. M. . . Mass. Institute of Technology, Boston, Mass.
Goodwin, Richard D. . . . . 28 Summer Street, Boston, Mass.
Gray, Joseph P. . . . . 31 Milk Street, Boston, Mass.
Greenleaf, Lewis S., Care Hudson River Telephone Co., Albany, N. Y.
Grover, Nathan C. . . . . . Maine State College, Orono, Me.
Guppy, Benjamin Wilder . . . . Melrose, Mass.

Hadaway, William S. . . . 107 Liberty Street, New York, N. Y.
Hadley, F. W . . . . Arlington Heights, Mass.
Hale, Richard A. . . . . Lawrence, Mass.
Hall, William T., Care Dr. Walker, Mass. Institute of Technology,
                                                        Boston, Mass.
Hamblett, George W. . . . 506 Lowell Street, Lawrence, Mass.
Hamilton, George Wyman . . . . 28 Court Square, Boston, Mass.
Hamlin, George H. . . . . . Maine State College, Orono, Me.
Hardy, W. B. . . . . . Care Diamond Rubber Co., Akron, Ohio.
Harrington, Francis . . . 5 Mount Vernon Street, Boston, Mass.
Harris, William A. . . . . . . Exchange, Liverpool, England.
Hart, Francis R., Old Colony Trust Co., Ames Building, Boston, Mass.
Hawkes, Levi G. . . . . Saugus, Mass.
Hayes, Hammond V. . 42 Farnsworth Street, South Boston, Mass.
Hazen, Allen . . . . 220 Broadway, New York, N. Y.
Herrick, Rufus F. . . . . Winchester, Mass.
Hicks, C. Atherton . . . . Needham, Mass.
Hobart, James C. . . . . 610 Baymuller Street, Cincinnati, Ohio.
Hobart, John D. . . . . Malden, Mass.
Hobbs, Franklin W. . . . . Brookline, Mass.
Hodgdon, Frank W. . . . . Arlington, Mass.
Hofman, H. O. . . Mass. Institute of Technology, Boston, Mass.
Hollingsworth, A. L. . . . . 35 Federal Street, Boston, Mass.
Hollingsworth, Sumner . . . . 44 Federal Street, Boston, Mass.
Hollis, Frederick S. . . . 3 Mount Vernon Street, Boston, Mass.
Holman, G. M. . . . . . 52 Pleasant Street, Fitchburg, Mass.
Holman, S. W. . . . . . The Abbottsford, Brookline, Mass.
Holton, Edward C. . . . . . 100 Canal Street, Cleveland, Ohio.
Holtzer, Charles W. . . . . Brookline, Mass.
Hood, George H. . . . . 9 Otis Street, Boston, Mass.

Hopkins, Arthur T. . . Mass. Institute of Technology, Boston, Mass.
Hopkins, Prescott A. . . . . . Drexel Institute, Philadelphia, Pa.
Hopkins, William J. . . . . . Drexel Institute, Philadelphia, Pa.
Horton, Theodore . . . . . 17 Everett Street, Melrose, Mass.
Howard, A. P. . . . . 13 Pearl Street, Boston, Mass.
Howard, L. Frederic . . . . . 142 P. O. Building, Boston, Mass.
Howe, Henry M. . . . . . 27 West 73d Street, New York, N. Y.
Howe, Horace J. . . . . . . Chestnut Street, Medford, Mass.
Hunt, Edward M. . . . . . . 22 Beckett Street, Portland, Me.
Hunt, Harry H. . . . . . 202 Equitable Building, Boston, Mass.
Hutchings, James H. . . 1672 Washington Street, Boston, Mass.

Jackson, Daniel D.   Mt. Prospect Laboratory, Flatbush Avenue,
     and Eastern Parkway, Brooklyn, N. Y.
Jackson, William . . . . 50 City Hall, Boston, Mass.
Jacques, W. W. . . . . 125 Milk Street, Boston, Mass.
James, Frank M. . . . . . 51 North Broadway, Haverhill, Mass.
Jaques, W. H. . . . . . . 227 Clarendon Street, Boston, Mass.
Jenkins, Charles D. . . . . . 32 Hawley Street, Boston, Mass.
Jenks, William J. . . . . . . 120 Broadway, New York, N. Y.
Johnson, Jesse F. . . Care Hamilton Powder Co., Montreal, Canada.
Johnston, William A. . . Mass. Institute of Technology, Boston, Mass.
Jones, Jerome . . . . 51 Federal Street, Boston, Mass.

Kales, William R. . . . 1119 Prospect Street, Cleveland, Ohio.
Kastner, Charles . . Mass. Institute of Technology, Boston, Mass.
Keith, Simeon C., Jr. . . . . Austin Street, Charlestown, Mass.
Kendall, Edward . . . 139 Magazine Street, Cambridgeport, Mass.
Kendall, Francis H. . . . . Court House, East Cambridge, Mass.
Killilea, James J. . . . . . . Tufts Wharf, East Boston, Mass.
Kimball, Fred M. . . . . . 58 Main Street, Winter Hill, Mass.
Kimball, Joseph H. . . . . West Newton, Mass.
Knapp, G. Frederick . . . 1110 Harrison Building, Philadelphia, Pa.
Knowles, Morris, 2d . . . . City Hall, Pittsburg, Pa.
Koehler, S. R. . . . . . Museum of Fine Arts, Boston, Mass.
Kunhardt, L. Henry . . . . . 31 Milk Street, Boston, Mass.

Lanza Gaetano . . Mass. Institute of Technology, Boston, Mass.
Laws, Frank A. . . Mass. Institute of Technology, Boston, Mass.

Leach, Albert E. . . . . Newtonville, Mass.
Lee, Elisha . . . . Tioga Centre, N. Y.
Lee, John C. . . . . Mountfort Street, Brookline, Mass.
Leeson, J. R. . . . . . . Post-Office Box 2221, Boston, Mass.
Lincoln, G. Russell . . . . Hingham, Mass.
Little, Samuel . . . . . 556 Warren Street, Roxbury, Mass.
Lodge, H. Ellerton . . . . 4 Post-Office Square, Boston, Mass.
Lodge, Richard W. . Mass. Institute of Technology, Boston, Mass.
Lothrop, Thomas M. . . . . . 13 Carlton Street, Brookline, Mass.
Loud, Joseph Prince . . . 135 Mount Vernon Street, Boston, Mass.
Lovejoy, Frank W. . . . . . . Kodak Park, Rochester, N. Y.
Low, John F. . . . . Chelsea, Mass.
Lowell, A. Lawrence . . 171 Marlborough Street, Boston, Mass.
Lowell, Percival . . . . 53 State Street, Boston, Mass.

Main, Charles T. . . . . 53 State Street, Boston, Mass.
Mandell, Samuel P. . . 302 Commonwealth Avenue, Boston, Mass.
March, Clement . . . . . 344 State Street, Bridgeport, Conn.
Martin, Henry . . . . South Gardiner, Me.
Melluish, James George . . . . . Eagle Block, Bloomington, Ill.
Metcalf, Frederick . . . . . 111 Elm Street, Cleveland, Ohio.
Metcalf, Leonard . . . . Concord, Mass.
Miller, Edward F. . Mass. Institute of Technology, Boston, Mass.
Miller, Franklin T. . . . . Auburndale, Mass.
Mixter, S. J. . . . . . 180 Marlborough Street, Boston, Mass.
Moody, Burdett . . Homestake Mining Co., Lead, South Dakota.
Moody, Frederick C. . . Care Bell Telephone Co., Philadelphia, Pa.
Moore, Alexander . . . . . . 3 School Street, Boston, Mass.
Moore, Fred F. . . . . South Framingham, Mass.
Morse, Henry C. . . . . . Post-Office Box 5285, Boston, Mass.
Morss, Everett . . . . 79 Cornhill, Boston, Mass.
Morton, Galloupe . . . 2 Westervelt Avenue, New Brighton,
                                        Staten Island, N. Y.
Moseley, Alexander W., Mass. Institute of Technology, Boston, Mass.
Mosman, Philip A. . . . . . Colorado Smelting Co., Pueblo, Col.
Mumford, Edgar H. . . . . 39 Cortlandt Street, New York, N. Y.
Munroe, James P. . . . . 179 Devonshire Street, Boston, Mass.

Newbegin, Parker C. . . Patten & Sherman Railroad Co., Patten, Me.
Niles, William H . . Mass. Institute of Technology, Boston, Mass.

Norman, George H. . . . . Newport, R. I.
Norris, Albert P. . . . 760 Massachusetts Avenue, Cambridge, Mass.
Noyes, Arthur A. . . . Mass. Institute of Technology, Boston, Mass.

Ober, Arthur J. . . . . Jamestown, R. I.
Osborne, George A. . . Mass. Institute of Technology, Boston, Mass.

Paine, Sidney B. . . . . Newton Centre, Mass.
Parce, Joseph Y., Jr. . . . . Denver, Col.
Patterson, George W., Jr. . . . 14 South University Avenue,
                                                    Ann Arbor, Mich.
Peabody, C. H. . . Mass. Institute of Technology, Boston, Mass.
Pickert, Leo W. . . American Sugar Refining Co., Boston, Mass.
Piper, Walter E.. . . . . . Boston Rubber Shoe Co., Fells, Mass.
Pitcher, Franklin W. . . . . 63 Franklin Street, Boston, Mass.
Pollock, Clarence D. . . . . . 333 State Street. Brooklyn, N. Y.
Pope, Macy S. . . . . 31 Milk Street, Boston, Mass.
Pope, T. E. . . . . Mass. Institute of Technology, Boston, Mass.
Porter, Dwight . . . Mass. Institute of Technology, Boston, Mass.
Prescott, Samuel C. . Mass. Institute of Technology, Boston, Mass.
Puffer, W. L. . . . Mass. Institute of Technology, Boston, Mass.

Read, Carleton A. . Mass. Institute of Technology, Boston, Mass.
Reed, Walter W. . . . . . . 38 Floyd Street, Waltham, Mass.
Reynolds, Howard S. Brockton Street Railway Co., Brockton, Mass.
Richards, Ellen H. . Mass. Institute of Technology, Boston, Mass.
Richards, R. H. . . Mass. Institute of Technology, Boston, Mass.
Ritchie, Thomas P. . 1057 Walnut Street, Newton Highlands, Mass.
Roberts, George L. . . . . . . 95 Milk Street, Boston, Mass.
Rolfe, George William, Mass. Institute of Technology, Boston, Mass.
Rollins, William Herbert . 250 Marlborough Street, Boston, Mass.
Rotch, A. Lawrence . . . . . 53 State Street, Boston, Mass.
Rowell, Henry K. . . . . . 141 Dale Street, Waltham, Mass.
Royce, Frederick P. . . . . . 256 Newbury Street, Boston, Mass.
Royce, H. A. . . . . 256 Newbury Street, Boston, Mass.

Safford, Arthur T. . . . . 66 Broadway, Lowell, Mass.
Sanborn, Frank E. . . . Ohio State University, Columbus, Ohio.
Sando, Will J. . . . . . 3 Mount Vernon Street, Boston, Mass.

Sawyer, Albert H. . . . . . . 19 Pearl Street, Boston, Mass.
Sawyer, Alfred H. . . . 237 West Newton Street, Boston, Mass.
Sawyer, Joseph . . . 31 Commonwealth Avenue, Boston, Mass.
Schwamb, Peter . . Mass. Institute of Technology, Boston, Mass.
Schwarz, F. H. . . . . . Lower Pacific Mills, Lawrence, Mass.
Sedgwick, W. T. . . Mass. Institute of Technology, Boston, Mass.
Shattuck, A. Forrest . . Care Solvay Process Co., Detroit, Mich.
Shaw, Henry S. . . . 339 Commonwealth Avenue, Boston, Mass.
Shepard, F. H. . . . . 227 East German Street, Baltimore, Md.
Sherman, George W. . Room 509, 53 State Street, Boston, Mass.
Shuman, A. . . . . . . 440 Washington Street, Boston, Mass.
Skinner, Joseph J. . Mass. Institute of Technology, Boston, Mass.
Slater, H. C. . . . . . . Post-Office Box 423, Milwaukee, Wis.
Slawson, Fred G., Burton Brewery, cor. Heath and Parker Streets,
Roxbury, Mass.
Smith, John W. . . Mass. Institute of Technology, Boston, Mass.
Smith, W. L. . . . Mass. Institute of Technology, Boston, Mass.
Snow, F. Herbert . . . . Brockton, Mass.
Snyder, Frederick T. . . . . . . Keewatin, Ontario, Canada.
Sondericker, Jerome . Mass. Institute of Technology, Boston, Mass.
Spear, Walter E. . . Care Metropolitan Water Board, Clinton, Mass.
Stantial, F. G. . . . Care Cochrane Chemical Co., Everett, Mass.
Stearns, Frederic P . . . 108 Cushing Avenue, Dorchester, Mass.
Stoddard, Arthur B. . . . . , . LaSalle, Ill.
Stone, Charles A. . . . . . . 4 Post-Offiee Square, Boston, Mass.
Sully, John M. . . Chickamauga Coal & Iron Co., Chickamauga, Ga.
Swain, George F. . . Mass. Institute of Technology, Boston, Mass.
Swan, Charles H. . . . . 25 Wabon Street, Boston, Mass.
Sweet, H. N. . . . . 4 Spruce Street, Boston, Mass.

Talbot, Henry P. . . Mass. Institute of Technology, Boston, Mass.
Talbot, Marion . . . . . . University of Chicago, Chicago, Ill.
Taylor, R. R. . . Normal and Industrial Institute, Tuskegee, Ala.
Tenney, Albert B. . . . . 35 Fremont Avenue, Everett, Mass.
Thompson, George K. . 42 Farnsworth Street, South Boston, Mass.
Thomson, A. C. . . . . . . Sumner Road, Brookline, Mass.
Thomson, Elihu . . . . . . 26 Henry Avenue, Lynn, Mass.
Thorndike, Sturgis H. . . . . . . 67 City Hall, Boston, Mass.
Thorp, Frank H. . . Mass. Institute of Technology, Boston, Mass.

Tinkham, S. Everett . . . . 65 City Hall, Boston, Mass.
Tolman, James P. . . . . . 115 Congress Street, Boston, Mass.
Towne, Walter I. . . . . 125 Milk Street, Boston, Mass.
Tucker, G. R. . . . . City Hospital, Boston, Mass.
Turner, E. K. . . . . 53 State Street, Boston, Mass.
Tuttle, Joseph H. . . . . . Post-Office Box 1185, Boston, Mass.
Tyler, Harry W. . . Mass. Institute of Technology, Boston, Mass.

Underwood, George R. . . . . Peabody, Mass.
Underwood, W. Lyman . . . . Belmont, Mass.

Vaillant, George W. . . . . . . 1 Broadway, New York, N. Y.
VanDaell, A. N. . . Mass. Institute of Technology, Boston, Mass.
VanEveren, Grace A. . . 841 Jefferson Avenue, Brooklyn, N. Y.
Very, Frank W. . . . . . 507 Morris Avenue, Providence, R. I.
Vogel, Frank. . . . Mass. Institute of Technology, Boston, Mass.

Walker, Charles R. . . 155 Western Avenue, Cambridgeport, Mass.
Walker, Elton D. . . . . 16 Gillespie Street, Schenectady, N. Y.
Warner, Charles F. . . 46-A Trowbridge Street, Cambridge, Mass.
Wason, Leonard C. . . . . 199 Harvard Street, Brookline, Mass.
Webster, Edwin S. . . . . 4 Post-Office Square, Boston, Mass.
Weeks, G. W. . . . . Clinton, Mass.
Wells, Webster . . . . Lexington, Mass.
Wendell, George V. . . 860 Massachusetts Avenue, Cambridge, Mass.
Wesson, Paul B. . . . . . . Lowell Machine Shop, Lowell, Mass.
Weston, David B. . . . . Watertown, Mass.
Whipple, George C. Mt. Prospect Laboratory, Flatbush Avenue
　　　　　　　　　　　and Eastern Parkway, Brooklyn, N. Y.
Whitaker, S. Edgar . . . . . 58 Oliver Street, Fitchburg, Mass.
White, J. Foster . . . . Box 76, Brookline, Mass.
Whitman, William . . . . . 78 Channey Street, Boston, Mass.
Whitney, Willis R. . Mass. Institute of Technology, Boston, Mass.
Wiggin, Thomas H. . . . . 154 Mountain Avenue, Malden, Mass.
Wigglesworth, George . . . . 89 State Street, Boston, Mass.
Willcutt, Levi L. . . . . . Post-Office Box 5239, Boston, Mass.
Williams, Francis H. . . . . 505 Beacon Street, Boston, Mass.
Williams, Henry J. . . . . . 161 Tremont Street, Boston, Mass.

Williams, Roger J. . . . . Canton, Mass.
Wilson, Fred A. . . . . Nahant, Mass.
Winkley, W. H. . . . . West Medford, Mass.
Winslow, Frederic I., City Engineer's Office, City Hall, Boston, Mass.
Winton, Henry D. . . . . Wellesley Hills, Mass.
Wood, Henry B. . . . . 138 State House, Boston, Mass.
Woodbridge, S. H. . Mass. Institute of Technology, Boston, Mass.
Woodbury, C. J. H. . . . . . 61 Commercial Street, Lynn, Mass.
Woodward, Edward O. . . . 29 Copeland Street, Roxbury, Mass.

TECHNOLOGY QUARTERLY

AND

PROCEEDINGS OF THE SOCIETY OF ARTS.

VOL. XI.	MARCH, 1898.	No. 1.

PROCEEDINGS OF THE SOCIETY OF ARTS.

THIRTY-SIXTH YEAR, 1897–98.

THE meeting of the SOCIETY OF ARTS which should have been held October 14 was omitted to allow members to attend a Memorial Meeting[1] held in Music Hall, under the auspices of the Corporation and the Faculty, in honor of the late President Francis Amasa Walker.

The meeting was called to order at 8 P.M. by Augustus Lowell, Esq., of the Corporation, who introduced His Excellency Governor Roger Wolcott as the presiding officer. The Governor upon taking the chair made a short but forceful address, and then introduced the principal speaker of the evening, Honorable George F. Hoar, United States Senator from Massachusetts. Senator Hoar delivered an oration, in which he spoke eloquently of the life and character of President Walker, and praised his achievements as a soldier, an administrator, and a scholar.

Invitations to this meeting had been sent to all members of the SOCIETY OF ARTS.

---

[1] For a complete report of the proceedings of this meeting, see Massachusetts Institute of Technology, Meetings Held in Commemoration of the Life and Services of Francis Amasa Walker. Boston, 1897. 39 pp. Portrait. 8°.

The 502d meeting of the Society of Arts was held at the Institute this day at 8 p.m.

The meeting was called to order by Mr. George W. Blodgett, Chairman of the Executive Committee, who announced that at a special meeting of the Corporation, held October 20, Professor James M. Crafts had been elected President of the Institute. President Crafts was then introduced, and took the chair.

The record of the previous meeting was read and approved. The Secretary announced the death of Mr. Thomas Doane, a distinguished engineer and friend of education, who in May had been elected a member of the Executive Committee of the Society of Arts. By unanimous consent the President was authorized to appoint the committee of five required to nominate a candidate to fill the vacancy in the Executive Committee.

The following were duly elected Associate Members of the Society: Messrs. George W. Bryden, of Portland, Maine; Orrin S. Doolittle, of Reading, Pennsylvania; W. H. Jaques, of Little Boar's Head, New Hampshire; Jesse F. Johnson, of Montreal, Canada; Marshall O. Leighton, of Montclair, New Jersey; George H. Norman, of Newport, Rhode Island; William T. Hall, of Boston; Walter E. Piper, of Fells, Massachusetts; J. C. Hobart, of Cincinnati, Ohio; and James H. Hutchings, of Boston.

In introducing the speaker, the President referred to the great difficulty of obtaining exact measures of heat, and described the methods employed at the International Bureau for the comparison of thermometers. Mr. Charles L. Norton then presented a paper on "Recent Work in Heat Measurement at the Institute," and began with a description of the apparatus used here for the comparison of thermometers. It contains an inner box, open at both ends. This is inclosed in an outer box provided with three propeller shafts, each bearing twelve propellers with blades bent so that when the inner box is in place the circulation is in a steady stream around and then through the inner box. There is a heating coil at one end. The whole box is lined with cork, which was found to be the best nonconductor. Between the box and its lining there is a layer of asbestos, with fine wires, forming an electric heater by which loss of heat is prevented. The apparatus may be kept at a constant temperature to within 0.005° C. for an hour.

An apparatus for testing the conductivity of steam pipe coverings was described next. This consists of an electric heater immersed in oil, with a stirrer in a section of pipe around which the substance to be tested is placed. Methods for determining the heat of combustion were then considered. Two kinds of apparatus are used in the Institute. A new form of bomb, suggested by Professor Holman, is made of aluminum bronze, and consists of two nearly hemispherical halves, with a nut to join them in the center. The manipulation is simplified by the shape, and the accuracy is about $\frac{1}{2}$ per cent. Equally good results are obtained by the apparatus for burning coal in a stream of oxygen, which consists of a combustion chamber, a cooling chamber, and a cooling coil. The coal is placed on a shelf in a platinum crucible with a perforated cover.

For measuring high temperatures the common thermo-electric pyrometer is employed. In determining the melting points of metals absolute control is secured by using an electric heater instead of a Bunsen flame. The paper was illustrated by an exhibition of apparatus and by lantern views.

While thanking Mr. Norton, the President pointed out that this work is of great value to the school, in that it makes it possible to place in the hands of the students instruments of precision possessing a degree of accuracy seldom found even in laboratories devoted entirely to research.

The Secretary announced that an amendment to the By-Laws would be brought up at the next meeting. The Society then adjourned.

THURSDAY, November 11, 1897.

On this day the 503d meeting of the SOCIETY OF ARTS was held at the Institute at 8 P.M., with Mr. Blodgett in the chair.

The record of the previous meeting was read and approved. There being no other business, the Chairman introduced Mr. Louis J. Hirt, of Brookline, who read a paper on "Cable and Underground Electric Roads." The paper dealt especially with the duplex system employed on the New York cable roads. In this system there are two cables, either of which may be used alone, or both of them may be used at once. The central station was described first, special attention being paid to the tension carriage for regulating the cable. Then views of cross sections of the duplex road were shown, exhibiting the details of construction, Provision is made for a signal system, consisting of electric bells and telephones, by means of which inspectors and conductors may communicate with the central station. The details of construction of grips, switches, curves, and the automatic gypsy used for keeping the cable in place at depressed curves, were described at length. Attention was then directed to underground electric conduits. The cost of running a road of this kind is one and one-half cents a car-mile higher than with the trolley. The method of operation and the plow used on the Lexington Avenue line were described, this being taken as the highest type of the underground electric road. A complete metallic circuit is used. The cost of construction of an underground electric road is one-third that of a cable road. The road with an underground conduit was declared to be the coming form of street railway. The paper was illustrated by a large number of lantern views. The Chairman thanked the speaker for his valuable contribution, and then the Society adjourned.

THURSDAY, December 9, 1897.

The 504th meeting of the SOCIETY OF ARTS was held at the Institute this day at 8 P.M., President Crafts in the chair. The record of the previous meeting was read and approved. Messrs. Arthur T. Hopkins, of Somerville, L. Frederic Howard, of Boston, and Professor Joseph J. Skinner, of the Institute, were duly elected Associate Members of the Society.

The following resolution, which, after having been approved by the corporation, had been presented in writing at a previous meeting, was adopted by unanimous vote of the Society:

Resolved, That the By-Laws of the SOCIETY OF ARTS be amended by striking out the word *four* in the last sentence of Section II (Duties of the Executive Committee), and by inserting in its place the word *three*, so that the sentence shall read, "Three members shall constitute a quorum for the transaction of business."

In introducing the speakers of the evening the President called attention to the improvement in the product and advantage, both to producer and consumer, that has resulted from the development of the canning industry. Mr. S. C. Prescott was the first speaker, presenting a paper written with Mr. W. Lyman Underwood, entitled " Contributions to Our Knowledge of the Micro-organisms and Sterilizing Processes in the Canning Industries," which dealt chiefly with the means employed to prevent the souring of corn.[1]   He said the first attempt to pack corn was made by Isaac Winslow, of Maine, in 1839. His first attempt to employ steam sterilization was a failure, and retorts did not come into general use until 1879.   Cookers were introduced in 1890, and enabled packers to do away with the preliminary boiling of the cans.   The losses due to sour corn and the means of detecting this condition were discussed in full.   Souring is always due to the action of bacteria, six species having been found in sour corn.   Intermittent sterilization is impracticable on a commercial scale, and retorts properly used will kill all bacteria in twenty minutes at a temperature of 120° C.   Mr. Underwood followed with an exhibition of lantern slides, illustrating the machinery employed in packing corn and the bacteria which are found in sour corn.

The paper was discussed by several speakers.   Professor Sedgwick criticised a report published by the Canadian Government, in which intermittent sterilization was advocated.   The President thanked the speakers, after which the Society adjourned.

ROBERT PAYNE BIGELOW, *Secretary.*

---

[1] The paper is published in full on pp. C-30 of this journal.

CONTRIBUTIONS TO OUR KNOWLEDGE OF MICRO-
ORGANISMS AND STERILIZING PROCESSES IN THE
CANNING INDUSTRIES.

II. THE SOURING OF CANNED SWEET CORN.

BY S. C. PRESCOTT AND W. LYMAN UNDERWOOD.[1]

Read December 9, 1897.

IN a paper read before the Society of Arts in October, 1896,[2] we showed the extent of the canning industry in this country, and the importance to it of accurate knowledge of the bacteriological principles of sterilization. In that paper we. dealt with the packing of clams and lobsters, and described some of the bacteria which are active in the deterioration of these products in case sterilization is not complete. It is interesting to notice that some of the results which we published at that time have since been confirmed by a specialist employed by the Canadian Government[3] to investigate the discoloration of canned lobsters.

We now desire to give an account of our more recent investigations in another branch of the industry, namely, the packing of sweet corn. This art constitutes a very large industry, as is shown by the fact that in 1895 seventy-two million 2-pound cans (72 thousand tons) were packed in the United States.

HISTORICAL.

The growth of the art has been rapid, for it was not until about 1853 that corn was packed at all with success. Maine has been generally acknowledged as the home of corn packing, and its claim to be so considered is probably just. In 1839 Isaac Winslow began experiments in canning corn at or near Portland. He was for a long time unsuccessful. He first attempted to cook the ears

---

[1] Mr. Prescott is Instructor in Biology, Massachusetts Institute of Technology, and Mr. Underwood is of the William Underwood Co., Boston.

[2] Technology Quarterly, Vol. X, No. 1, March, 1897, pages 183–199.

[3] Canadian Department of Marine and Fisheries, 29th Annual Report, Supplement No. 2, Ottawa, 1897.

of corn whole, but this proved unsatisfactory on account of their bulk, and it was also thought that the cobs absorbed the sweetness. He next tried to remove the kernels whole by means of a fork, but this was soon abandoned, and the corn was afterward cut from the cob. His first experiments were made in a common household wash boiler, and in a very limited way. Small quantities were treated by various methods, but nearly all the corn spoiled. Some kept, however, and gave promise of ultimate success. In 1843 he built a small copper steam boiler of about two barrels capacity, and carrying ten or twelve pounds of steam. To this he connected wooden tanks lined with zinc and made steam tight. In these crude retorts he "processed" the corn, subjecting it to the direct action of live steam. Nearly the whole lot spoiled, and in consequence of this failure, steam apparatus was abandoned. The next year he returned to open boilers, and continued his experiments with varying success for ten years. In 1853 he applied for a patent, but this was not allowed until 1862.

An abstract from the patent may be of interest: "After a great variety of experiments I have overcome the difficulties of preserving Indian corn in the green state without drying the same, thus retaining the milk and other juices, and the full flavor of fresh green corn until the latter is desired for use. Instead of a hard, insipid, or otherwise unpalatable article, I have finally succeeded in producing an entirely satisfactory article of manufacture, in which my invention consists. I have employed several methods of treatment. My first success was obtained by the following process: The kernels being removed from the cob were immediately packed in cans, and the latter hermetically sealed so as to prevent escape of the natural aroma of the corn, or the evaporation of the milk or other juices of the same. Then I submitted the sealed cans and their contents to boiling or steam heat for about four hours. In this way the milk and other juices of the corn are coagulated as far as may be, boiling thus preventing the putrefaction of these most easily destructible constituents. At the same time the milk is not washed away or diluted, as would be more or less the case if the kernels were mixed with water and then boiled. By this method of cooking green corn the ends of the cans are bulged out, as though putrefaction and the escape of the resultant gases had commenced within the cans. Consequently strong cans are required.

"I recommend the following method: Select a superior quality of the green corn in the green state, and remove the kernels from the

cob by means of a curved or gauged knife or other suitable means.
Then pack these kernels in cans and hermetically seal the latter so as
to prevent the evaporation under heat or the escape of the aroma of
the corn.   Now expose these cans of corn to steam or boiling heat
for about one hour and a half, and then puncture the cans and immediately seal the same while hot, and continue to heat for about two
and one-half hours longer.   Afterwards the can may be slowly cooled
in a room at a temperature of 70° to 100° F."

For nearly twenty years this method was in use, the only change
being that the time of processing was somewhat shortened.   About
1879 retorts were introduced in corn packing, and the second heating
was done in them, the time being reduced from two and one-half hours
to one hour.   The advent of cookers about 1890 did away with the
first heating in the water bath, so that now this is abandoned as an
agency of sterilization.   Many of the processes formerly carried on
by hand are now carried on by machinery.   Maine leads in the packing of sweet corn, but large quantities are packed in New York and
Maryland and in the West, particularly in Iowa, Illinois, and Michigan.

THE SOURING OF CANNED SWEET CORN.

Sweet corn, when properly prepared, is one of the most valuable
of all canned foods, as it retains much of its original flavor, is popular,
and is sold at a price within the reach of all.   If, however, the sterilizing has not been done thoroughly, there may result fermentations
caused by bacteria which have not been killed, producing what is
known as "sour" corn.   It is not definitely known when sour corn
first appeared.   In the experiments of Isaac Winslow, spoiling of some
kind resulted, but so far as we have been able to ascertain, its nature
has never been described.   In a Massachusetts factory, however,
where corn has been packed with success for nearly twenty years,
souring suddenly occurred in 1878.   Maine was also somewhat affected
at the same period.   Until this time corn had been processed for five
hours at a boiling temperature with no loss, but in the year just mentioned, with exactly the same treatment, this manufacturer experienced
a total loss.   Some of this corn was sent to chemists for analysis with
the hope that a remedy might be found at once.   It was reported by
them to be due to "fungus consisting of little globules that boiling
heat did not dissolve."

Early in the following year (1879) the Massachusetts packer who owned the factory referred to attempted to continue with the old process, but the corn spoiled. Retorts were procured, and with their higher heat satisfactory results were obtained. For sixteen successive years he experimented with the old process with the intention of returning to it if possible. The corn so packed and kept at a temperature of 90 to 100° F. invariably spoiled, swelling on the third or fourth day. It was thought that the trouble might be local, and to decide this he visited distant sections, carefully selected and gathered corn, and, returning at once to his factory, packed it in the old way with the least possible delay, working sometimes all night that this might be done. The results were always the same — the corn could not be successfully packed by the old method. Had any locality been found where this could have been done, he intended removing his factory to that neighborhood.

The exact chemical changes which take place when sour corn is produced are difficult to state, and vary under different conditions. The sugar and starch in the corn are fermented for the most part to lactic, acetic, and butyric acids, thus giving rise to the souring. There are also other products of decomposition. Gases are frequently evolved, but being dissolved by the liquid in the can at the ordinary temperature, in the majority of cases no swelling results.

The loss resulting from sour corn during the last eight or ten years has been enormous, in some years being much more than in others. Thousands of dollars have been lost in a single year by individual manufacturers who have experienced this trouble. Moreover, the uncertainty and the possibility that losses may be incurred are constant sources of worry and uneasiness to those engaged in this industry.

DETECTION OF SPOILED CANS.

Spoiling in canned goods is generally indicated by bulging of the ends of the cans, caused by the pressure of the gases produced within. Thus a packer may generally detect any unsoundness before the goods are put upon the market, as all are overhauled and inspected before ultimate shipment. In the case of sour corn, however, at least in its first stages of deterioration, there is no outward indication of trouble. It is only under rather exceptional conditions that swelling occurs. If the temperature and other conditions are favorable for the rapid development of germs which can produce fermentation with the formation of gas, swelling will result.   Since, however, the latter conditions rarely prevail in the factories, the detection of sour corn becomes difficult.   Corn which is sweet when shipped may become sour many months afterward.   To illustrate this fact an instance may be cited where from the same day's packing two lots of corn were shipped, one to the northern and one to the southern part of the United States. That which was sent to the North was in perfect condition at the end of a year, while that which went to the warmer climate became sour in a short time.   Many instances of the same nature have been noticed by different packers, and similar results may be obtained by laboratory experiments.   The explanation of this fact probably is that all the bacteria were not killed by the heating to which these cans were subjected, and that the conditions for growth of the micro-organisms became favorable only in the warmer locality.   Provided sterilization is not complete, there seems little reason to doubt that climatic condition is a most important factor in the souring of corn.   It should always be borne in mind that if processing or sterilization is complete sour corn cannot result, because the germs of fermentation are destroyed.   When souring occurs the percentage of bad cans may be small, but often runs from 10 to 40 per cent., or even higher.   Such goods are generally returned, and an attempt is made to separate the sweet from the sour cans.   To do this there are two methods in common use.

According to the first method, the cans are put into a tank of water at a temperature of 80° F., where they stand for from six to twelve hours in order that the contents may be heated uniformly throughout.   They are then removed and their ends just submerged in water at 190° F.   Here they remain for not more than thirteen minutes.   At the end of that time those cans which are swelled are rejected as sour.   The other method is to boil the cans for one hour. This causes all the ends to bulge.   They are then cooled, and those whose ends remain bulged for more than eight hours are rejected, while those which "snap back" within this time are considered satisfactory.   Both these methods depend for their success upon the fact that at certain temperatures gas is produced rapidly by bacteria within the cans.

BACTERIOLOGY OF SOUR CORN.

Our investigations commenced in February, 1897, with the examination of a large number of cans of sour corn. On opening the cans no change was noticeable to the eye, the corn appearing fresh and of a natural color. In some cases a sour odor could be detected, but in others this was not observed. It was to the taste that the trouble was most apparent, the corn being sour and of a peculiar astringent quality. Bacteriological examination showed sound cans to be sterile, while spoiled cans invariably gave evidence of bacterial action. Pure cultures of twelve species of bacteria were obtained, of which eleven were bacilli, and one was a micrococcus. It must not be supposed that these bacteria are disease-producing; they probably act merely upon the saccharine and starchy matter, transforming it to organic acids and other substances of more or less disagreeable taste and odor, and so make the corn unpalatable and destroy its commercial value.

By inoculating sterile cans of corn with these organisms we have been able to produce souring in all respects similar to that of the spoiled cans from which they were originally taken. Our experiments were conducted in the laboratory in the following manner: A number of cans were selected and all of them were punctured, this operation being done in a sterile glass chamber. A part of the cans were inoculated with cultures obtained from sour corn, and all the cans were then sealed and put in an incubator kept at the blood heat. The cans which had been inoculated commenced to swell in from twelve to twenty-four hours, while those not inoculated remained as sound as when put in the incubator. Thus we easily proved that a vacuum is not necessary for keeping canned corn, and that air may be admitted to a sound can and spoiling will not result, provided proper precautions are taken that the air so admitted be free from germs. This statement will undoubtedly be regarded with incredulity in some quarters, so strong is the popular belief among packers in the indispensability of a vacuum, yet a long line of experiments from the days of Tyndall to the present time prove the validity of this assertion. Moreover, there are bacteria which can develop in a vacuum, and which could find favorable conditions within cans from which the air has been expelled. Sterilization, not the driving out of air, is the important factor in keeping all kinds of canned foods; and although, as we have shown in our earlier paper, the vacuum is necessary in testing the cans, no preserving power can be rightfully ascribed to it. These experiments have been made repeatedly, and always with the result that souring takes place only when living bacteria are present. The presence and activity of the bacteria in sour corn have also been shown by inoculating various kinds of culture media with material from spoiled cans. Active fermentations, of the various kinds previously mentioned, have been brought about in this way.

In order to study these fermentations more thoroughly, and to ascertain, if possible, the source of the bacteria causing them, we spent nearly the whole of the corn-packing season of 1897 at an establishment in Oxford County, Maine, where every convenience for scientific study of the process was kindly put at our disposal by the proprietors. We were thus enabled to investigate thoroughly the methods of procedure, from the harvesting of the green corn to its ultimate shipment in cans.

THE PROCESS OF PACKING.

It is very important that the utmost cleanliness and dispatch should be observed in all the operations, so that the chances of infection from bacteria may be reduced to a minimum. In this factory the strictest caution was exercised in these respects, everything being kept scrupulously clean. The corn is generally picked in the morning, and is delivered to the cannery as early as possible. One or two men make it their special duty to visit the farms once or twice a week during the season to keep informed as to the condition of the crop, and to "order in" the corn as it becomes sufficiently matured. As the ears are delivered at the factory they are arranged in low piles on the ground in an open shed to protect them from the sun. The husks and the silk are taken off by hand, and the corn is then quickly carried to the cutting machines, in which, by a series of knives and scrapers, the kernels are quickly and cleanly separated from the cob. Any stray bits of cob or silk which may be mixed with the corn are now taken out as it passes through the "silker," a machine arranged somewhat on the plan of a gravel-sifter; that is, with two cylindrical wire screens one inside the other, placed on an incline, and rotating in opposite directions. The corn drops through the meshes of the screens, while the refuse passes out at the lower (open) end.

The corn is now weighed, mixed with water in the proper proportions, and is then ready for the cooker. There are several varieties of these machines in use, all of which are alike in principle, but differ somewhat in the details of construction. Their object is to heat the corn evenly and quickly to a temperature of 82–88° C. (180–190° F.) and to deliver it automatically into the cans. A single machine fills about thirty cans a minute. The duty of the cooker is threefold: First, in the heating to which the corn is here subjected some of the bacteria, particularly those in the vegetative state, are killed. Second, the corn being filled into the cans while hot expands the air, so that after sealing and cooling a partial vacuum is produced, which, as before stated, is essential for the detection of unsound cans. Finally, this cooking heats the corn to such a temperature that the subsequent sterilization in the retorts is brought about more quickly, and the danger of browning or scorching of the corn next to the tin is minimized.

The cans are next capped, soldered, and tested for leaks. Sterilization, the final and most important step in the whole process, now follows, and is done in retorts, by steam under pressure. The length of heating or processing, and the pressure which is given, vary somewhat in different factories. As we have shown in our previous paper, in practice, in order to insure sterilization it is necessary to obtain and maintain a temperature in excess of 100° C. (212° F.) *throughout the contents of the can*, and for a period of time varying with the substance to be sterilized.

METHOD OF STERILIZATION.

It is thought by some that intermittent sterilization might be employed in packing, but we consider this entirely impracticable upon a commercial scale. Intermittent sterilization consists in heating to the temperature of boiling water for a length of time varying from thirty minutes to one hour, on three or four successive days, the substance to be sterilized being cooled and kept cool between the heatings. It is supposed that in the first heating all the active bacteria, the so-called vegetative cells, are killed, while the more resistant forms, spores, retain their vitality. According to the theory, the majority of the spores germinate and become active before the second heating, and in turn are killed, while by the third heating all the remaining spores will have developed into active bacteria, and will then be destroyed.

To insure success by this method of sterilization, apparatus and means must be employed which, while practicable in a small way, are in our opinion absolutely impracticable on such an extensive scale as would be demanded commercially. To use this method would necessitate at least three times as much sterilizing apparatus, much more room, a greater amount of labor, and a great loss of time.

To show the resistance of bacteria to the continuous action of a boiling temperature, we have found that certain species isolated from sour corn will survive actual boiling for more than five hours, and other species of bacteria which are met with in spoiled canned goods have been boiled for eight hours without being killed. These facts serve to show conclusively the impracticability of the ordinary water bath. On the other hand, the retort with its high temperature will, if properly used, kill all forms of bacteria at a single heating, without injury to the food substance, the length of time required varying, as has already been said, with the conductivity of the medium for heat. We have found by experiment that sixty minutes at 121° C. (250° F.), as indicated by the thermometer on the outside of the retort, is sufficient time for sterilizing corn in two-pound cans, and it seems probable that this can be shortened somewhat, or the temperature reduced. Further experiments are in progress to decide this question.

WHITENESS OF CANNED CORN.

Through a demand that canned corn shall be very light in color, there has been, apparently, a pressure put upon the packer to shorten the time of heating or to reduce the temperature in his retorts. The large losses which have resulted in recent years from sour corn have, it is claimed, been due principally to this demand. Instances are known where the desired result has been brought about by the use of some bleaching reagent, generally sulphite of sodium. While this may not be unwholesome, it greatly injures the flavor of the corn, as a comparison of such corn with that without bleachers will show. Although such cases sometimes occur, it cannot be said to be the fault of the packer; for if the dealers demand very white corn the packer must resort to some unusual means in order to render his product saleable. In this connection a statement in a recent trade journal is noteworthy: " The volume of poor corn which has found its way to market in the last few years has had, and is still having, a considerable effect upon the consumption of that article, and there are a good many families who never buy canned corn nowadays because they have found little but disappointment in their corn purchases of the last few years." [1]

It is much to be doubted if the consumer demands that the corn be very white in color. What he desires is a palatable article with a natural flavor. It seems evident that in the near future the dealers must regard this very white corn with disfavor, and reject any in this condition.

MAXIMUM TEMPERATURE WITHIN THE CANS.

By the use of small registering thermometers which can be sealed up within the cans, and which record the maximum temperature reached, we proved, in an extended study of the process as it is actually carried on at the factory, that corn is a very poor conductor of heat, and that the time necessary to bring all portions of the center of the can to the requisite temperature is a factor whose importance cannot be overestimated. Corn as it comes in cans from the cooker is at a temperature of 82 to 88° C. (180 to 190° F.). At the end of thirty minutes in a retort with a pressure of thirteen pounds, the corresponding temperature of which is 118.8° C. (246° F.), a thermometer in the center of a can placed in the middle of the retort, which was full of corn, registered 108.3° C. (227° F.). At the end of forty-five minutes under the same conditions a temperature of 114° C. (237.2° F.) was reached, and at the end of fifty-five minutes the retort temperature of 118.8° C. (246° F.) was registered by the thermometer in the can. From this it is evident that if a packer were giving his corn an hour in the retort at this pressure, the central portions of the can would in reality be subjected to the full effect of the heat for only five minutes. Thus it is evident that with the present methods any reduction of time of heating is attended by considerable risk. If any means could be devised by which the heat could reach more quickly the center of the cans, it might be safe to shorten the time of heating. There is a prospect that before long some such modifications may be possible.

BACTERIOLOGY OF SWEET CORN.

The source of the bacteria producing the fermentations described was also a problem, the solution of which we sought with great care.

[1] Canner and Dried Fruit Packer, Vol. V, No. 19.

Every step of the process was investigated bacteriologically, and all channels of infection, the water supply for example, were studied. The general cleanliness and the liberal use of water and steam throughout the factory which we visited reduced the liability of infection from dust to a minimum. We examined the green corn on the cob, the corn as it came from the cutting machines, as it went to the cooker, as it came from the cooker, and as it came from the retorts after the usual processing and after some periods of heating given for experimental purposes. Living bacteria were found ou the raw corn, and at all stages of the process before the final sterilization. The corn as it went to the cooker was found to contain many germs, but in the short heating to which it was subjected there, some of the organisms were destroyed. Cans which had been retorted for thirty minutes or less were found to contain living bacteria, and cans so treated spoiled and became much distended within four days. No living bacteria were found in cans which had received the full time of processing at this factory. By culture methods and by microscopical examination we have found that the bacteria living upon the kernels of corn and those which are found in the later stages of the process are undoubtedly of the same species. They also correspond in all respects with species which we obtained from cans of sour corn in the laboratory experiments carried on in the early part of our investigations.

All these organisms are characterized by great rapidity of growth when allowed to develop at a temperature of 37° C. (98.6° F.). In evidence of this fact we need only to state that of the large number of cans incubated at this temperature many swelled within twenty-four hours, while in several cases the cans exploded within that time. Agar streak-cultures of these bacteria frequently showed well-marked growth within six hours, and in some cases in four hours. The growth is much retarded at a temperature of 20° C. (70° F.). None of the organisms which we have obtained correspond closely to the published descriptions of the lactic or butyric acid organisms, or to that of the *Bacillus maidis* of Cuboni.

If sour corn is the result of bacterial action, the question naturally arises, Why should a packer have trouble in a certain year, when he is using presumably the same methods of treatment that he has employed without loss in former years ? A number of conditions might exist that would account for this. In the first place, it is a well-known fact that diseases which are caused by bacteria may be much more prevalent in some years than in others. The same is probably true in the case of the bacteria which attack corn. The weather may be much more favorable for the growth of these germs in certain years than in others, and there is good reason to believe that a warm moist season is more apt to give sour corn than a cool dry one. Is the packer entirely sure that the conditions prevailing within the factory are always the same from year to year? Other things being equal, if exactly the same methods are used, similar results should be obtained. But to all outward appearances the conditions may be the same, when in reality they are quite different. Differences in the steam gauges or thermometers, or a little carelessness on the part of some operative, may be sufficient to turn the scale and give rise to sour corn where before none had existed. That trouble might be caused by such slightly changed conditions can be seen readily when we realize that, as we have already shown, in being processed in the retort for an hour at a temperature of 240° C., or over, the corn at the center of the can is in reality only receiving this intensity of heat for five minutes.

Believing that, in order to be of practical value, all laboratory experiments must be carried on under conditions as nearly as possible like those existing in the factory, we have recorded only such results as have been obtained under these conditions. There are still some facts to be determined which cannot be settled by laboratory experimentation, and which, owing to the shortness of the packing season, we were unable to push to completeness last year. We hope another year to investigate these points more fully.

We wish to express our gratitude and indebtedness to all those who have so kindly helped us, and particularly to Professor Sedgwick, without whose coöperation this work would have been long delayed.

In conclusion we would again affirm :

1. That sour corn appears to be always the result of bacterial action, and due to imperfect sterilization.

2. That in case of insufficient processing souring does not always result unless the cans are subjected to conditions favorable to the growth of the bacteria within.

3. That the bacteria which produce sour corn are found on the kernels and beneath the husks of the corn as it comes from the field.

4. That the bacteria found on the ears of corn correspond in all respects to those originally found by us in cans of sour corn.

5. That swelling may be caused by bacteria other than those which produce sour corn, but it is also a natural consequence and a further development of this process of souring, provided the cans be subjected to a favorable temperature.

6. That so far as we have been able to discover, the organisms present in sour corn are capable of producing serious commercial damage and an unpleasant taste, but are otherwise harmless.

7 That a vacuum is not necessary for the preservation of canned foods, but is a valuable factor in the detection of unsound cans.

8. That the use of bleachers is not to be recommended, and is unnecessary if proper methods of sterilization be employed.

9. That the utmost cleanliness at every step is absolutely essential.

10. That intermittent sterilization is not practicable on a commercial scale.

11. That the open water bath is inefficient as a means of sterilization.

12. That with the present methods of retorting it takes fifty-five minutes for the temperature which is indicated on the outside thermometer to be registered at the center of a two-pound can of corn previously heated in the cooker to 82 to 88° C. (180 to 190° F.).

13. That heating for ten minutes with a temperature of 126° C. (250° F.) throughout the whole contents of such a can of sweet corn, appears to be sufficient to produce perfect sterilization.

BIOLOGICAL LABORATORY,
MASSACHUSETTS INSTITUTE OF TECHNOLOGY, BOSTON,
January 4, 1898.

APPENDIX.

DESCRIPTIONS OF BACTERIA FOUND.
BACILLUS A.

OCCURRENCE.	Found in cans of sour corn.
GENERAL CHARACTERS.	*Shape and arrangement:* Bacillus, occurring singly and in short chains. *Size:* Generally 2–4μ long by 1μ broad. Many cells are very long, and vary from 10–50μ in length. *Motility:* Rapid serpentine and spinning movements. *Spore formation:* Oval, centrally located spores. *Relation to temperature:* Develops rapidly at 37$\frac{1}{2}$° C.; more slowly at 20° C. *Relation to air:* Aërobe and facultative anaërobe. *Relation to gelatin:* Liquefies. *Color:* Non-chromogenic.
GELATIN.	*Stick culture:* Develops rapidly throughout whole length of puncture. Liquefaction begins within twenty-four hours, and at the end of two days a horn-shaped liquefied portion is observed. *Plate culture:* Surface colonies: Very small. Liquefaction begins almost as soon as colonies are visible; in two days the plate culture is entirely liquid. Submerged colonies: Apparently same as on surface.
AGAR.	*Streak culture:* A thin, smooth layer, covering nearly the whole surface. Edges dissected and bluish in color. In two days lower part of culture becomes dryer, white, and finely wrinkled. *Plate culture:* Surface colonies: Vary much in size and shape. Young colonies are very small, oval or circular. Spreading soon begins, giving irregularly branched or sullate colonies. Submerged colonies: Very small, oval or spherical.
POTATO.	Potato much darkened. A thin film of growth covers the surface. This film is at first moist, but at end of three days becomes dry and finely wrinkled.
MILK.	Not coagulated. Acidity: strong.
SMITH SOLUTION.	No gas produced. Thin film on surface. Sediment at bend of tube. Turbid throughout. Strongly acid.
NITRATE.	Is reduced to nitrite. Solution clear.
BOUILLON.	Slightly turbid at end of twenty-four hours at room temperature. Film develops in twenty-four hours in incubator at 37$\frac{1}{2}$°. No sediment.

BACILLUS B.

OCCURRENCE.	Found in cans of sour corn.
GENERAL CHARACTERS.	*Shape and arrangement:* Bacilli, occurring singly and in short chains. *Size:* 2–5µ long by .6µ broad. *Motility:* Slightly motile. *Spore formation:* Very small oval spores. *Relation to temperature:* Develop more rapidly at 37½° C. than at 20° C. *Relation to air:* Aerobic, facultatively anaerobic. *Relation to gelatin:* Liquefy slowly. *Color:* Non-chromogenic.
GELATIN.	*Stick culture:* Growth well marked throughout entire length of line of inoculation. A small cup-shaped depression is observed on second day. This increases in size slowly as liquefaction occurs. *Plate culture:* Surface colonies: First appear as small, translucent blue dots, which later become white or gray, and slowly liquefy the plate. Colonies from ¼″ to ⅟₁₆″ in diameter. Submerged colonies: Small and blue when seen by transmitted light.
AGAR.	*Streak culture:* A thick, white, milky layer, covering the whole surface of the agar. The lower portion becomes somewhat wrinkled. *Plate culture:* Surface colonies: Shiny, almost porcelain white in color. When about ⅟₁₂″ in diameter often send out little branches or processes on one side, giving a very characteristic appearance. Submerged colonies: Small, spherical dots.
POTATO.	Drab growth spreading over whole surface of potato, and moist and somewhat wrinkled in appearance. Potato much darkened.
MILK.	Not coagulated in two days. Ac'd in reaction.
SMITH SOLUTION.	No gas produced. Solution acid in reaction. Growth throughout tube. Sediment at bend and film on surface.
NITRATE.	Not reduced to nitrite.
BOUILLON.	Faintly turbid at end of twenty-four hours at 20° C. At 37½° C. a dry looking film appears on surface in from twelve to twenty-four hours.

BACILLUS C.

OCCURRENCE.	Found in cans of sour corn.
GENERAL CHARACTERS.	*Shape and arrangement:* Bacilli, with rounded ends, occurring in chains. *Size:* 3–9µ x 1.8µ. *Motility:* The chains swim with slow, steady, undulating motion. *Spore formation:* Large, oval, centrally located spores. *Relation to temperature:* Develop rapidly at 37½° C.; more slowly at 20° C. *Relation to air:* Aërobic and facultatively anaërobic. *Relation to gelatin:* Liquefy readily. *Color:* Non-chromogenic.
GELATIN.	*Stick culture:* Growth throughout, but most abundant at surface. A trumpet-shaped, liquefied portion is quickly formed, with flocculent material in suspension and precipitate at bottom. Film on surface. *Plate culture:* Surface colonies: At first white and small. As soon as they break through the surface liquefaction commences, and colonies rapidly become large and of a homogeneous gray color. At end of a week colonies are 1″ in diameter and covered by a thin film, with concentric markings and fluted edges. Submerged colonies: Rounded and white. Soon break surface of gelatin and begin to liquefy.
AGAR.	*Streak culture:* Thick granular layer, with dull luster. Edges sharply defined and scalloped. At end of two or three days wrinkles appear on older portion. *Plate culture:* Surface colonies: Smooth and somewhat waxy in appearance. Often spread to form irregularly shaped patches with thickened edges. Submerged colonies: Small when separated, but often unite, forming a thin film on lower surface of the agar.
POTATO.	A rather abundant yellowish layer of dull, waxy luster. Potato darkened.
MILK.	Is coagulated with clear separation of whey. Acidity: Slight.
SMITH SOLUTION.	No gas produced. Very heavy sediment. Solution comparatively clear. Strongly acid.
NITRATE.	Is reduced to nitrite strongly. Slight turbidity.
BOUILLON.	Strongly turbid throughout at end of twenty-four hours at 20° C. After twenty-four hours at 37½° C. a slight amount of sediment is formed, and a ring of growth appears at the surface.

BACILLUS D.

OCCURRENCE.	Found in cans of sour corn.
GENERAL CHARACTERS.	*Shape and arrangement:* Bacilli, occurring singly and in chains. *Size:* 3.5–10μ x 2μ. *Motility:* Chains not motile. Single cells move with slow serpentine motion. *Spore formation:* Spores formed in center or near one end. *Relation to temperature:* Develops rapidly at 37½° C. *Relation to air:* Aërobic and facultatively anaerobic. *Relation to gelatin:* Liquefy. *Color:* Non-chromogenic.
GELATIN.	*Stick culture:* Growth slight, but noticed throughout. Liquefaction soon begins. *Plate culture:* Surface colonies: First appear as white dots. Liquefaction begins quickly, and a liquefied saucer-shaped depression, with a white dot at center, is soon formed. Colonies rapidly become large, and have flocculent precipitate near center and finger-like processes projecting inward from edges. At end of a week the plate is nearly all liquefied and a thin film is developed at surface. Submerged colonies: Few and small.
AGAR.	*Streak culture:* Thick, slimy growth readily removed. It occurs in form of scalloped patches, with smooth edges. *Plate culture:* Surface colonies: White or gray, regular in outline, and smooth and shiny when young. Later become somewhat irregular in shape. Submerged colonies: First appear like woolly or burr-like rounded masses, which soon break through surface and become shiny and smooth, like surface colonies.
POTATO.	At end of three days a thick, spreading growth, dull white in color, and looking like a piece of wet cracker. Irregular edges. Potato not discolored.
MILK.	Partially coagulated in two days. Acidity: Marked.
SMITH SOLUTION.	No gas produced. Turbid throughout. Considerable sediment. Very strongly acid.
NITRATE.	Is reduced to nitrite faintly. Slight turbidity and sediment.
BOUILLON.	Turbid throughout at end of first day. A thin film is formed after several days.

BACILLUS E.

OCCURRENCE.	Found in sour corn.
GENERAL CHARACTERS.	*Shape and arrangement:* Long, narrow bacilli, generally occurring singly. Very variable in size. *Size:* 2-50µ x .6µ. *Motility:* Move rapidly, with eccentric darting and twisting movements. *Spore formation:* Small, oval, centrally located. *Relation to temperature:* Develops rapidly at 37½° C.; slower at 20° C. *Relation to air:* Aérobic and facultatively anaërobic. *Relation to gelatin:* Non-liquefying. *Color:* Non-chromogenic.
GELATIN.	*Stick culture:* Slight spreading growth at surface, and growth all along line of inoculation. Filmy, ragged surface. Growth at end of second day. Transparent. *Plate culture:* Surface colonies: Circular; bluish by transmitted light. Grow to about ¼" in diameter. Submerged colonies: Small white dots, developing more slowly than surface colonies.
AGAR.	*Streak culture:* In growth very similar to A. Bluish edges, finely dissected. Surface of agar covered with a thin, white layer, finely wrinkled at the base. *Plate culture:* Surface colonies: At first small, rounded masses, which on third or fourth day show a thin surrounding outgrowth, appearing bluish by transmitted light. Submerged colonies: Many small colonies about size of pin points.
POTATO.	A yellowish growth along line of inoculation. On lower part of surface a thin, wrinkled film extends. Potato slightly darkened. Young growth moist in appearance.
MILK.	Not coagulated. Acidity: strong.
SMITH SOLUTION.	No gas produced. Solution turbid throughout. Sediment at bend. Thick film on surface. Slightly acid.
NITRATE.	Not reduced. Considerable turbidity.
BOUILLON.	Faintly turbid throughout at end of twenty-four hours at room temperature. Tough, thin, wrinkled film at end of twenty-four hours in incubator.

BACILLUS S.

OCCURRENCE.	Found on ears of green corn and in cans of sour corn.
GENERAL CHARACTERS.	*Shape and arrangement:* Bacilli, generally occurring singly, but frequently in chains of three or four. *Size:* 2–10μ x 1–1.8μ. Average: 3–4μ x 1.5μ. *Motility:* Quick swimming motion; chains also motile. *Spore formation:* Small, oval, centrally located spores. *Relation to temperature:* Develops more rapidly at 37$\frac{1}{2}$° C. than at 20° C. *Relation to air:* Aerobic and facultatively anaérobic. *Relation to gelatin:* Liquefies rapidly. *Color:* Non-chromogenic.
GELATIN.	*Stick culture:* Growth throughout in twenty-four hours. Liquefaction at surface. Thick film, marked with concentric rings on surface. At end of a week liquefaction extends to walls and $\frac{1}{4}''$ down from surface. *Plate culture:* Surface colonies at end of two days are small, and white or bluish in color. Liquefaction begins about the third day, and proceeds slowly until the whole plate is liquid. Colonies form saucer-shaped depressions with a central disk of gray color and sharply defined edges. Submerged colonies: Small, irregular, and hazy in outline.
AGAR.	*Streak culture:* A thin, smooth, shiny, transparent layer, with bluish color and scalloped edges, covering nearly the whole surface. *Plate culture:* Surface colonies first appear as small, round, white spots. Spreading soon begins, and stellate or branched colonies, with bluish fluorescence, are formed. If many colonies are present on the plate the branching is less conspicuous, and the plate soon becomes covered with a thin layer.
POTATO.	A yellowish, moist, slimy layer extends over the whole surface. The potato is wet in appearance and very much darkened in color.
MILK.	Not coagulated. Acidity: strong.
SMITH SOLUTION.	No gas produced. Slightly turbid throughout. Film on surface. Neutral reaction.
NITRATE.	Is reduced to nitrite slightly. Solution turbid.
BOUILLON.	Turbid throughout at end of twenty-four hours at 20° C. At end of a week slight surface growth, considerable cloudiness, but no sediment.

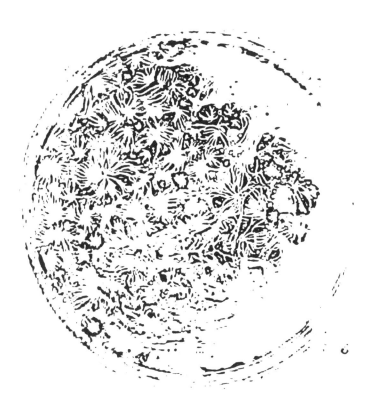

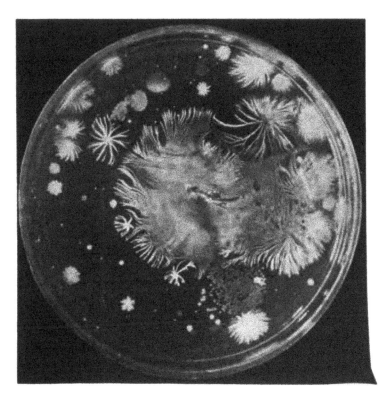

BACILLUS T.

OCCURRENCE.	Found in cans of sour corn.
GENERAL CHARACTERS.	*Shape and arrangement:* Rods occurring singly in short chains. *Size:* 2–4μ x 1μ. *Motility:* Slightly motile. *Spore formation:* Oval spores, filling nearly the whole cell. *Relation to temperature:* Develops more rapidly at 37.5° C. than at 20° C. *Relation to air:* Aërobe and facultative anaërobe. *Relation to gelatin:* Liquefies. *Color:* Non-chromogenic.
GELATIN.	*Stick culture:* Development all along line of inoculation in twenty-four hours. Liquefaction takes place, forming a trumpet-shaped mass, somewhat depressed at surface. A film develops on surface, and a flocculent substance is held in suspension. *Plate culture:* Surface colonies first appear as small spots. Soon liquefaction begins, forming a cup-shaped depression, with a central, whitish mass. At end of a week a thick, waxy scum, marked by concentric rings, covers the entire surface.
AGAR.	*Streak culture:* Thick, gray film, with irregular edges. Dull, granular mat surface on lower portion, and smooth and·lustrous above. *Plate culture:* Surface colonies: Grayish or brownish in color, irregular in outline. Thickened edges; sometimes a dot is seen at center. Submerged colonies: Like surface colonies in general appearance.
POTATO.	Thick, white growth, with dull luster spreading over surface. Potato darkened.
MILK.	Partially coagulated in two days. Acidity: slight.
SMITH SOLUTION.	No gas produced. Very turbid throughout. Ring of growth at surface. Strongly acid.
NITRATE.	Is reduced to nitrite strongly. Turbid.
BOUILLON.	Becomes turbid within twenty-four hours. Thin, flaky surface growth, which breaks up and settles, forming a deposit. Waxy film.

BACILLUS U.

OCCURRENCE.	Found on ears of green corn and in cans of spoiled corn.
GENERAL CHARACTERS.	*Shape and arrangement:* Rods occurring singly and in chains of two or three elements. *Size:* Variable — 2 to 16μ x 1.5μ. *Motility:* Rapid, serpentine, and spinning motion. *Spore formation:* Oval spores formed. *Relation to temperature:* Develop rapidly at 37½° C., more slowly at 20° C. *Relation to air:* Aërobe and facultative anaërobe. *Relation to gelatin:* Liquefy. *Color:* Non-chromogenic.
GELATIN.	*Stick culture:* At end of first day faint growth along needle track. On second day slightly liquefied at surface. Liquefaction spreads rapidly to wall of tube, and whole upper portion soon becomes liquid. *Plate culture:* Surface colonies: When very small show slight branching, but as soon as liquefaction begins colonies become circular and form depressions in the gelatin. Plates become entirely liquid in a few days. Submerged colonies: Small, spherical, and inconspicuous.
AGAR.	*Streak culture:* Smooth, white, shiny layer, with branched or serrated edges, and extending over nearly the whole surface. *Plate culture:* Surface colonies: Circular when very young, but branching takes place as colonies develop, producing stellate forms. The fewer the colonies the more marked the branching.
POTATO.	Growth rapid, and extends over the whole surface of the potato. In early stages appears drab and slimy, but darkens with age, and becomes somewhat dry and wrinkled, like a paint skin. Potato darkened.
MILK.	Not coagulated. Acidity: strong.
SMITH SOLUTION.	Much gas produced. Heavy sediment. Turbid throughout. Very strongly acid.
NITRATE.	Is reduced to nitrite faintly. Solution clear.
BOUILLON.	Turbid throughout at end of twenty-four hours in incubator. Sediment and film.

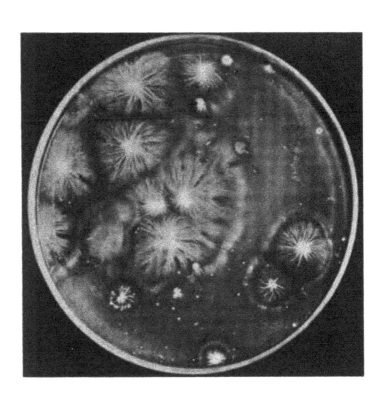

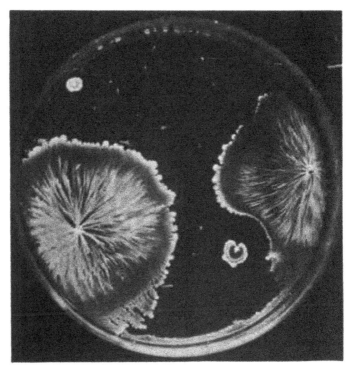

BACILLUS W.

OCCURRENCE.	Found on ears of green corn and in cans of corn.
GENERAL CHARACTERS.	*Shape and arrangement:* Stout rods, rounded ends, generally in chains. *Size:* 4–6μ x 1.8–2μ. *Motility:* Swim rapidly with undulating motion. *Spore formation:* Oval spores centrally located. *Relation to temperature:* Develops rapidly at 37½° C., slowly at 20° C. *Relation to air:* Aërobe and facultative anaërobe. *Relation to gelatin:* Liquefies rapidly. *Color:* Non-chromogenic.
GELATIN.	*Stick culture:* Development well marked at end of twenty-four hours. On second day a large trumpet-shaped mass of liquefied gelatin is formed, in which is suspended a heavy flocculent precipitate. *Plate culture:* Surface colonies: Circular, rapidly growing, and containing a gray or brown precipitate at center, surrounded by a broad ring of clear, liquefied gelatin. On long standing surface becomes covered with a thin, scaly film or incrustation. Submerged colonies: Small, circular or oval.
AGAR.	*Streak culture:* A thick, finely-granular layer, with bluish irregular and indistinct edges covering nearly the whole surface of the agar. *Plate culture:* Surface colonies: Granular, brownish-gray colonies, of rather dull luster, irregular in shape and thickened at the edges. Young colonies appear somewhat finely branched or woolly.
POTATO.	An abundant white, spreading layer, with dull, somewhat waxy luster. Edges somewhat thickened and whiter than central portion. Potato slightly darkened.
MILK.	Totally coagulated. Acidity: strong.
SMITH SOLUTION.	No gas produced. Slightly turbulent. Heavy sediment. Strongly acid.
NITRATE.	Is reduced to nitrite strongly. Turbid throughout.
BOUILLON.	Turbid throughout. Ring of growth at surface and slight amount of sediment.

MICROCOCCUS X.

OCCURRENCE.	Found in sour corn and on ears of green corn.
GENERAL CHARACTERS.	*Shape and arrangement:* Micrococci, occurring singly and in irregular clusters. *Size:* 1u in diameter. *Motility:* Not motile. *Spore formation:* Not observed. *Relation to temperature:* Develop well at 37½° C., slowly at 20° C. *Relation to air:* Aërobic and facultatively anaërobic. *Relation to gelatin:* Does not liquefy. *Color:* Non-chromogenic.
GELATIN.	*Stick culture:* Development slow; growth throughout somewhat raised at surface, forming a small button-like mass. *Plate culture:* Surface colonies: Sharp outline, raised above surface, concentric markings, bluish white in color, and of somewhat waxy luster. Submerged colonies: Small, spherical.
AGAR.	*Streak culture:* Growth closely follows line of inoculation. Bluish white, semi-translucent, lustrous, and moist. *Plate culture:* Surface colonies: Circular and somewhat dome-shaped. White in color. Develop in about three days. Submerged colonies: Small, oval or rounded.
POTATO.	Very meager growth at end of three days. White and shiny. After two weeks becomes dry and almost exact color of potato.
MILK.	Not coagulated. Acidity: slight.
SMITH SOLUTION.	No gas produced. Slight turbidity. Some sediment. Slightly acid reaction.
NITRATE.	Not reduced to nitrite.
BOUILLON.	Slight turbidity at end of twenty-four hours at 37½° C. On standing a heavy sediment is deposited and solution becomes clear.

BACILLUS U. VEGETATIVE STATE FROM BOUILLON. (Magnified 1,000 times.)

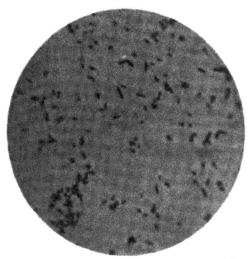

BACILLUS U. VEGETATIVE STATE FROM AGAR. (Magnified 1,000 times.)

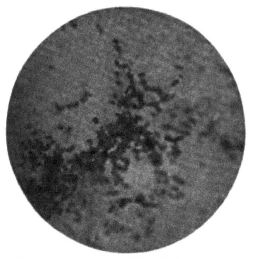

BACILLUS U. SPORES. (Magnified 1,000 times.)

PLATE 4.

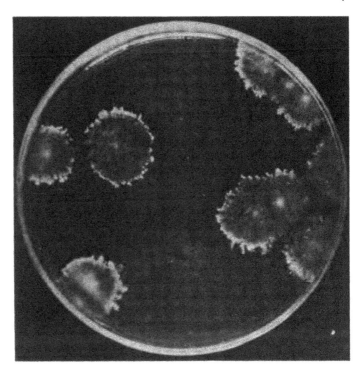

PLATE CULTURE, SHOWING COLONIES OF BACILLUS W
AT END OF 48 HOURS. (Actual size.)
Found under Husks of Green Corn.

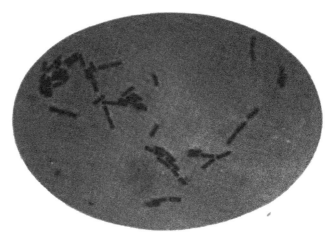

BACILLUS W.
(Magnified 1,000 times.)

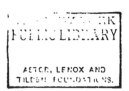

BACILLUS Y.

OCCURRENCE.	Found in cans of sour corn.
GENERAL CHARACTERS.	*Shape and arrangement:* Stout, thick rods, occurring singly and in chains. *Size:* 4–5μ x 2.5μ. *Motility:* Slow, serpentine motion; chains also motile. *Spore formation:* Oval spores. *Relation to temperature:* Develops more rapidly at 37½° C. than at 20° C. *Relation to air:* Aërobic and facultative anaërobe. *Relation to gelatin:* Liquefies rapidly. *Color:* Non-chromogenic.
GELATIN.	*Stick culture:* Growth throughout in one day, and liquefaction already begun at surface; much increased on second day, and flocculent precipitate in lower part of liquefied portion. Gelatin finally becomes entirely liquid, and heavy sediment formed. *Plate culture:* Surface colonies: Circular and rapidly liquefying; soon become covered with film. Saucer-shaped depression formed, at center of which is a flocculent, suspended mass, surrounded by ring of clear liquid. Submerged colonies: Small and inconspicuous.
AGAR.	*Streak culture:* Very thick, much-wrinkled layer, white and somewhat shiny. Edges finely scalloped. *Plate culture:* Surface colonies first appear as small dots; later form irregular spreading growths of varying thickness. Submerged colonies: Thin, blue, irregular in outline.
POTATO.	Thick, white layer, somewhat granular in appearance. On thickest portion of potato growth has a faint luster.
MILK.	Totally coagulated. Acidity: strong.
SMITH SOLUTION.	No gas produced. Turbid throughout. Strongly acid.
NITRATE.	Is reduced to nitrite strongly. Some turbidity.
BOUILLON.	Development slow. Liquid remains clear, but a film is formed on surface and a heavy, flaky sediment produced by sinking of film.

BACILLUS Z.

OCCURRENCE.	Found in cans of sour corn.
GENERAL CHARACTERS.	*Shape and arrangement :* Bacilli, occurring singly and in long chains. *Size :* 3.5–7μ x 2μ. *Motility :* Moves with slow serpentine motion. *Spore formation :* Small oval, centrally located spores. *Relation to temperature :* Develops rapidly at 37½° C., more slowly at 20° C. *Relation to air :* Aerobic and facultatively anaërobic. *Relation to gelatin :* Liquefying. *Color :* Non-chromogenic.
GELATIN.	*Stick culture :* Growth throughout and liquefaction at surface at end of first day. On second day liquefaction had spread to walls of tube, a wrinkled film was present on surface, and flocculent precipitate in suspension. *Plate culture :* Surface colonies develop rapidly. When about $\frac{1}{16}''$ in diameter liquefaction begins. A central area of flocculent material is present. At end of a week colonies are large, and are covered by a thick film of waxy appearance and showing concentric rings. Submerged colonies : Small.
AGAR.	*Streak culture :* A thick, white layer, of rather dull luster and finely granular appearance, covers the whole surface of the agar. *Plate culture :* Surface colonies : Irregular in outline, slightly thickened at the center. Brown and somewhat shiny. Some colonies show irregular outgrowths and appear woolly. Submerged colonies : Spread on lower surface of agar, forming a thin layer, which appears bluish by transmitted light.
POTATO.	Abundant white growth covering nearly the whole surface. Somewhat dull and granular in appearance.
MILK.	Totally coagulated in two days. Acidity : strong.
SMITH SOLUTION.	No gas produced. Turbid throughout. Sediment at bend ; ring of growth at surface. Strongly acid.
NITRATE.	Is reduced to nitrite strongly. Slight turbidity.
BOUILLON.	Strongly turbid throughout at end of twenty-four hours at 20° C. Growth at surface and flaky sediment when grown at 37½° C.

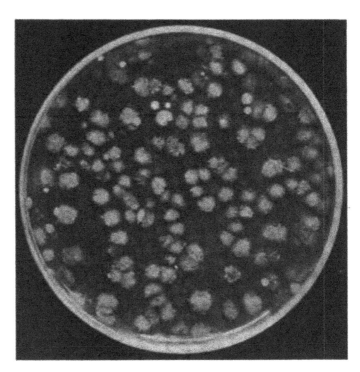

PLATE CULTURE, SHOWING COLONIES OF BACILLUS W
AT END OF 24 HOURS. (Actual size.)
Found in Cans of Sour Corn.

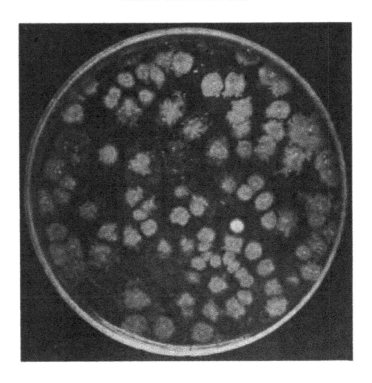

PLATE 6.

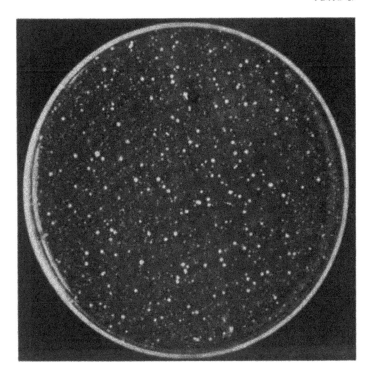

PLATE CULTURE, SHOWING COLONIES OF MICROCOCCUS X.
(Actual size.)
Found under Husks of Green Corn.

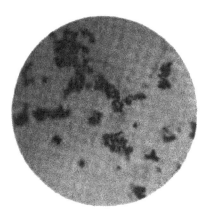

MICROCOCCUS X.
Taken from Above Colonies.
(Magnified 1,000 times.)

RESULTS OF TESTS MADE IN THE ENGINEERING LABORATORIES.

IX.

STEAM AND HYDRAULICS.

DESCRIPTION AND RESULTS OF TWO TESTS ON NO. 1 BABCOCK & WILCOX BOILER WITH HAWLEY DOWN DRAUGHT FURNACE, AT THE MASSACHUSETTS INSTITUTE OF TECHNOLOGY, JANUARY 13, 14, 15 AND 16, 1897.

THESE tests, which served as a part of the regular work of the Engineering Laboratories, were made by the students under the direction of the regular instructing force of the laboratories.

The tests were divided up into watches of about eight hours each, six men working in each watch.

The different stations were as follows :

STATIONS.

(1) Coal, ash, log. Temperature of room. Temperature of outside air. Boiler pressure every fifteen minutes. Barometer.

(2) Weight of water fed.

(3) Calorimeter every thirty minutes, ten-minute test.

(4) Gas analysis base of stack. Sample taken on the hour. Temperature of feed water every fifteen minutes.

(5) Gas analysis every hour, taken on the half-hour. Draught and temperature of flue at base of stack every thirty minutes.

(6) Draught and temperatures at different landings above basement every hour.

ARRANGEMENTS FOR THE TEST.

With the exception of the feed pipe, all pipes entering the water space of the boiler were disconnected and blanked, thus preventing all chance of leakage. The boiler was fed by a duplex pump connected with this boiler only.   The suction of this feed pump was taken from a barrel to which water was supplied from a second barrel placed on scales directly over the first barrel.   An auxiliary pump served to feed boiler No. 3, and also to pump the returns from the heating system into the barrel on scales.   The temperature of the feed water was taken from a thermometer inserted in a cup screwed into the feed pipe close to the boiler.   A throttling calorimeter was connected to the steam main about 7 feet from the boiler, the quality of the steam being determined every thirty minutes.   Samples of flue gases were taken at frequent intervals from the chimney about 8 feet above the grate.

Seven hundred and fifty pounds of coal were weighed out every hour and dumped in front of the boiler.

A sample of coal was taken from each barrow as it left the pocket, and the per cent. of moisture in the coal was determined from these samples.   The coal was thoroughly wetted before firing.

All gauges and thermometers used were tested, and their errors determined.   The weighing scales were verified with sealed weights.

In connection with the test on the boiler, readings of temperature and draught were taken at five different levels in the chimney.   The chimney is unlined 3' x 3' inside with 8" walls, and is 102 feet in height above the grate.   All but about 15 feet of the chimney is inside the building.

Observations as to the amount and density of the smoke issuing from the chimney were taken by one of the instructing staff.

Starting and Ending the Test.

The tests were started with the boiler running at its usual rate and with good fires on the grate.   At the time of starting the pressure and water level were observed, and the thickness of the fires both on the upper and lower grates noted.   At the end of the tests the fires were brought, as near as could be judged, to the same condition and to the same thickness as at starting.

Remarks.

The test was divided into two parts.   During the first thirty-one hours New River coal was used; during the last forty-eight hours Pocahontas coal was used.   The New River was of the same grade that had been furnished to the Institute. The Pocahontas was of good quality, the greater part being in lumps.

During the day watches, the firing was done by an expert fireman furnished by the Hawley Furnace Company. During the night, the firing was done by the regular fireman.

Fires were cleaned three hours before starting each test and three hours before ending each test, and at such other times as seemed advisable. Tubes were blown each night at about midnight.

It was shown by the gas analysis that a great excess of air was admitted, the per cents of O and CO_2 being reversed from what they should have been. Probably a large part of this excess air leaked in around the tiles supported on the boiler tubes directly over the upper grate.

The greatest possible error of the first test is not greater than $\frac{8}{10}$ of 1 per cent. The greatest possible error of the second test is not over $\frac{6}{10}$ of 1 per cent.

BOILER TEST.

Date, January 13-14, 1897.

Duration of test	31 hrs.
Average pressure of air	14.7 lbs.
Average pressure of steam	109.4 lbs.
Average temperature of feed water	123.26° F.
Coal used was New River	New River.
Moisture in coal (2.9 %)	2.9 %
Boiler was 208 H. P. (rated) B. & W., with Hawley furnace	B. & W.
Grate surface of upper grate	34.22 sq. ft.
Heating surface	2291.0 sq. ft.
Heating surface to grate surface	67 to 1
Pounds coal weighed out	23,250 lbs.
Dry coal burned	22,576 lbs.
Refuse from coal	1,438 lbs.
Combustible	21,138 lbs.
Quality of steam (dry steam = 1)	.992
Total water pumped into the boiler	221,203 lbs.
Equivalent water evaporated from and at 212°	249,296 lbs.
Equivalent water evaporated into dry steam from and at 212° F. per pound of dry coal	11.04 lbs.
From and at 212° per pound of combustible	11.79 lbs.
Coal per square foot upper grate per hour	21.9 lbs.
Water from and at 212° per square foot heating surface per hour	3.51 lbs.
Boiler H. P. developed (A. S. M. E. rating)	233 H. P.
Per cent. excess over rating (208)	12. %

BOILER TEST.

Date, January 14, 15, 16, 1897.

Duration of test	48 hrs.
Average pressure of air	14.7 lbs.
Average gauge pressure	113.4 lbs.
Average temperature of feed water	120.74° F.
Coal used was Pocahontas	Pocahontas.
Moisture in coal (1.9 %)	1.9 %
Boiler was 208 II. P., B. & W., with Hawley furnace	B. & W.
Grate surface of upper grate	34.22 sq. ft.
Heating surface	2291 sq. ft.
Heating surface to grate surface	67 to 1
Pounds coal weighed out	36,000 lbs.
Dry coal burned	35,316 lbs.
Refuse from coal	1,350 lbs.
Combustible	33,966 lbs.
Quality of steam (dry steam = 1)	.992
Total water pumped into boiler	349,767 lbs.
Equivalent water evaporated from and at 212° F.	395,326 lbs.
Equivalent water evaporated into dry steam from and at 212° per pound of dry coal	11.19 lbs.
From and at 212° per pound of combustible	11.64 lbs.
Coal per square foot of upper grate per hour	21.9 lbs.
Water from and at 212° per square foot heating surface per hour	3.60 lbs.
Boiler H. P. developed (A. S. M. E. rating)	239 H. P.
Per cent. excess over rating (208)	14.9 %

SMOKE TESTS.

These observations were taken on January 14 and 15, 1897, to determine the grade of smoke issuing from the chimney of No. 1 Babcock & Wilcox boiler, with Hawley furnace.

On account of the difficulty of estimating the grade of smoke by the eye alone, it seemed best to rate the smoke by means of some device which would eliminate personal judgment as far as possible.

The apparatus used was made up of pieces of smoked glass so arranged in a fixed frame that the observer could, with but little motion of his head, observe the smoke through one, two, three, four, five or six thicknesses of the glass.

The observations, which were taken from the top of the Walker Building at the Massachusetts Institute of Technology about 200 feet from the chimney, consisted in noting the number of thicknesses of the glass which just obscured the smoke, and the period of time that the smoke was covered by any given number of thicknesses.

Although the method is rather unsatisfactory, and the apparatus crude, yet it gives us a means of telling at any time, more accurately than could be told by the eye alone, how the amount of smoke compares with that noted during the test.

A smoke visible through four thicknesses of glass, but not visible through five, would be recorded as in grade No. 5. No. 6 denotes very black smoke; No. 4 corresponds to what is indefinitely called 10 per cent. smoke. A few of the observations taken are given in the following table :

NEW RIVER COAL.

January 14, 1897. Sky, light gray.

Time smoke noticed.	No. of glasses.	Duration.		Time smoke noticed.	No. of glasses.	Duration.	
h. min. sec.		min.	sec.	h. min. sec.		min.	sec.
1 32 0	3	..		38 30	6	..	15
32 30	2	..	30	38 40	5	..	10
33 20	3	..	50	38 50	4	..	10
33 35	0	..	15	39 5	3	..	15
34 5	5	..	30	39 20	2	..	15
34 15	4	..	10	40 10	0	..	50
34 25	1	..	10	40 45	3	..	35
34 35	5	..	10	41 05	2	..	20
34 45	4	..	10	41 15	4	..	10
35 5	0	..	20	41 30	2	..	15
38 5	4	3	0	41 40	0	..	10
38 15	5	..	10				

SUMMARY OF SMOKE RESULTS. NEW RIVER COAL. January 14, 15, 1897.

Time, 1.32 to 4.02 P. M., or 150 minutes interval. Sky, light gray.

PER CENTS. OF TOTAL TIME THAT THE SMOKE WAS COVERED BY EACH OF THE DIFFERENT NUMBERS.

No. 0   23.4 %
1   1.9 %
15.1 %
16.6 %
24.5 %
14.8 %
3.7 %
100.0 %

SUMMARY OF SMOKE RESULTS. POCAHONTAS COAL.

January 15, 1897.

Observations were taken from 9.26 A. M. to 11.56 A. M.		Observations were taken from 1.35 P. M. to 4.05 P. M.	
No. o	41.0	No. o	47.2
1	14.0	1	16.1
-	10.0	-	7.8
	8.4		12.8
4	14.7	4	10.4
5	10.2	5	4.9
6	1.7	6	0.8

BRIEF SUMMARY OF BOILER TEST RESULTS NEEDED FOR SMOKE COMPARISON.

Area upper grate (square feet)	34.22	34.22	
Area lower grate (square feet)	35.0	35.0	
Heating surface (square feet).	2291	2291	
Kind of coal used	New River.	Pocahontas.	
Coal fired each hour (pounds)	750	750	
Boiler rating (A. S. M. E.)	208	208	
Boiler H. P. developed	233	239	
Draught at base of stack (inches of water)	.614"	.599"	
Temperature of gases at base of stack	467° F.	453° F.	
Carbonic acid, per cent. by vol. in flue gases	7.30	7.92	
Carbonic oxide, per cent. by vol. in flue gases	0.10	0.12	
Free oxygen, per cent. by vol. in flue gases	12.32	11.47	
Time smoke observations taken	{ 1.32 to 4.02	{ 9 26 to 11.56	1.35 to 4.05
Condition of sky	Light gray.	Light gray.	Blue.
Per cent. of total time covered by No. o	23.4	41.	47.2
Per cent. of total time covered by No. 1	1.9	14.	16.1
Per cent. of total time covered by No. 2	15.1	10.	7.8
Per cent. of total time covered by No. 3	16.6	8.4	12.8
Per cent. of total time covered by No. 4	24.5	14.7	10.4
Per cent. of total time covered by No. 5	14.8	10.2	4.9
Per cent. of total time covered by No. 6	3.7	1.7	0.8

TESTS ON A RIDER COMPRESSION HOT AIR ENGINE.

Date.	Revolutions per minute.	M. E. P. working cylinder.	M. E. P. compression cylinder.	Total lift.	Apparent capacity. (Gals. per minute.)	Real capacity. (Gals. per minute.)	Horse power indicated.	Horse power by pump.	Mechanical Efficiency.
1896. Dec. 11	138.0	13.6	7.50	88.84	12.03	10.51	.812	.236	29.0
Dec. 8	128.3	13.1	8.02	96.0	11.18	9.67	.666	.234	35.2
Dec. 7	118.6	12.7	7.76	107.0	10.34	7.25	.582	.196	33.6
Dec. 9	128.3	12.3	7.51	108.0	11.18	9.42	.614	.257	41.8
Dec. 10	143.1	13.3	7.65	111.6	12.47	10.65	.786	.300	38.2
Dec. 9	135.8	13.0	7.57	115.4	11.84	10.09	.679	.294	43.3
Dec. 11	74.7	13.9	7.62	166.9	6.51	4.24	.454	.179	39.4

```
Diameter of working cylinder . . . . . . . . . . . . . . .  6.75″
Stroke of working cylinder . . . . . . . . . . . . . . .  9.53″
Diameter of compression cylinder . . . . . . . . . . . .  6.75″
Stroke of compression cylinder . . . . . . . . . . . . .  8.59″
```

TESTS ON AN ERICSSON HOT AIR ENGINE.

Date.	Revolutions per minute.	M. E. P.	Capacity. (Gallons per minute.)	Total lift.	Horse power indicated.	Horse power by pump.	Mechanical Efficiency.
1896. Dec. 1	100.8	4.56	5.53	44.3	.229	.062	27.0
Nov 24	93.9	4.30	4.97	47.9	.201	.060	30.0
Dec. 2	110.6	4.95	6.14	51.7	.272	.080	29.4

```
Diameter of cylinder . . . . . . . . . . . . . . . . .  8″
Stroke of piston . . . . . . . . . . . . . . . . . . .  3.92″
```

TESTS ON THE FLOW OF STEAM.

Experiments on the Flow of Steam Through an Orifice $\frac{1}{4}''$ in Diameter.

Date.	Boiler pressure.	Back pressure.	Barometer pressure.	Weight of condensed steam.	Steam per hour by experiment.	By Napier's formula.	Coefficient of flow by Napier's equation.	Ratio of absolute back pressure to absolute boiler pressure.
1896. Mar. 11	112.4	15.0	14.5	971.5	1,296	1,281	1.012	.2325
Dec. 2	70.3	7.8	14.8	391.5	870	859	1.013	.2656
Dec. 9	72.27	9.83	14.3	445.2	890	874	1.018	.2783
Dec. 9	71.59	10.96	14.3	353	883	867	1.019	.2934
Mar. 12	111.9	23.4	14.5	966	1,288	1,276	1.009	.2998
Mar. 12	111.4	23.4	14.5	968.5	1,291	1,272	1.015	.3009
Nov. 9	68.20	10.77	14.5	56	844	835	1.012	.3060
Nov. 4	51.9	8 5	14.7	343	686	672	1.021	.3483
Oct. 28	49.3	7.8	14.8	330	661	647	1.022	.3526
Oct. 21	51.7	9.15	14.6	339	678	669	1.014	.3575
Mar. 2	109.6	35.6	14.6	955	1,273 5	1,254	1.016	.4042
Mar. 4	111.9	38.2	14.6	975	1,300	1,277	1.017	.4174
Oct. 28	51.4	14.0	14.7	238	680	667	1.020	.4327
Mar. 15	93.8	35.2	14.7	560	1,120	1,095	1.023	.4550
Feb. 11	95.7	37.2	14.7	567	1,134	1,115	1.017	.4701
Feb. 12	94.8	40.2	14.6	563	1,126	1,105	1.019	.5008
Feb. 19	101.0	44.8	14.5	896	1,195	1,168	1.024	.5135
Nov. 4	48.23	18.07	14.7	324	649	635	1.022	.5214
Oct. 27	48.66	20.03	14.6	328	657	639	1.028	.5467
Nov. 4	49.2	21.1	14.7	330	661	645	1.024	.5603
Mar. 15	111.3	56.7	14.6	129.0	1,290	1,272	1.015	.5662
Feb. 23	91.6	48.4	14.4	821	1,095	1,070	1.023	.5926
Mar. 8	111.7	51.7	14.8	975	1,300	1,277	1.017	.6047
Feb. 24	100.4	55.3	14.6	592	1,184	1,161	1.020	.6081
Oct. 19	50.8	27.3	14.5	669	669	646	1.035	.6400
Dec. 9	70.27	44.52	14.3	330	826	801	1.031	.6958
Oct. 28	51.0	33.0	14.7	313	626	603	1.038	.7259
Oct. 21	51.9	33.8	14.5	324	648	609	1.064	.7273
Oct. 28	41.1	26.5	14.8	267	535	506	1.058	.7391
Dec. 2	64.7	44.1	14.8	380	760	718	1.059	.7408
Nov. 18	71.67	50.89	14.6	403	806	760	1.059	.7590
Oct. 21	52.1	37.2	14.6	306	613	572	1.070	.7766
Nov. 18	73.95	63.94	14.6	314	628	580	1.082	.8861

The apparatus used was made of 3″ pipe arranged as shown by the cut on page 359 of *Technology Quarterly*, Vol. V, No. 4, December, 1892.

RESULTS OF TESTS ON A 9″–24″ x 30″ COMPOUND ENGINE.

Number.	Horse power.	Steam per horse power per hour.	Steam through cylinders per hour.	Water by Jackets.			Revolutions per minute.	Boiler pressure by gauge.	Vacuum in condenser. (Inches of mercury.)	Barometer. (Ins.)
				High jacket.	Second receiver jacket.	Low jacket.				
1	69.11	19.31	1,334	....	....	....	80.87	105.8	26.5	28.88
2	75.33	19.56	1,477	....	....	....	80.25	105.8	26.5	29.84
3	77.49	19.42	1,512	....	....	....	80.22	105.8	26.2	29.20
4	51.87	19.95	1,035	....	....	....	82.72	119.3	24.7	29.56
5	54.89	19.34	1,061	....	....	....	83.25	119.6	25.4	29.80
6	57.00	19.27	1,098	....	....	....	82.75	119.6	24.9	29.50
7	63.89	18.49	1,181	....	....	....	81.88	119.2	26.0	29.50
8	65.36	18.30	1,196	....	....	....	81.62	119.6	25.4	29.48
9	66.24	18.07	1,197	....	....	....	82.40	121.0	26.4	29.32
10	70.02	18.17	1,272	....	....	....	81.75	119.3	26.1	29.46
11	71.27	18.14	1,293	....	....	....	81.30	119.2	25.1	29.61
12	81.41	17.46	1,421	....	....	....	81.28	123.9	26.4	29.95
13	81.78	17.62	1,443	....	....	....	81.00	125.5	26.3	29.74
14	82.92	17.40	1,469	....	....	....	81.38	124.8	26.5	29.63
15	87.07	18.01	1,568	....	....	....	80.89	119.6	24.7	29.62
16	69.2	17.24	1,016	....	....	177.1	82.83	120.6	26.3	30.17
17	72.08	16.11	1,006	....	....	155.1	82.90	120.7	26.3	30.00
18	72.61	16.21	1,019	....	....	163.4	83.20	118.9	26.9	29.40
19	93.68	16.24	1,354	....	....	167.3	81.48	119.2	26.0	29.50
20	98.32	16.73	1,436	....	....	208.6	81.12	119.5	26.3	29.39
21	73.14	16.16	990	46.0	....	145.7	82.57	119.2	25.7	29.32
22	63.07	15.96	844.5	42.0	37.5	82.8	82.73	100.2	26.6	29.49
23	63.19	15.59	820.0	39.2	51.5	74.5	82.90	101.0	26.6	29.49
24	78.13	16.16	1,059.0	42.6	72.8	88.5	81.72	100.9	26.3	29.49
25	80.47	16.14	1,086.0	44.2	80.1	87.9	81.37	101.4	26.3	29.49

RESULTS OF TESTS ON A 9″–24″ x 30″ COMPOUND ENGINE — *Continued.*

HIGH.

Number.	Initial pressure.	Per cent. of cut-off.	Pressure at cut-off.	Pressure at release.	Pressure at compression.	Per cent. of steam in cylinder at cut-off.	Per cent. of steam in cylinder at release.	M. E. P. crank end.	M. E. P. head end.	Horse power.
1	94.6	32.6	86.4	26.2	3.8	70.92	77.51	60.9	57.0	44.45
2	94.6	39.0	89.8	32.2	7.3	76.11	79.61	64.5	61.1	47.00
3	94.0	40.8	87.2	33.5	6.8	74.76	79.45	62.8	63.5	47.04
4	117.0	14.4	105.3	15.1	2.0	60.46	72.14	53.1	43.1	37.02
5	117.8	13 7	110.0	15.5	1.2	61.08	75.58	52.6	49.5	39.63
6	117.4	16.1	108.9	17.0	1.9	62.28	75.76	53.5	51.3	40.46
7	117.5	20.0	108.0	21.1	3.4	66.55	77.24	58.8	55.6	43.68
8	116.1	21.5	112.2	23.0	3.8	71.03	80.90	61.1	56.3	44.66
9	118.1	20.4	111.1	21.4	2.6	70.02	78.33	59.4	59.3	45.65
10	117.8	23.0	107.1	25.1	3.2	67.97	80.63	64.9	59.8	47.52
11	118.5	24.7	114.3	26.7	4.7	73.60	81.63	64.3	62.0	47.88
12	121.6	28.4	113.8	30.9	8.1	74.83	81.00	69.5	68.9	52.50
13	121.8	28.1	114.5	30.1	7.5	76.37	78.39	70.0	66.9	51.71
14	122.0	30.0	110.9	31.3	6.7	74.40	79.20	71.6	69.8	53.68
15	118.0	33.5	108.8	36.1	3.4	73.85	82.05	76.6	74.3	56.92
16	118.4	15.2	110.4	16.7	8.4	65.97	79.59	44.9	44.4	34.51
17	117.9	14.5	110.7	16.3	7.2	65.88	80.44	47.2	46.5	36.25
18	116.4	15.0	107.7	15.8	7.4	64.78	77.18	46.3	42.9	34.60
19	115.3	25.6	108.3	27.6	3.0	69.79	80.42	68.0	64.9	50.49
20	116.9	29.9	110.1	31.5	4.1	74.71	81.95	71.3	69.4	53.24
21	117.9	15.8	113.2	18.3	5.4	69.58	85.10	53.5	46.8	38.58
22	96.5	16.3	92.3	12.8	3.0	70.18	83.98	43.4	37.9	31.41
23	97.9	15.6	94.1	12.6	4.1	71.11	85.59	42.2	37.5	30.77
24	97.4	24.0	93.0	20.6	4.5	73.60	84.97	52.2	49.2	38.63
25	98.3	25.4	93.7	21.4	5.5	76.24	83.73	53.2	50.4	39.30

RESULTS OF TESTS ON A 9″-24″ x 30″ COMPOUND ENGINE—*Concluded.*

					Low.							
Number.	Initial pressure.	Per cent. of cut-off.	Pressure at cut-off.	Pressure at release.	Pressure at compression.	Per cent. of steam in cylinder at cut-off.	Per cent. of steam in cylinder at release.	M. E. P. crank end.	M. E. P. head end.	Horse power.	B. T. U. per horse power per minute actual.	B. T. U. per horse power per minute reduced to a like standard.
1	+0.9	11.5	−0.9	−10.3	−11.4	43.59	64.40	4.44	4.13	24.66	354.6	372.2
2	+3.5	11.5	+0.8	−10.2	−11.6	42.20	65.61	5.30	4.98	28.33	356.3	361.9
3	+4.2	22.2	1.1	− 9.9	−11.3	45.14	63.48	5.92	5.14	30.45	355.5	362.4
4	−4.8	20.7	−4.0	−10.7	−10.8	41.38	78.68	2.75	2.48	14.85	358.8	345.7
5	−2.7	13.7	−5.2	−11.3	−11.4	40.49	68.25	2.90	2.44	15.26	349.3	346.3
6	−1.5	11.0	−3.6	−11.1	−11.6	42.89	66.42	3.03	2.79	16.54	347.1	338.5
7	−0.1	11.7	−4.0	−11.2	−12.0	39.17	62.32	3.97	3.22	20.21	336.7	343.4
8	+0.5	11.0	−3.1	−10.7	−11.7	39.91	69.97	4.05	3.34	20.70	331.1	331.1
9	−1.3	14.0	−4.4	−10.8	−11.9	41.17	67.19	3.80	3.48	20.59	331.2	344.7
10	0.0	13.7	−3.7	−10.8	−11.8	40.45	64.62	4.18	3.84	22.50	331.0	336.7
11	+2.2	11.0	−2.0	−10.6	−11.5	40.41	66.18	4.58	3.80	23.39	327.2	333.4
12	+3.4	14.0	−1.2	−10.3	−11.7	45.26	67.20	5.35	5.01	28.91	317.4	321.5
13	+3.8	14.0	−0.8	−10.3	−11.7	41.85	66.15	5.53	5.27	30.07	320.5	325.3
14	+3.3	14.0	−1.4	−10.2	−11.5	43.01	64.27	5.59	4.88	29.24	323.0	328.9
15	+2.3	22.0	−1.4	− 9.8	−10.7	48.69	64.71	5.43	5.16	30.15	323.9	317.6
16	+4.9	25.0	+1.0	−10.3	−11.7	76.44	95.41	6.37	5.83	34.60	302.8	304.8
17	+2.2	31.0	−1.6	−11.7	−12.3	80.44	79.80	6.58	6.01	35.83	284.3	308.6
18	+4.2	37.0	−0.3	−10.7	−12.2	78.99	81.75	6.93	6.38	38.01	289.2	304.4
19	+2.2	35.7	−2.0	− 9.4	−11.6	81.21	82.35	7.74	7.70	43.19	288.8	292.5
20	+2.2	36.3	−1.5	− 8.9	−11.6	79.97	85.00	8.00	8.18	45.08	297.5	303.1
21	+2.4	31.5	−2.2	−10.6	−11.6	69.79	84.05	6.37	5.82	34.56	285.7	289.0
22	+1.5	18.0	−0.5	−11.1	−12.4	91.94	87.62	5.95	5.17	31.66	282.1	294.7
23	+1.8	18.0	−0.3	−11.2	−12.3	95.20	88.19	6.01	5.38	32.42	275.1	287.4
24	+2.8	24.7	−0.1	−10.2	−12.0	92.13	88.57	7.38	6.70	39.90	284.4	290.4
25	+2.5	24.7	+0.5	−10.2	−12.0	92.38	87.86	7.65	7.09	41.17	283.8	289.7

TESTS ON HANCOCK LOCOMOTIVE INSPIRATOR NO. 4, C.

Number of test.	Duration. (Minutes.)	Lift. (Feet.)	Boiler pressure. (Lbs. per sq. in.)	Delivery pressure. (Lbs. per sq. in.)	Barometer. (Lbs. per sq. in.)	Suction temperature. C°.	Delivery temperature. C°.	Water pumped. (Lbs. per hour.)	Water delivered. (Lbs. per hour.)	Steam used. (Lbs. per hour.)	Water delivered. (Per lb. of steam.)	Delivery capacity. (Gals. per hour.)
71	30	3.94	90.7	99.6	14.4	5.8	55.8	6,420	6,958	538	12.9	846
75	30	4.02	95.6	99.6	14.6	13.3	64.9	6,430	6,984	554	12.6	853
76	30	4.01	100.6	106.0	14.5	5.8	58.4	6,642	7,326	584	12.4	870
77	30	4.01	100.6	110.0	14.6	6.1	58.7	6,676	7,276	600	12.1	886
78	30	4.03	101.7	114.0	14.5	6.2	59.2	6,652	7,340	588	12.3	882
79	30	4.01	105.2	116.2	14.8	5.5	59.2	6,759	7,365	606	12.1	897
80	30	4.03	105.6	112.0	14.2	11.0	65.1	6,581	7,181	600	12.0	878
81	30	4.04	109.9	113.2	14.7	5.7	60.7	6,840	7,456	616	12.1	909
83	30	4.04	110.5	118.8	14.5	7.4	62.4	6,781	7,424	643	11.6	905
84	30	4.05	110.5	119.3	14.5	14.0	69.3	6,676	7,312	636	11.5	896
86	29	4.04	110.6	119.0	14.7	6.6	61.3	6,829	7,465	636	11.7	910
87	30	4.02	110.6	120.5	14.5	11.9	66.5	6,743	7,382	639	11.5	903
88	30	4.03	110.6	120.5	14.5	10.5	65.3	6,746	7,392	646	11.4	903
89	30	4.	110.6	120.5	14.5	9.9	64.8	6,474	7,408	634	11.7	905
90	30	4.	110.7	118.7	14.8	9.4	64.1	6,804	7,444	640	11.6	909
91	30	4.	110.7	118.8	14.8	9.8	64.7	6,812	7,459	647	11.5	911
94	30	4.	115.2	123.8	14.3	6.1	62.4	6,891	7,538	648	11.7	919
95	30	4.02	115.6	128.7	14.6	14.2	69.8	6,796	7,431	635	11.7	910
96	30	4.02	115.6	123.0	14.7	11.4	67.4	6,856	7,522	666	11.3	921
97	30	4.02	115.6	124.9	14.6	9.0	65.2	6,885	7,567	682	11.1	924
99	30	4.04	115.7	125.0	14.3	7.1	63.6	6,892	7,552	660	11.4	922
101	30	4.02	115.7	127.6	14.7	9.3	65.3	6,908	7,584	676	11.2	927
102	30	4.03	115.7	125.4	14.7	9.9	65.8	6,874	7,538	664	11.4	921
103	30	4.01	115.8	124.7	14.2	7.2	63.3	6,840	7,504	664	11.3	916
107	30	4.02	115.8	126.9	14.7	8.9	65.1	6,914	7,587	673	11.0	927
108	30	4.02	115.9	125.9	14.5	9.9	66.0	6,862	7,521	659	11.4	919
109	30	4.03	116.0	125.7	14.6	6.2	62.6	6,923	7,589	666	11.4	926

Hirn's Analysis Applied to a Receiver Jacketed Cross Compound Engine.

By the application of Hirn's analysis to an engine, the interchange of heat between the steam and the walls of the cylinder can be calculated for different parts of the stroke.

To carry on such a test, it is necessary to make what may be called a heat balance test. This requires, in addition to the observations usu-

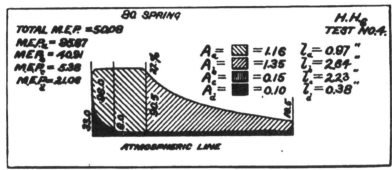

FIG. 1.

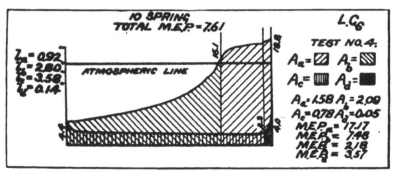

FIG. 2.

ally taken for simple engine tests, a knowledge of the quality of the steam entering the cylinder, the weight and gain in temperature of the condensing water, and the weight and temperature of the condensed steam leaving the condenser. The work as shown by the indicator cards is broken up into four parts, that during admission, during exhaust, during expansion, and during compression.

Figures 1 and 2 represent cards taken during test No. 4; Figure 1 from the head end of the high-pressure cylinder, and Figure 2 from the crank end of the low-pressure cylinder.   The different cross-hatchings show the areas which have to be planimetered in order to calculate the work for these several parts of the stroke.

The heat equivalents of the work done per stroke during admission, expansion, exhaust, and compression are represented by $A W_a$, $A W_b$, $A W_c$, and $A W_d$ respectively, where W represents the foot-pounds of work done, and A is the heat equivalent of 1 foot-pound, or $\frac{1}{778}$ B. T. U. In calculating the different values of W for the high-pressure cylinder cards, the mean effective pressures above the absolute zero must be .used.   It is more convenient to add the pressure of the atmosphere in pounds to the M. E. P. calculated above the atmospheric line, than to draw in the absolute zero line, as has been done on the low-pressure cards.

The following equations are made use of in the calculations.   These equations are quoted from an article printed in the *Technology Quarterly*, Vol. IV, No. 3, October, 1891, by Professor C. H. Peabody.

' The heat taken inside the cylinder per stroke is $M (xr + q)$, M being the weight of condensed steam per stroke, x being the quality of the entering steam (or 1 — priming), q and r the heat of the liquid and the total latent heat respectively. at the absolute steam pressure at the throttle.   The heats Q_J, Q_{JR}, and Q'_J supplied by the jackets on the high-pressure cylinder, intermediate receiver, and low-pressure cylinder, are calculated by multiplying the weight per stroke condensed in each by $x r$; the engine being jacketed with full boiler pressure.   The heats Q_e, Q_{eR}, and Q'_e, lost from the jackets by radiation to the air, are figured from the rates of condensation in the jackets noted immediately after stopping the engine.

The heat equivalent of the intrinsic energy of the mixture in the cylinder at the different points of the stroke is

at beginning of admission, $I_0 = M_0 (q_0 + x_0 \rho_0)$
at cut-off . . . . . . $I_1 = (M + M_0) (q_1 + x_1 \rho_1)$
at release . . . . . . $I_2 = (M + M_0) (q_2 + x_2 \rho_2)$
at compression . . . . $I_3 = M_3 (q_3 + x_3 \rho_3)$

where M_0 is the average weight of steam in the cylinder at compression or admission, figured by assuming the steam to be dry at compression.

The different values of q and ρ, the heat of the liquid and the internal latent heat, are taken at the average absolute pressure measured at the proper points on a pair of cards.

The per cent. of the weight of mixture present accounted for as steam at cut-off or at release, denoted by x_1 or x_2, is found as follows: The average volume at cut-off is $V_1 + V_0$ where V_1 is the mean piston displacement up to cut-off and V_0 is the average clearance. This volume is occupied by $(M + M_0)$ pounds of the mixture of steam and water. The volume of one pound of a mixture is $x s + (1 - x) \frac{1}{62.4}$ where s is the volume of a pound of steam. This may be reduced to $x (s - .016) + .016$, or $x u + .016$.

Then $(M + M_0) (x_1 u_1 + .016) = V_1 + V_0$ from which x_1 can be calculated. The value u_1 or $(s_1 - .016)$ is taken from the tables at the mean absolute pressure at cut-off. x_0 and x_3 are assumed to be unity. A considerable error in this assumption would affect the results but slightly. The heat carried away per stroke by the condensed steam and the condensing water is $M q_4 + G (q_4 - q_i)$; where q_4, q_4 and q_i are the heats of the liquid corresponding to the temperatures of the condensed steam, the hot and the cold condensing water, and where G is the weight of cooling water per stroke.

The accuracy of the test may be checked by noting the agreement between the radiation loss as determined by experiment, and the heat unaccounted for by the following equation: "Heat unaccounted for"
$$= Q + Q_J + Q_{JR} + Q'_J - M q_4 - G (q_4 - q_i) - A W_{high} - A W_{low}$$
In all tests excepting No. 4 this agreement is very good.

The interchanges of heat for the high-pressure cylinder may be written as follows:

during admission $\quad Q_a = Q + I_0 - I_1 - A W_a$

during expansion $\quad Q_b = I_1 - I_2 - A W_b$

during exhaust $\quad Q_c = I_2 - I_3 - Q - Q_J + Q_e + A W_{high} + A W_c$

during compression $Q_d = I_3 - I_0 + A W_d$

It is evident if there was no interchange during admission that $Q + I_0 = I_1 + A W_a$. On account of the interchange, the term Q_a, necessary to balance the two sides, may be given either a positive or a negative sign. As applied in these equations, the positive sign indicates heat given up by the walls, and the negative sign heat absorbed by the walls.

In the equation for Q_e, the interchange during exhaust, the expression $- (Q + Q_f - Q_e - A W_{high})$ represents the heat carried away by the exhaust.

The interchanges of heat for the low-pressure cylinder are:

during admission $Q'_a = Q + Q_f - Q_e - A W_{high} + Q_{JR} - Q_{eR} + I'_0 - I'_1 - A W'_a$

during expansion $Q'_b = I'_1 - I'_2 - A W'_b$

during exhaust $Q'_c = I'_2 - I'_3 - M q_4 - G(q_4 - q_2) + A W'_c$

during compression $Q'_d = I'_3 - I'_0 + A W'_d$

The following tests were conducted and calculated by the students as a part of the regular work of the Engineering Laboratories.

CRANKS SET WITH LOW LEADING THE HIGH BY 120°.

High-Pressure Cylinder.

Dia. of piston = 8.99″.   Dia. of piston rod = 2.19″.   Stroke = 30″.
Piston displacement: H. E. = 1.102 cu. ft.; C. E. = 1.037 cu. ft.
Clearance in % of P. D.: H. E. = 8.83; C. E. = 9.76.
Engine constant: H. E. = .004809; C. E. = .004524.

Low-Pressure Cylinder.

Dia. of piston = 24.063″.   Dia. of piston rod = 2.16″.   Stroke 30″.
Piston displacement: H. E. = 7.894 cu. ft.; C. E. = 7.831 cu. ft.
Clearance in % P. D.: H. E. = 12.18; C. E. = 12.27.
Engine constant: H. E. = .0345; C. E. = .03417.

Test Number.	I.	II.	III.	IV.
Duration of test, minutes	60	60	60	60
Total number of revolutions	4,976	4,974	4,903	4,882
Revolutions per minute	82.93	82.90	81.72	81.37
Steam consumption during test, pounds :				
Passing through cylinders	844.5	820.0	1059.0	1086.0
Condensation in h. p. Jacket	42.0	39.0	42.5	44.0
" in receiver Jacket	37.5	51.5	73.0	80.0
" in l. p. Jacket	82.8	74.5	88.5	88.0
Total	1006.8	985.0	1263.0	1298.0
Condensing water for test, pounds . . .	· 16,066	16,287	21,800	21,960
Priming, by calorimeter	1.0%	0.9%	1.13%	0.9%
Temperatures, Fahrenheit :				
Condensed steam	93.4	91.4	86.0	89.2
Condensing water, cold	52.2	52.2	52.5	52.7
Condensing water, hot	105.8	104.2	102.6	104.0
Pressure of the atmosphere, by the barometer, pounds per square inch	14.5	14.5	14.5	14.5
Boiler pressure, pounds per square inch, absolute	114.7	115.5	115.4	115.9
Vacuum in condenser, inches of mercury .	26 6	26.6	26.3	26.3
Events of the stroke, per cent. :				
High-pressure cylinder —				
Cut-off, crank end	16.8	15.8	23.8	25.2
head end	15.7	15.4	24.2	25.6
Release, both ends	100	100.0	100.0	100.0
Compression, crank end	7.5	7.5	7.5	7.5
head end	11.5	11.5	10.5	10.5
Low-pressure cylinder —				
Cut-off, crank end	18.0	18.0	24.5	24.5
head end	18.0	18.0	25.0	25.0
Release, both ends	100.0	100.0	100.0	100.0
Compression, crank end	3.5	3.5	3.5	3.5
head end	5.5	5.5	4.0	4.0
Absolute pressures in the cylinder, pounds per square inch :				
High-pressure cylinder —				
Cut-off, crank end	106.5	109.5	107.5	108.3
head end	107.1	107.7	107.5	108.1
Release, crank end	29.9	29.5	37.0	38.2
head end	24.7	24.8	33.2	33.7
Compression, crank end	17.2	17 7	18.0	18.7
head end	17.8	19.5	20.1	21.3
Admission, crank end	32.7	35.5	35.5	38.2
head end	42.2	42.5	45.2	46.9

Test Number.	I.	II.	III.	IV.
Low-pressure cylinder —				
Cut-off, crank end	14.4	14.5	14.7	15.1
head end	13.6	13.9	14.2	14.9
Release, crank end	3.4	3.3	4.2	4.2
head end	3.3	3.3	4.4	4.5
Compression, crank end	1.9	2.0	2.2	2.2
head end	2.4	2.5	2.8	2.9
Admission, crank end	3.4	3.5	4.1	3.7
head end	4.4	4.7	5.8	4.9
Heat equivalents of external work, B. T. U. from areas on indicator diagram to line of absolute vacuum :				
High-pressure cylinder —				
During admission, AW_a, crank end	3.59	3.38	5.13	5.45
head end	3.49	3.47	5.43	5.83
During expansion, AW_b, crank end	7.94	8.05	8.38	8.43
head end	7.81	7.97	8.48	8.57
During exhaust, AW_c, crank end .	2.86	2.96	3.11	3.28
head end .	2.91	3.03	3.18	3.54
During compres'n, AW_d, crank end	0.35	0.37	0.38	0.40
head end	0.67	0.72	0.72	0.72
Low-pressure cylinder —				
During admission, AW_a, crank end	4.13	4.15	5.93	6.11
head end .	4.07	4.13	5.99	6.25
During expansion, AW_b, crank end	7.19	7.25	7.78	7.99
head end .	7.05	7.12	7.88	8.16
During exhaust, AW_c, crank end .	2.64	2.55	2 87	2.98
head end .	3.25	3.10	3.98	4.00
During compres'n, AW_d, crank end	0.12	0 13	0.17	0.17
head end .	0.33	0.32	0.28	0.29
Quality of the steam in the cylinder. At admission and at compression the steam was assumed to be dry and saturated :				
High-pressure cylinder —				
At cut-off . . . . . x_1 . . .	70.18	71.11	73.60	76.24
At release . . . . . x_2 . . .	83.98	85.59	84.97	83.73
Low-pressure cylinder —				
At cut-off . . . . . x_1 . . .	91.94	95.20	92.13	92.58
At release . . . . . x_2 . . .	87.62	88.19	88.57	87.86

Test Number.	I.	II.	III.	IV.
Interchanges of heat between the steam and the walls of the cylinders, in B. T. U. Quantities affected by the positive sign are absorbed by the cylinder walls; quantities affected by the negative sign are yielded by the walls:				
High-pressure cylinder —				
Brought in by steam . Q . . .	99.82	97.10	126.90	131.30
During admission . . Q_a . . .	25.22	23.85	27.14	25.83
During expansion . . Q_b . . .	−15.01	−15.84	−16.13	−12.60
During exhaust . . . Q_c . . .	−11.13	−10.04	−12.99	−15.47
During compression . Q_d . . .	0.98	0.45	0.46	0.47
Supplied by jacket . . Q_f . . .	3.70	3.45	3.81	3.97
Lost by radiation . . Q_e . . .	1.87	1.87	2.20	2.20
Second intermediate receiver —				
Supplied by jacket . . Q'_{fR} . .	3.30	4.53	6.50	7.19
Lost by radiation . . Q'_{eR} . .	0.69	0.69	1.60	1.60
Low-pressure cylinder —				
Brought in by steam . Q'' . .	96.24	94.26	123.40	128.53
During admission . . Q''_a . .	7.63	5.44	9.63	10.57
During expansion . . Q''_b . .	− 1.75	− 1.12	− 2.03	− 0.97
During exhaust . . . Q''_c . .	−11.11	− 7.53	− 9.92	−16.53
During compression . Q''_d . .	0.33	2.21	0.11	0.64
Supplied by jacket . . Q''_f . .	7.30	6.56	7.91	7.85
Lost by radiation . . Q''_e . .	4.44	4.44	5.70	5.70
Total loss by radiation:				
By preliminary test . . ΣQ_e . .	7.39	7.39	9.52	9.52
By equation	7.00	5.78	7.83	4.42
Power and economy:				
Heat equivalents of works per stroke:				
High-pressure cylinder . AW . .	8.02	7.89	10.01	10 17
Low-pressure cylinder . AW'' .	8.10	8.27	10.04	10.53
Totals	16.12	16.16	20.05	20.70
Total heat furnished by jackets . . .	14.30	14.54	18.22	19.01
Distribution of work:				
High-pressure cylinder	1.00	1.00	1.00	1.00
Low-pressure cylinder	1.01	1.05	1.00	1.04
Total horse-power	63.07	63.19	78.13	80 47
Steam per horse-power per hour . . .	15.96	15.59	16 16	16.14
B. T. U. per horse-power per minute .	282.1	275.1	284.4	283.8
B. T. U. per horse-power per minute, 2 pounds absolute	294.7	287.4	290.4	289.7

TESTS ON DUPLEX PUMP AT MASSACHUSETTS INSTITUTE OF TECHNOLOGY.

Date.	Single strokes per minute.	Length of stroke (West).	Length of stroke (East).	Steam pressure by gauge.	Weight of condensed steam per hour (Lbs.)	Total head (Feet.)	M. E. P. Steam Cylinder West (Head end.)	West (Pump end.)	East (Head end.)	East (Pump end.)	Horse-power (Steam cyl. Indrs.)	Horse-power (Output.)	Efficiency (Per cent.)	Apparent capacity of pump (Gallons.)	Capacity of pump by lbs.	Steam per horse-power per hour.	B. T. U. per horse-power per minute.	Duty (Foot-pounds per 1,000,000 B. T. U.)
Nov. 10, 1896	181.3	11.4	11.2	40.6	2,544	162.3	33.30	30.88	34.50	33.25	34.03	28.6	84.0	753	698	74.8	1,237	22,400,000
Nov. 11, 1896	186.2	10.87	10.98	40.5	2,599	163.10	33.29	32.10	31.88	32.39	33.26	28.57	86	790	694	78.1	1,288	22,010,000
Nov. 11, 1896	190.6	11.11	11.79	45.3	2,781	184.7	37.2	37.7	36.8	38.1	41.2	34.7	84.3	809	745	67.6	1,117	24,910,000
Nov. 12, 1896	194.1	11.07	11.29	44.9	2,802	183.6	36.9	37.4	36.3	37.4	40.5	34.4	84.89	843	741	64.2	1,145	24,540,000
Nov. 16, 1896	108.8	11.19	11	49.11	1,978	229.9	45.03	45.02	46.09	45.5	27.6	23.2	84	444	767	71.6	1,176	33,599,000
Nov. 19, 1896	215.9	10.4	10.2	46.1	2,891	177.8	30.24	37.21	37.60	37.71	42.5	34.2	80.4	819	764	67.9	1,123	33,620,000
Nov. 20, 1896	191.2	10.6	10.5	34.9	2,275	154.5	32.3	29	29.8	28.9	30.4	26.7	87.9	742	687	74.8	1,231	33,570,000
Nov. 23, 1896	185.8	10.7	10.2	35.0	2,368	147.1	31.01	27.19	29.73	27.72	28.4	24.8	87.3	715	668	83.4	1,314	21,940,000
Nov. 30, 1896	106	11.78	11.86	53.4	2,008	261.7	49.60	45.5	47.7	49.1	30.3	26.8	88.6	643	666	66.3	1,097	26,660,000
Nov. 30, 1896	188.3	10.75	10.82	36.42	2,355	156	31.91	27.75	30.26	29.60	30.5	27.6	90.6	748	702	77.1	1,271	33,540,000
Dec. 2, 1896	190.5	11.15	11.05	40.8	2,718.8	172.9	34.68	34.25	34.59	31.61	35.8	32.1	89.8	778	737	75.9	1,251	33,449,000
Dec. 2, 1896	193.7	11.11	11.02	40.5	2,588.3	174.5	35.7	32.6	34.8	33.0	36.8	34.8	89.1	787	745	70.4	1,160	35,380,000
Dec. 3, 1896	221.0	10.7	10.5	50	3,120	204.3	42.2	40.0	41.5	38.6	47.9	41.1	84.2	861	801	67.02	1,109	35,670,000
Dec. 4, 1896	101.6	11.6	11.2	45.56	1,719	228.9	43.4	42.5	43.0	42.6	24.8	21.8	88.0	426	378	69.39	1,145	35,390,000
Dec. 4, 1896	94.5	11.5	11.5	40.11	1,526	204.9	38.31	37.41	36.80	37.52	20.6	18.2	88.5	400	353	73.1	1,204	44,240,000
Dec. 7, 1896	187.7	11.4	11.25	40.53	2,540.3	172.9	34.51	33.94	33.49	33.57	36.1	32.3	89.5	783	741	70.60	1,163	35,390,000
Dec. 9, 1896	152.1	11.11	10.93	25.0	1,583	116.2	21.8	21.1	22.0	21.6	18.2	17.2	94.5	617	586	86.98	1,425	21,870,000
Dec. 9, 1896	162	11.5	11.1	30.01	1,966	132.1	25.5	25.3	26.0	24.6	23.3	21.1	90.6	674	633	84.3	1,385	21,590,000
Dec. 10, 1896	152.0	11.12	11.01	24.91	1,617	115.2	22.67	21.57	22.24	21.93	18.8	16.8	89.4	619	578	108.2	1,408	20,950,000

TESTS ON PULSOMETER.

Date.	Total lift. (Feet.)	Steam pressure at pulsometer. (Gauge.)	Steam used per hour. (Pounds.)	Water pumped per hour including steam. (Pounds.)	Temperature at suction. °C.	Temperature at discharge. °C.	Work done by pulsometer. (Foot-pounds.)	Total heat given up by steam. (B.T.U.)	Efficiency. (Per cent.)	Duty, foot-pounds per 1,000,000 B.T.U. (Delivery.)	Duty, foot-pounds per 1,000,000 B.T.U. (Suction.)	Capacity. (Gallons per minute.)
1896												
Oct. 5	28.56	26.5	394.9	37,420	27.5	32.5	1,063,000	326,000	.42	3,260,000	3,229,000	74.6
Oct. 6	30.41	30.54	386.6	42,280	30.3	35.8	1,302,000	419,100	.40	3,107,000	3,070,000	85.2
Oct. 7	29.74	30.6	370.2	42,770	27.1	32.2	1,269,000	410,000	.39	3,095,000	3,070,000	91.7
Oct. 8	31.07	31.0	370.7	45,360	28.1	33.2	1,406,000	410,000	.44	3,439,000	3,408,000	90.8
Oct. 12	31.32	30.4	359.0	46,380	24.94	29.76	1,449,000	399,200	.44	3,631,000	3,594,000	92.8
Oct. 13	30.20	32.3	370.6	42,070	27.19	32.62	1,303,000	410,600	.41	3,173,000	3,140,000	86.0
Oct. 14	30.43	30.2	369.0	42,740	28.8	34.1	1,308,000	407,000	.41	3,186,000	3,150,000	85.5
Oct. 14	31.26	27.8	380.9	43,610	29.0	34.6	1,360,000	419,800	.40	3,240,000	3,210,000	87.3
Oct. 15	30.18	21.8	431.4	43,270	30.3	34.8	1,372,000	474,200	.34	2,682,000	2,650,000	84.6
Oct. 19	23.18	29.0	241.5	31,540	24.4	29.0	729,500	268,700	.35	2,714,000	2,695,000	63.1
Oct. 21	23.97	28.6	244.5	31,270	25.3	30.2	748,000	271,100	.35	2,758,000	2,730,000	62.6
Oct. 22	23.25	22.6	241.0	29,810	27.8	32.5	601,400	266,100	.33	2,508,000	2,570,000	59.6
Oct. 22	24.09	21.6	251.7	31,550	25.5	30.3	785,800	278,500	.36	2,821,000	2,799,000	63.1
Oct. 26	24.62	26.9	234.6	29,770	23.2	28.2	737,300	261,000	.36	2,885,000	2,802,000	59.6
Oct. 26	24.59	23.7	254.4	23,470	26.1	30.8	831,000	283,900	.37	2,892,000	2,909,000	66.9
Oct. 27	24.18	30.1	229.4	30,980	26.3	31.1	746,200	266,000	.36	2,805,000	2,789,000	62.0
Oct. 28	23.89	30.3	278.2	30,060	26.8	31.6	716,400	261,700	.35	2,737,000	2,715,000	60.1
Oct. 28	22.76	24.1	201.8	28,860	26.2	30.2	655,300	224,100	.38	2,922,000	2,900,000	57.7
Oct. 29	23.91	24.6	244.1	33,330	27.2	31.5	795,300	270,400	.38	2,943,000	2,920,000	66.7
Nov. 2	26.73	28.6	241.7	34,530	25.2	28.6	887,100	271,300	.42	3,269,000	3,295,000	69.1
Nov. 3	23.01	27.7	231.2	32,310	27.3	31.7	800,100	258,000	.40	3,128,000	3,104,000	64.6
Nov. 3	22.92	29.0	228.8	30,400	31.9	27.5	695,100	255,100	.35	2,744,000	2,726,000	60.8
Nov. 4	24.35	22.4	242.2	32,920	25.9	30.4	799,400	272,200	.38	2,956,000	2,937,000	65.9
Nov. 4	24.57	28.0	242.4	32,630	27.6	32.1	800,300	270,200	.38	2,984,000	2,962,000	65.3
Nov. 5	30.91	29.0	400.1	43,430	28.9	34.4	1,339,000	446,800	.40	3,160,000	3,132,000	86.9
Nov. 9	28.58	20.8	324.3	41,860	26.7	31.7	1,194,000	361,000	.43	3,334,000	3,308,000	83.7
Nov. 9	28.88	20.8	326.8	39,990	29.1	34.1	1,152,800	359,400	.41	3,207,000	3,183,000	80.0
Nov. 11	28.86	20.3	323.2	39,620	27.1	32.2	1,141,000	356,500	.41	3,201,000	3,178,000	79.3
Nov. 11	28.52	20.2	326.8	39,270	30.2	34.9	1,118,000	358,800	.40	3,115,000	3,093,000	78.6
Nov. 12	27.42	19.9	323.6	37,280	30.2	35.4	1,020,000	355,000	.37	2,873,000	2,850,000	74.6
Nov. 16	28.78	20.14	326.5	38,580	25.2	30.2	1,031,000	364,200	.36	2,854,000	2,831,000	77.2
Nov. 16	28.73	20.5	331.2	37,640	27.81	33.15	1,074,000	356,900	.39	3,038,000	3,010,000	75.7
Nov. 17	28.73	20.5	328.7	38,870	28.9	33.9	1,114,000	364,400	.39	3,080,000	3,056,000	77.8
Nov. 18	28.90	20.17	32.05	39,680	28.9	32.8	1,165,000	355,300	.42	3,290,000	3,270,000	79.4
Nov. 18	28.88	20.1	326.0	38,560	28.6	33.7	1,100,000	361,400	.39	3,068,000	3,044,000	77.2
Nov. 19	29.00	20.8	329.1	30,930	29.4	34.5	115,900	364,500	.41	3,304,000	3,180,000	79.9
Nov. 20	28.97	23.63	323.2	39,280	31.94	36.9	1,136,000	357,500	.41	3,203,000	3,178,000	78.6
Nov. 20	28.00	20.1	325.2	38,630	16.0	41.2	1,080,000	353,500	.39	3,055,000	3,030,000	77.3
Nov. 23	28.69	26.14	331.1	34,610	27.9	33.5	1,136,000	368,700	.47	3,656,000	3,623,000	69.2
Nov. 24	26.94	23.2	330.7	32,880	30.0	35.9	1,211,000	366,300	.42	3,334,000	3,306,000	65.7
Nov. 30	26.40	25.4	330.7	35,500	23.9	29.5	1,260,000	367,000	.44	3,434,000	3,402,000	71.0
Dec. 2	26.11	31.19	322.8	32,750	24.65	29.99	1,180,000	358,900	.42	3,288,000	3,261,000	65.5
Dec. 3	27.24	27.9	326.8	32,100	28.1	34	1,193,000	360,300	.43	3,311,000	3,278,000	64.2
Dec. 3	30.71	26.5	330.0	31,890	26.6	32.4	1,168,000	353,400	.42	3,301,000	3,271,000	63.8
Dec. 7	30.01	13.16	200.5	26,480	26.2	30.6	793,100	221,000	.46	3,589,000	3,563,000	53.0
Dec. 7	28.38	12.55	184.9	26,180	24.31	28.42	741,700	204,200	.47	3,631,000	3,603,000	52.4
Dec. 8	05.35	11.4	146.0	21,440	26.8	31.9	546,400	202,600	.35	2,696,000	2,675,000	42.9
Dec. 9	25.09	10.4	184.3	21,000	27.0	32.2	526,400	202,200	.33	2,603,000	2,583,000	42.2
Dec. 9	24.96	10.8	188.3	20,880	28.3	33.7	519,800	206,100	.32	2,523,000	2,501,000	41.8
Dec. 10	25.08	10.6	186.6	21,130	26.9	31.6	528,700	206,000	.33	2,566,000	2,548,000	42.3
Dec. 11	24.20	10.7	178.2	21,510	27.2	31.8	521,400	195,700	.34	2,664,000	2,645,000	43.0
Dec. 11	24.65	11.0	186.8	22,520	27.8	32.5	553,800	204,900	.35	2,702,000	2,683,000	45.1

TESTS ON DAVIS PUMP.

Date.	Revolutions per minute of pump shaft.	Suction head in feet.	Discharge head in feet.	Velocity head in feet.	Total head in feet.	Power given to pump. (H. P.)	Work done by pump. (H. P.)	Efficiency. (Per cent.)	Capacity. (Apparent.)	Capacity. (Real.)
1896.										
Oct. 5	20.5	7 0	35.8	0.1	43.9	1.06	.54	50.7	65.4	48.6
Oct. 5	20.5	6 9	43.6	0.1	51.6	1.06	.55	51.9	65.4	42.3
Oct. 6	20 4	6.8	48.5	0.1	53.4	1.10	.60	54.5	65.0	42.3
Oct. 7	14.1	6.6	49.9	0.0	57.5	.68	.39	56.9	44.8	26.6
Oct. 7	20.4	6.5	39.7	0.0	47.2	.97	.50	51.5	65.0	41.8
Oct. 12	20.5	6.3	70.4	0.1	77.8	1.24	.86	69.2	65.3	43.6
Oct. 14	20.5	6.6	59.1	0.1	66.8	1.58	.97	61.2	64.9	57.5
Oct. 14	20.3	7.4	20.1	0.1	28.6	.71	.45	63.7	64.7	62.4
Oct. 15	20.3	6 3	82.2	0.2	89.7	1.99	1.39	70.0	64.7	61.5
Oct. 19	20.5	6.6	62.1	0.1	69.8	1.51	1.11	73.8	65.2	63.0
Oct. 19	20.3	6.2	68.8	0.2	76.2	1.58	1.19	75.4	64.6	61.8
Oct. 21	20.4	6.7	40.2	0.1	48.0	.88	.76	86.3	65.0	63.1
Oct. 28	20.2	6 2	64.4	0.2	71.8	1.26	1.12	88.9	64.4	61.9
Oct. 29	20.5	5.6	57.3	0.1	64.0	1.11	1.02	91.9	65.4	63.6
Nov. 2	20.5	5.8	57.7	0 1	58.6	1.11	.94	84.9	65.4	63.6
Nov. 2	20.5	5.9	50.4	0.1	57.4	1.09	.93	84.7	65.3	64.0
Nov. 3	20.5	6.1	57.5	0.1	64.7	1.11	1.03	92.7	65.4	63.1
Nov. 4	20.5	6.1	51.0	0.1	58.2	1.05	.91	86.5	65.2	61.5
Nov. 4	20.3	5.6	56.6	0.1	63.3	1.27	.99	77 7	64.6	61.8
Nov. 5	25.2	6.0	50.6	0.2	57.8	1.46	1.13	77.6	80.1	77.6
Nov. 9	25.8	5.6	98.8	0.2	105.6	2.74	2.11	76.9	82.0	79.0
Nov. 9	24.8	5.7	92.2	0.2	99.1	2.47	1.90	77.2	79.1	76.1
Nov. 11	25.0	6.5	87.5	0.2	95.2	2.37	1.81	76.7	79.6	75.5
Nov. 11	24 8	6.5	92.4	0.2	100.1	2.47	1.89	76.3	78.9	74.8
Nov. 12	25.0	6.7	78.1	0.2	86.0	2.04	1.64	80.6	79.7	75.7
Nov. 18	25.2	5.8	79.0	0.2	86.0	2.09	1.68	80.2	80.2	77.3
Nov. 18	25.0	5.8	70.9	0.2	77 9	2.01	1.44	71.7	79.5	73.9
Nov. 19	25.2	6.6	74.4	0.2	82.2	1.97	1.59	80.6	80.2	76.6
Nov. 23	24.9	7.4	88.5	0.2	97.1	2.26	1.86	82.4	79.3	76.0
Nov. 24	25.0	6.8	64.0	0.2	72.0	1.79	1.40	78.0	79.7	76.8
Nov. 30	16.0	7.7	155.4	0.1	164.2	2.59	1.94	74.8	50.8	46.7
Nov. 30	15.5	7 6	117.0	0.1	125.7	1.97	1.49	75.7	49.5	47.1
Dec. 1	18.4	7.6	161.6	0.1	170 3	3.21	2.38	74.0	58.5	55.3
Dec. 2	20.5	7.6	127 1	0.1	135.8	2.59	2 11	81.4	65.4	61 6
Dec. 3	18.9	7.2	165.8	0.1	174.1	3.18	2.48	78.0	60.1	56.5
Dec. 4	20.5	7 5	155 0	0.1	163.6	3.16	2 55	80.8	64.7	61.8
Dec. 4	17.3	7.3	150.5	0.1	158 9	2.67	2.05	77.0	55.0	51.2
Dec. 7	21.1	7.0	150.5	0.1	158.6	3.27	2.48	76.0	67.3	62 0
Dec. 7	21.1	7.1	144.5	0.1	152.7	3.15	2.50	79.4	67.5	64.9
Dec. 8	22.3	7.3	152.4	0.2	161.0	3.40	2.68	78.8	71.0	66.0
Dec. 9	22.1	6.5	152.2	0.2	159.9	3.43	2.55	74.4	70.2	63.2
Dec. 9	25.0	6 9	101.6	0.2	109.7	2.66	2.08	78.1	79.6	75.0
Dec. 10	24.6	6.8	64.9	0.2	72.9	1.76	1.38	78.6	78.1	75.6
Dec. 11	25.3	6.8	95.2	0.2	103.2	2.31	2.03	88.2	80.6	78.1
Dec. 11	26.6	7.0	107.4	0.3	115.7	2.73	2.38	87.2	84 6	81 6

There was a distance of one foot between the points where the suction head and the discharge heads were measured. This is included in the values of total head.

TESTS ON ROTARY PUMP.

Date.	Revolutions of pump per minute.	Pressure of discharge. (Pounds per sq. in.)	Total head. (Feet.)	Capacity. (Gallons per minute.)	H. P. of engine. (Pumping.)	H. P. of engine. (Not pumping.)	H. P. delivered to pump.	H. P. delivered by pump.	Efficiency of pump.	Friction of pump.
1897. Mar. 2	235	30.4	85.8	836	68.9	31	37.9	18.2	48.2	51.8
Mar. 4	237	31.1	91.3	843	70.8	31.1	39.7	19.5	49	51
Mar. 9	236.9	30.4	88.8	831	70.9	35.1	35.8	18.6	52	48
Mar. 9	235.5	30.6	90.3	823	74.3	36.2	38.1	18.8	49.2	50.8
Mar. 15	236	31.2	90.7	833	75.5	35.9	39.6	19.1	48.1	51.9
Mar. 16	236	31.2	92.3	828	73.3	33.7	39.6	19.3	48.7	51.3
Mar. 18	235.7	30.9	90.4	823	69.4	33.4	36	18.8	52.1	47.9
Mar. 18	236	29.7	85.5	807	66.5	32.3	34.2	17.4	50.9	49.1
Mar. 19	236	28.4	82.9	788	66.6	33.3	33.3	16.5	49.5	50.5
Mar. 19	235.9	29.9	87.5	818	68.1	33.1	35	18.1	51.6	48.4
Mar. 22	201.7	30.36	89.6	818	67.7	31	36.7	18.5	50.4	49.6
Mar. 23	235.6	30.5	90.9	820	70.8	32.7	38.1	18.8	49.4	50.6
Mar. 23	222.4	30.02	88.3	818	69.3	34.6	34.7	18.2	52.5	47.5
Mar. 25	233	30.53	90.6	818	71.1	34	37.1	18.7	50.4	49 6
Mar. 29	236	30	88.3	818	66.6	30.3	36.3	18.2	50.2	49.8
Apr. 1	237.2	30.41	88.3	820	69.2	31.5	37.7	18.3	48.5	51.5
Apr. 1	236.5	30.32	89.4	820	68.4	30.5	37.9	18.5	48.8	51.2
Apr. 2	237.2	30.5	89.3	825	68.5	31.2	37.3	18.6	49.9	50.1
Apr. 5	236.8	30.6	90.2	820	70.9	33.5	37.4	18.7	50	50
Apr. 5	243.6	29.5	86.5	807	66.2	30	36.2	17.6	48.6	51.4
Apr. 6	238.1	30.7	90.1	825	66	30	36	18.8	52.1	47.9
Apr. 8	238	30.9	89.7	825	61.5	23.4	38.1	18.7	49	51
Apr. 9	237.2	29.8	88.5	813	67.1	30.5	36.6	18.1	49 6	50.4
Apr. 12	237.8	29.81	90.6	807	59.7	23.3	36.4	18.4	50.7	49.3
Apr. 13	238.6	30.5	89.9	828	59.5	23.8	35.7	18.8	52.6	47.4

THE CONCENTRATION OF ORES.

By ROBERT H. RICHARDS.

Read January 13, 1898.

QUARTZ ores, which are mined from veins in the solid rock, contain valuable minerals which are usually of high specific gravity; for example, galena carrying lead, or copper pyrites carrying copper, associated with light-weight minerals; for example, quartz, calcite, etc., which are of little or no value.

Since the smelting process is expensive, ores that are smelted at once without concentration may cost more than the metal contained in them is worth, leaving a balance of indebtedness as the result of the transaction. For example, suppose lead smelting costs $10.00 per ton, and the lead ore contains 8 per cent. of lead, 10 per cent. of which is lost in smelting, then the account for treating 100 tons would stand as follows:

Cr.	100 tons ore at $10.00 smelting charges	$1,000.00
Dr.	100 tons ore at 8% lead, 10% lost in smelting, yields 14,400 pounds lead at 3 cents	432.00
	Balance of loss	$568 00

To overcome this difficulty and to substitute profit for loss, the processes of concentration of ores have been developed. By them, the values contained in 2 tons, 10 tons, or even 20 tons may be concentrated into 1 ton, which is proportionally enriched.

The large quantity of mine ore can afford to pay the low price of concentration, while the small quantity of much enriched concentrates is in the best condition to pay the high smelting charges and to leave a good margin for profit.

For example, suppose, in addition to the figures above, that concentration costs 60 cents, and that 15 per cent. is lost in concentrating, then we have:

Cr.	100 tons ore at $0.60 concentrating expense	$60.00
	13 tons 1,200 pounds concentrates at $10.00 per ton smelting charges	136.00
		$196.00
Dr.	13 tons 1,200 pounds concentrates at 50% lead, and 10% loss in smelting, yields 12,240 pounds lead at 3 cents per pound,	$367.20
	Balance of profit	$171.20

The methods of concentration depend upon certain properties of minerals, of which the following are the most important: *Structure and cleavage* influence the shapes of the particles resulting from crushing, some breaking into rounded or cubical forms, and others into flat shapes. These different shapes have an important bearing on the power of the grains to settle in water. *Mineral aggregation.* — When the minerals occur in large crystals, it is comparatively easy to sever the valuable from the waste by breaking, and then to separate them; but if the valuable minerals are finely disseminated in the waste rock the separation becomes more difficult. *Hardness, tenacity, and brittleness.* — The harder the minerals the greater the wear on the crushing machinery. Certain minerals, such as talc, gypsum, and native copper, though soft are very tough, and are, therefore, not easily broken. Other minerals, such as certain varieties of quartz, though hard are very brittle. These tend to form a large amount of slimes from which it is difficult to extract the values, because they do not readily settle either in water or in air. *Specific gravity.* — Differences in specific gravity furnish the most valuable means of separating minerals. Nearly all of the concentrating machines are based primarily upon this principle. *Adhesion.* — A clean particle of gold that has become coated with mercury will adhere to an amalgamated copper plate, and is thus separated from the quartz with which it was associated. *Magnetism* is utilized for the enrichment of some kinds of iron ore, and also for certain other special cases. *Porosity.* — Certain minerals, such as pyrites, become porous by roasting. Their specific gravities are thus decreased, and they can then be separated from minerals from which they could not be separated before. *Greasiness.* — When minerals are finely divided they are apt to act as though they were greasy, and float on the surface of water. This is frequently a cause of loss, especially in the treatment of cassiterite, of native copper, and of native gold. *Decrepitation.* — Certain minerals, such as calcite, barite, and fluorite, when heated fly to pieces, while other minerals are not so affected. This treatment, applied to a carefully sized product, may permit a separation by subsequent screening.

There are two chief steps in concentration : the breaking up of the rock to sever the valuable minerals from the quartz, and the subsequent separation of these "values" from the worthless portions.

For coarse breaking the Blake, Dodge, and Gates crushers are used. Each machine has special qualities which adapt it to its work, but they all act by approaching and receding crusher plates, between which the ore is broken.

For fine crushing a great variety of machines are employed, among which may be mentioned steam stamps and gravity stamps, which break the ore by a blow; and rolls, Huntington roller mill, and Bryan roller mill, which break the ore by rolling pressure. All of these machines find appropriate places in one or another of the mills, according to their fitness for the work to be done, and the preference of the manager. For example, the gravity stamp is almost universally used for crushing gold ores, the steam stamp is the only machine now in use for crushing the native copper of the Lake Superior region, rolls are almost universally employed for lead ores and sulphide copper ores, and the Huntington and Bryan mills are much used for the finest crushing of the various ores.

The concentration proper, or the separation of the "values" from the refuse in the crushed material, is divided into two kinds of work, namely: the preliminary work, which separates the materials into groups or classes, which may or may not be enriched thereby, and final work, which separates the "values" from the refuse of each class, the former going to the smelter, the latter to the dump.

The preliminary separators include the grizzly or bar screen, the drum screens (trommels), the hydraulic classifiers, and the Spitzkasten. The screens separate the materials which come to them into coarse and fine, and a series of screens, with sizes ranging from very large spaces in the first screen down to very small in the last, will give a series of products ranging from the largest down through all the sizes to the finest. Each of these sizes is kept separate from all the others and sent to its proper place in the treatment.

The hydraulic classifier subjects the sands which pass through the finest screen to an upward current of water in a confined space, and the particles which have sufficient weight to settle down through the current can go out through a discharge spigot below, while those which cannot do so, go over and try their luck in the next pocket of the classifier, where an upward stream of less force awaits them, and so on to a third and a fourth pocket. The overflow passes on to the " Spitzkasten," which has three or more pointed boxes, each larger than the last. In this the particles settle from the carrying current of water which passes over the surface to the discharge end of the box, and no upward current is used. The second is wider than the first and so drops finer particles; thus here, too, a series of graded sizes is made. Looking in review over the products of the preliminary separators, it will be seen that the ore has been divided up into a series of sizes ranging from the coarsest, obtained from the first screen, down to the finest size of the Spitzkasten. The products from the classifier and the Spitzkasten, however, are both what are called "water sorted products," which differ from the "sized products" obtained by the screens, in that the quartz and galena in any one of the latter are of the same size, while the quartz particles are much larger than the galena in the former products.

Having examined into the preliminary machines, we will next take up the final concentrators which deliver the dressed ore ready for shipment to the smelter. These include picking tables, jigs, vanners, shaking tables, and slime tables. They all turn out finished concentrates for the smelter. The waste from the coarser machines, however, may require recrushing to recover certain included or attached particles of the valuable minerals called middlings, which failed to be severed from the quartz in the first crushing. These recrushed middlings are then washed again. Waste from the finer sizes is refuse.

On picking tables, rich minerals above 1 inch in size may be picked out by hand for the smelter; the rest goes back to crushers.

The jigs take the sizes next below that treated by the picking table, each of the sizes having its own jig down to nearly the last of the classifier products. The jig treats the ore on a sieve by the action of an up and down movement of water. When screen-sized quartz and galena are subjected on an immersed sieve to an upward current of water which loosens up the ore, the galena being the heavier drops below the quartz and a separation is effected. As the ore is fed continuously to the jig, the light quartz passes off by the overflow, while means are provided for drawing off from the sieve the accumulation of galena as fast as it forms.

When a water-sorted product from the classifier is fed to the jig, the downward movement which follows each lift is quite as important as the upward movement, for the downward current draws down the small particles of galena through the interstices between the large grains of quartz. The galena passes down through the sieve in this case, and the quartz passes off at the overflow as before.

Vanners are used to make the final separation of the last products of the classifier or the first of the Spitzkasten, or both. The Frue vanner is an endless belt 4 feet wide, with flanges or ribs on the edges to keep the pulp from running off the sides. The belt runs upon end rollers 12 feet apart, with little supporting rollers all along. It has a gentle slope; the belt moves slowly up hill, and it is given rapid vibrations sidewise. When water and ore are distributed across this belt they are subjected to three actions: at first the vibrations settle the rich minerals upon the belt with the quartz floating above it; secondly, the slow upward travel of the belt draws the former towards the upper end of the machine, where they are discharged; thirdly, the downward flow of the water carries the quartz with it to be discharged at the lower end of the machine, and thus separation is effected. The fact that the quartz is larger and galena smaller in size is of less consequence for this machine than for the slime tables.

The slime tables, which treat the two or three last products of the Spitzkasten, including the finest mud, are generally circular tables of about 17 feet diameter, with a very slight inclination from the center toward the circumference; they revolve continuously and very slowly in one direction, so that if the fine slimes from the Spitzkasten are fed on one side, at the center, and clean water on the other side, the water will in both cases flow down and over the margin of the table, carrying with it the grains that are most affected by the current. The large grains of quartz are more affected than the small grains of galena, for two reasons: first, they project up higher into the water film where the water is moving faster; and secondly, the quartz is of less specific gravity, and, when other things are equal, it will be moved more rapidly than galena. These conditions, then, serve admirably, first, by the aid of the water which brings the ore, and secondly, by the clean wash water on the other side of the table, to push the quartz off the margin of the table and to retain the galena until it reaches the wash-off jet, which forcibly removes it, preparing the table to treat the next coming ore.

A number of shaking tables have been designed which may take the place of vanners or slime tables, or of both. They all have a bump or a jerk given to the tables, which acts either at right angles to the flow of the water film or diametrically opposite to it, and in every case the heavy minerals are settled below the quartz by the jerk, and are thrown farther in the direction of the jerk than the quartz. The quartz, on the other hand, as it is floating in the upper layer, is more affected by the water current than are the heavy minerals. By adjusting the slope of the table and the quantity of water these two opposing forces are made to deliver the valuable minerals at one point and the quartz at another.

A great deal of study by mill men and engineers has been put upon the question as to how the machines shall be arranged in a mill to turn the good qualities of each machine to the best account, and much difference of opinion still exists in regard to it. The more we study the principles of action of these machines, the better are we qualified to discuss the matter intelligently and to contribute suggestions of real value towards the solution of the question.

The principles which govern the action of some of the machines have been stated in the descriptions which have just been given. Some results of investigations into the laws which regulate those principles will now be given. For example: In jigging, the first question which arises is, "Does suction help or hinder the jigging of screen-sized products; also of water-sorted products?" Tests were made to decide these questions by varying the amount of suction used, and also by varying the size of the valuable minerals while the quartz remained the same. The results are shown in the table, and the ease or difficulty of jigging is shown by the number of pulsions needed to separate the per cent. indicated.

The figures marked with (*b*) clearly show that screen-sized products should be jigged with as little suction as possible, for, with much suction 2,129 pulsions were needed, against 147 pulsions with no suction. On the other hand, figures (*a*) show that to jig water-sorted products much suction greatly helps the process, for, with much suction, only 297 pulsions were needed to effect an almost perfect separation, while with no suction separation was impossible, the finer sizes of valuable minerals floating on top of the quartz.

This table shows further that when the valuable mineral is so heavy that the diameter of the quartz is 3.52 $\left(=\dfrac{.0683}{.0195}\right)$ or more times that of the former, suction acts with great ease and rapidity; but when the figure is much less than 3.52 it is difficult jigging, unless the jig has a very coarse sieve upon it.

Diameter of quartz	.0683 ins.	.0683 ins.	.0683 ins.
Diameter of valuable minerals	.0683 ins.	.0262 ins.	.0195 ins.
WITH STRONG SUCTION.			
Pulsions needed	2,129 *b*	.1759	297 *a*
Per cent. separated	96	95	95
WITH MILD SUCTION.			
Pulsions needed	306 *b*	846	1,382 *a*
Per cent. separated	99	100	98
WITH NO SUCTION.			
Pulsions needed	147 *b*	496	0 *a*
Per cent. separated	98	50	0

The ultimate effect also of pulsion, or the upward current of a jig, has been studied in a vertical tube, in which mixed sizes of quartz and galena from $\frac{1}{10}$ inch in diameter down to dust, were subjected to an upward current of water, which was tried both as a pulsating intermittent current, and also as a steady current, with identically the same results. It was found that after all the heaviest of the galena had settled beneath the quartz, and equilibrium of the particles in the quicksand had been attained, alongside the quartz small grains of galena were balanced in equilibrium, and when these were taken out and measured, the diameter of the quartz was found to be about 5.8 times that of the galena. This test was tried and figures obtained for a number of minerals of different specific gravities, among which are the following: The specific gravity of the quartz being 2.640 —

Specific gravity.

Quartz is 8.6 times the diameter of copper 8.479
Quartz is 5.8 times the diameter of galena 7.586
Quartz is 3 7 times the diameter of arsenopyrite 5.627
Quartz is 2 8 times the diameter of pyrrhotite 4.508
Quartz is 2 1 times the diameter of sphalerite (blende) 4.046

These are called the "hindered settling" ratios, because the particles are bumping against and hindering one another.

This table, taken in connection with the previous one, shows that classifier products of quartz and pyrrhotite, and still more so of quartz and blende, will be difficult to work by suction unless a very coarse sieve is used on the jig.

The next question which occurs affects both the jigging and the slime table work. It is, "What is the ratio of quartz to valuable mineral in the products of the hydraulic classifier and the Spitzkasten?" Measurements have been made of the diameter of particles of galena and quartz, when settling freely and with the same velocity. The following table gives some of the results of these measurements:

DIAMETERS OF THE QUARTZ AND GALENA PARTICLES WHICH ARE EQUAL-SETTLING IN THE UPWARD CURRENTS SPECIFIED WHEN TREATED UNDER FREE-SETTLING CONDITIONS, TOGETHER WITH THE OBSERVED AND CALCULATED DIAMETER-RATIOS.

Particles fall in current of inches per second.	Particles rise in current of inches per second.	Diameter of particles in inches.		Ratio between diameters of particles actually obtained.	Diameter of particles in millimeters.		Particles fall in currents of millimeters per second.	Particles rise in currents of millimeters per second.	Ratio between diameters of particles averaged by a curve.
		Quartz.	Galena.		Quartz.	Galena.			
.000	.050	.00119*	.00076*	1.55	.0301*	.0194*	0.00	1.26	1.54
.050	.099	.00132	.00078	1.69	.0335	.0198	1.26	2.51	1.68
.099	.199	.00224	.00115	1.95	.0568	.0292	2.51	5.05	1.82
.394	.577	.00561	.00242	2.32	.1423	.0613	10.01	14.68	2.23
.780	1.186	.0089	.0041	2.18	.2254	.1032	19.80	30.12	2.48
1.186	1.589	.0135	.0051	2.62	.3416	.1305	30.12	40.37	2.61
1.589	1.972	.0153	.0055	2.76	.3880	.1404	40.37	50.08	2.72
2.366	2.769	.0232	.0079	2.95	.5892	.1997	60.09	70.34	2.92
3.552	3.919	.0403	.0135	2.99	1.0234	.3428	90.21	99.54	3.21
5.526	5.918	.0632	.0180	3.52	1.6032	.4560	140.37	150.31	3.54
7.106	7.826	.0778	.0228	3.42	1.9744	.5776	180.51	198.78	3.70

EXAMPLE. — Mixed particles of galena and quartz treated in free space with rising currents of water. Those that fall in 30.12 and rise in 40.37 millimeters per second current, will have diameters .3416 mm. for quartz, .1305 mm. for galena approximately, and the diameter of the quartz will be 2.6 times that of the galena.

* These averages have less value than the others, for the diameters in these cases range from the figures given down to zero.

A hydraulic classifier probably gives products with nearly these ratios. The grains may hinder one another to some extent, and if they do the ratios will be larger, but not so large as the hindered settling ratios. These ratios have also been obtained for a number of minerals, among which are the following:

FREE-SETTLING FACTORS, OR MULTIPLIERS, FOR OBTAINING THE DIAMETERS OF QUARTZ WHICH WILL BE EQUAL-SETTLING WITH THE MINERAL SPECIFIED WHEN SETTLING FREELY IN AMPLE WATER.

	VELOCITY IN INCHES PER SECOND.								
	1	2	3	4	5	6	7	8	9
	MULTIPLIERS.								
Sphalerite	....	1.46	1.05	1.17	1.62	1.64	1.68	1.66	1.56
Pyrrhotite	....	1.73	1.29	1.48	2 00	2.22	2.26	2.13	2.08
Arsenopyrite . . .	....	1.90	1.57	1.89	2.42	2.56	2.72	2.84	2.94
Galena	....	2.71	1.83	2.26	3.00	3.42	3.65	3.76	3.75
Copper	....	2.71	2.00	2.36	3.00	3.20	3.58	3.76	3.75

EXAMPLE. — If a compact particle of galena, falling freely in water, settles 7 inches per second, the particle of quartz of the same shape that will settle at the same rate will be approximately 3.65 times the diameter of the galena.

In regard to the Spitzkasten, the evidence thus far obtained indicates that while the above ratios ("free-settling" ratios) are probably true for the coarser grains of each product, the earlier products will always be contaminated with grains that belong in the latter, a fault which causes loss of rich mineral upon the tables.

The last question which will be discussed in this paper affects the slime tables: "At what angle of slope should a slime table be built, and with what water quantity should it be fed?" To gain facts which will throw light upon these questions, measurements have been made of the lowest angles at which all the grains of quartz in any given free-settling product will roll, the water quantity varying from 1 pound up to 25 pounds per minute on every 2 feet of width of table. A similar set of angles have been measured for the galena, and they are found to be much larger than those of the quartz; in fact, at the angle where all the quartz rolls, scarcely any of the galena moves. These measurements have all been repeated for a number of different free settled products, giving a range from the finer to the coarser sizes.

The following two tables give these facts for galena and quartz:

GALENA. — ANGLES AT WHICH ALL THE GRAINS MOVE.

Pounds of water per minute on a ft. of width.	1.26 o	Millimeters per second of Current which Lifts the Particles 2.51 5.05 14.68 40 27 50.08					70.34
		Millimeters per second of Current in which the Particles Fall. 1.26 2.51 10.01 30.12 40.37					60.09
		Galena Finish-Angles (Minimum of Three Trials in Most Cases)					
1	15°20′	. . . . . .	9°00′	12°45′	10°40′	. . . .	. . . .
2	12°20′	. . . . . .	11°00′	9°30′	10°00′	. . . .	. . . .
4	7°50′	10°50′	6°15′	7°10′	5°30′	7°30′	9°00′
6	3°05′	6°00′	4°15′	5°15′	4°40′	5°55′	7°30′
8	. . . . . .	4.30′	5°00′	4°45′	3°55′	4°55′	6°10′
10	0°50′	. . . . . .	. . . . . .	4°15′	4°00′	4°20′	5°05′
12	. . . . . .	. . . . . .	. . . . . .	4°00′	3°50′	4°05′	4°15′
14	1°05′	. . . . . .	. . . . . .	4°15′	4°10′	4°20′	4°50′
16	. . . . . .	. . . . . .	. . . . . .	4°20′	4°05′	4°10′	4°25′
18	0°50′	. . . . . .	. . . . . .	4°30′	4°40′	4°35′	4°15′
20	. . . . . .	. . . . . .	. . . . . .	4°30′	4°55′	5°05′	4°15′
22	. . . . . .	. . . . . .	. . . . . .	5°05′	5°05′	5°00′	4°50′
25	. . . . . .	. . . . . .	. . . . . .	5°35′	7°20′	5°30′	4°55′

QUARTZ. — ANGLES AT WHICH ALL THE GRAINS MOVE.

Pounds of water per minute on 2 ft. of width.	1.26 / 0	Millimeters per second of Current which Lifts the Particles. 2.51 / 1.26	5.05 / 2.51	14.68 / 10.01	40 37 / 30 12	50.08 / 40.37	70.34 / 60.09
		Millimeters per second of Current in which the Particles Fall.					
		Quartz Finish-Angles (Maximum of Three Trials in Most Cases).					
1	4°	3°20′	1°55′	3°45′	11°05′	7°30′	8°40′
2	3°40′	2°35′	2°25′	3°25′	7°30′	5°20′	5°25′
4	2°40′	2°30′	2°20′	3°00′	3°50′	3°10′	3°15′
6	1°05′	2°35′	2°00′	2°40′	3°10′	3°10′	2°45′
8	......	2°15′	2°10′	2°40′	2°50′	2°50′	2°35′
10	0°50′	2°15′	2°00′	2°25′	2°50′	2°45′	2°55′
12	......	1°35′	2°20′	2°05′	3°05′	3°00′	2°40′
14	1°05′	2°05′	2°00′	2°40′	3°30′	2°55′	3°15′
16	......	1°40′	1°35′	2°00′	3°30′	2°55′	3°10′
18	0°50′	1°20′	1°40′	1°40′	3°15′	2°45′	3°25′
20	......	2°15′	1°45′	2°30′	3°10′	2°25′	2°45′
22	......	1°30′	1°20′	1°40′	2°05′	2°45′	4°00′
25	......	1°30′	1°55′	1°50′	2°10′	2°20′	2°40′

Mill practice of this country generally puts about 10 pounds of water on a slime table for every two feet of periphery. Admitting that this has been settled by practice, we examine the tables and find that 3° of slope will cause all the quartz to roll off, while practically none of the galena goes. These adjustments will work satisfactorily for all sizes except the very finest slimes, and of these the whole of the galena may be lost. To catch this galena, we turn again to the tables and see very favorable figures, if perhaps, 6° or even 7° slope was 'used and only 2 pounds of water on 2 feet of width; for under these conditions the quartz will all roll at 3°40′, while the galena does not roll until 12°20′ is reached.

INVESTIGATION OF THE THEORY OF THE SÓLUBILITY EFFECT IN THE CASE OF TRI-IONIC SALTS.

BY ARTHUR A. NOYES AND E. HAROLD WOODWORTH.

Received January 20, 1898.

THE theory of the effect of salts on the solubility of one another has been quite thoroughly tested and confirmed in the case of di-ionic salts having one ion in common.[1] The solubility of tri-ionic salts in the presence of other salts has, however, been much less investigated. To be sure, the theory of the phenomenon has been developed already, and has been partially tested by experiments with lead chloride in the presence of other salts.[2] There was found, however, to be only an approximate agreement between the theory and the facts, probably because the dissociation-values involved are uncertain. Moreover, the theory could not be tested in the case where a salt with a common bivalent ion was added, probably owing to the fact that a double salt was formed. Further investigation of the subject seems, therefore, to be desirable.

To this end we have determined the solubility of lead iodide in pure water, and in solutions of potassium iodide and of lead nitrate of varying strengths. We made use of lead iodide because of its slight solubility (one molecular weight in about 600 liters); for in such dilute solutions the influence of the two substances on the dissociation of one another is hardly appreciable, and the tendency to the formation of double salts is ordinarily very slight — two phenomena which often have a disturbing influence in concentrated solutions. In order to determine the solubility, we have measured the electrical conductivity of the saturated solutions, and subtracted from it the conductivity of the water or of the solutions before treating with lead iodide. This method has the advantage of greater convenience over the analytical

---

[1] Compare especially, Noyes and Abbot: Ztschr. phys. Chem., 16, 1, 125.

[2] Noyes: Ztschr. phys. Chem., 9, 626.

determination; and in this case it is especially well adapted, as it furnishes directly a knowledge of the concentration of the ions, so that it is not necessary to consider dissociation-values.

Two samples of each of the three salts were prepared by different methods. These samples were in all cases shown to be exactly alike so far as the conductivity of their solutions was concerned, which indicates that the substances were in all probability pure. One sample of the potassium iodide was prepared by treating the commercial chemically pure salt with alcohol until it was about one-half dissolved, and by subsequent crystallization from this solution by evaporation. The other sample was obtained by crystallizing from water the residue left undissolved by the alcohol. The lead nitrate was obtained in one case by crystallization of a commercial sample from water; in the other, by precipitation with nitric acid followed by crystallization from water. One sample of lead iodide was prepared by metathesis from lead acetate and potassium iodide; the other, by dissolving a commercial preparation in a strong potassium iodide solution, and then diluting with water. In both cases the salt was purified by crystallizing twice from water.

The determinations of solubility were carried out as follows: Small 40 c.c. glass-stoppered bottles were charged with an excess of the solid lead iodide and with pure water or with solutions of lead nitrate or potassium iodide; the stoppers were coated with paraffin, and the bottles were rotated for five hours in a thermostat by means of an apparatus previously described.[1] The solutions were then allowed to settle for a short time, and blown out by means of a wash bottle arrangement into a resistance cell of the Arrhenius type; and the conductivity was measured in the usual way.[2] All the determinations of solubility and measurements of conductivity were carried out at 25° C. In all cases duplicate determinations of the solubility were carried out in such a way that the condition of saturation was approached from both sides — that of supersaturation and that of undersaturation. Moreover, in order to make certain of the purity of the lead iodide, not only were the two different samples used, but each sample was treated also with successive amounts of water, and the conductivity of the corresponding solutions determined. A complete agreement was found to exist.

[1] Ztschr. phys. Chem., 9, 606.
[2] Ostwald: Ztschr. phys. Chem., 2, 561.

The following tables contain the specific conductivities (multiplied by 10^{-7}) expressed in Siemen's units. In all cases the conductivity of the water used (which was equal to 9×10^{-7}) has been subtracted.

CONDUCTIVITY OF WATER SATURATED WITH LEAD IODIDE.

	I.	II.	III.	IV.	Mean.
Undersaturated	4016	4016	4023	4030	4021
Supersaturated	4039	4041	4042	4041	4041

CONDUCTIVITY OF SOLUTIONS OF POTASSIUM IODIDE SATURATED WITH LEAD IODIDE.

CONCENTRATION OF THE POTASSIUM IODIDE.	0.003077 EQUIVALENTS PER LITER.		0.002000 EQUIVALENTS PER LITER.	
Condition of the Solution.	Undersaturated.	Supersaturated.	Undersaturated.	Supersaturated.
I	6174	6184	5316	5343
II	6172	6200	5310	5335
Mean	6173	6192	5313	5339

CONDUCTIVITY OF SOLUTIONS OF LEAD NITRATE SATURATED WITH LEAD IODIDE.

CONCENTRATION OF THE LEAD NITRATE.	0.003077 EQUIVALENTS PER LITER.		0.002000 EQUIVALENTS PER LITER.	
Condition of the Solution.	Undersaturated.	Supersaturated.	Undersaturated.	Supersaturated.
I	6807	6783	5798	5800
II	6791	6799	5796	5784
III	6799	6802	....	....
Mean	6799	6795	5797	5792

A consideration of the results in these tables shows that the comparable values agree closely with one another, although the determinations were carried out with different salts and at different times. The supersaturated solutions are, to be sure, about half of one per cent. higher than the undersaturated, except in those experiments where lead nitrate was used, in which case they were practically equal. In the calculation of the corresponding concentrations of the ions described below, the mean of the supersaturated and of the undersaturated values was used always.

The specific conductivities of the pure potassium iodide and lead nitrate solutions are given in the following tables. In the upper half are given the conductivities of the freshly-made solutions, and in the lower half, those of the same solutions after they had been rotated for the same time and under the same conditions as prevailed in the solubility experiments.

CONDUCTIVITY OF THE SOLUTIONS OF POTASSIUM IODIDE AND LEAD NITRATE BEFORE TREATMENT WITH LEAD IODIDE.

	Potassium Iodide Solution.		Lead Nitrate Solution.	
Equivalents per liter . . .	0.003077	0.002000	0.003077	0.002000
First solution	4247	2809	3782	2498
Second solution	4257	2807	3785	2497
Mean	4252	2808	3784	2498
First solution	4288	2836	3771	2487
Second solution	4288	2849	3774	2487
Mean	4288	2843	3773	2487

In order to ascertain the conductivity of the dissolved lead iodide in the presence of the two other salts, we have subtracted the mean values in the lower half of this table from those of the saturated solutions in the two previous tables; it being assumed that the conductivity at these dilutions is an additive property. That this assumption is approximately correct may be concluded from the fact that the lead nitrate and the potassium iodide at a dilution of 500 liters are dissociated 93.6 and 98.5 per cent. respectively, while at a dilution of 325 liters they are 92.0 and 96.9 per cent. dissociated respectively; in other words, the degree of dissociation is so great, and changes so little with increasing concentration, that it cannot be appreciably affected by the presence of the relatively small amount of lead iodide ions. In order, now, to derive from the conductivity-values the corresponding concentrations of the ions, it is only necessary to divide them by the molecular conductivities of the respective salts at infinite dilution. We take as the values of the rates of migration those calculated by Bredig, namely, $K = 70.6$, $I = 72.0$, and $NO_3 = 65.1$. The value for the lead ion has, so far as we know, not as yet been derived. Franke[1] has measured the conductivity of lead ·nitrate in dilute solution, from which the rate of migration can be calculated. Since, however, this value is for our purpose of fundamental importance, we thought it advisable to confirm the accuracy of his determinations by measurements of our own. We have, therefore, determined the conductivity of both of our samples of lead nitrate at different dilutions. Our values (represented by μ_1 and μ_2, the mean value by μ), together with those of Franke (μ_{Fr}), are given in the following tables:

EQUIVALENT CONDUCTIVITY OF LEAD NITRATE.

v	μ_1	μ_2	μ	μ_{Fr}
256	120.8	120.9	120.9	122.2
325	122.6	122.8	122.7	....
500	124.9	124.8	124.9	....
512	125.2	125.3	125.3	125.9
1024	128.2	127.7	128.0	127.1

As is seen, the conductivities of our two samples agree almost completely with each other, and also fairly well with those of Franke. To obtain the value at infinite dilution we make use of Bredigs'[2] extrapolation values, which have been derived from a consideration of a large number of salts. The values extrapolated in this manner from our measurements are 132.9, 133.3, and 134.0 for the three dilutions, 256, 512, and 1024 liters respectively. We, therefore, take the mean, 133.4, as the value of the conductivity of the lead nitrate at infinite dilution.

Since the rate of migration of the NO_3-ions is equal to 65.1, that of the lead ion must be 68.3, from which it follows that lead iodide has for its value at infinite dilution 140.3. Finally, the value derived for potassium iodide at infinite dilution from the above given rates of migration is 142.6. Now if the corresponding specific conductivities

[1] Ztschr. phys. Chem., 16, 471.
[2] Ztschr. phys. Chem., 13, 198.

are divided by these limiting values, we obtain the concentrations of the ions which correspond to the measurements of conductivity given above. The results of these divisions are presented in the following table, those in the case of the lead iodide being placed under the heading "found."

CONCENTRATION OF THE IONS IN THE SATURATED SOLUTIONS.

Potassium iodide ions.	Lead Iodide Ions.		Lead nitrate ions.	Lead Iodide Ions.	
	Found.	Calculated.		Found.	Calculated.
0.00	0.002873	......	0	0.002873	......
0.001969	0.001770	0.001731	0.001872	0.002358	0.002366
0.002982	0.001351	0.001296	0.002837	0.002155	0.002175

It is readily seen from these numbers that the solubility in the solutions of potassium iodide, as well as in those of the lead nitrate, is considerably less than in pure water, and further, that the diminution caused by equivalent amounts of the two salts is quite different. We will now compare quantitatively the experimental results with the requirements of the principle of solubility effect.

The laws of mass action require that in the case of a saturated solution of a tri-ionic salt, the product of the concentration of the bivalent ion into the square of the concentration of the monovalent ion shall be a constant quantity, whatever other salts may be present at the same time.[1] In this case the equation

$$Pb \times (I)^2 = \text{a constant}$$

should hold, where the chemical symbols denote the concentrations of the respective ions.

If m_0 is the solubility of lead iodide in pure water, m its solubility in a solution of potassium iodide or lead nitrate whose concentration is n (all expressed in equivalents), and a_0, a, and a_1, the corresponding dissociation-values, the following equation applies in the case where lead nitrate is present :

$$(ma + na_1)\, m^2a^2 = m_0{}^3a_0{}^3$$

and where potassium iodide is present :

$$ma\, (ma + na_1)^2 = m_0{}^3a_0{}^3.$$

---

[1] Ztschr. phys. Chem., 9, 627.

The value $m_0 a_0$, the concentration of the ions in the saturated solution in pure water, is found to be 0.002873; and the values of $n a_1$, the concentrations of the ions in the solutions of lead nitrate and potassium iodide in the four solubility experiments are given in the first and fourth columns of the last table. From these we have calculated the values of ma (the concentration of the lead iodide ions), by the two formulæ given above.

This cubic equation is easily solved by substituting an estimated value of ma, and calculating the value of the left member of the equations, repeating this process until the value thus found becomes equal to $m_0^3 a_0^3$. The theoretical values obtained in this way are found in the last table under the headings "calculated."

The agreement between the experimental and the theoretical values, although not complete, is, however, entirely sufficient to prove the essential correctness of the solubility principle. The deviations are in the case of the lead nitrate 0.9 and 0.3 per cent., and in that of the potassium iodide 4.1 and 2.2 per cent. It is to be considered, however, that the values of ma were not directly measured, but were obtained as the differences of two larger experimentally determined values, whereby the percentage error is greatly increased.

In order to give an idea of the existing error-relations, it may be mentioned that an error of half of one per cent. in the conductivity of the solution of lead nitrate saturated with lead iodide, and an equal error in the subtracted conductivity of the pure solution of lead nitrate, would cause in the value ma an error which would probably amount to 1.3 per cent. in the 0.003077 normal solution, and 0.9 per cent. in the 0.002000 normal solution. Equal errors in the experiments with potassium iodide would cause errors of 2.0 and 1.0 per cent., respectively, in ma. It is, therefore, seen that the method cannot furnish very accurate results, and it is not improbable that the differences between the actual and the calculated values are due to experimental errors.

It has, therefore, been established by this investigation within the somewhat wide limits of experimental error, that *the solubility of lead iodide is diminished both by potassium iodide and by lead nitrate, in such a way that the product of the concentration of the lead ions into the square of the concentration of the iodine ions remains constant.*

MASSACHUSETTS INSTITUTE OF TECHNOLOGY,
January, 1898.

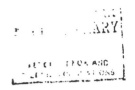

TECHNOLOGY QUARTERLY

AND

PROCEEDINGS OF THE SOCIETY OF ARTS.

VOL. XI.	JUNE, 1898.	No. 2.

PROCEEDINGS OF THE SOCIETY OF ARTS.

THIRTY-SIXTH YEAR, 1897-98.

THURSDAY, January 13, 1898.

THE 505th regular meeting of the SOCIETY OF ARTS was called to order by the President at 8 P.M.

The records of the previous meeting were read and approved. Professor R. H. Richards read a paper[1] on the "Concentration of Ores, with Special Reference to Recent Investigations in this Subject." The separation of gold was the first process to be described. Next the concentration of lead, zinc, and copper ores was taken up, and the process of crushing, preliminary separating, sizing, and final separation were discussed at length. Turning to the theoretical principles involved in water sizing, Professor Richards described the difference in the proportions of ore and quartz at the various stages of free settling and of hindered settling. The relative advantages of these modes of settling for various kinds of ores were discussed, and principles used in determining the sieve scale and the best angle of slope for slime tables were explained.

After the speaker had received the thanks of the Society, the meeting was adjourned.

---

[1] Printed in full in *Technology Quarterly*, Vol. xi, No. 1, p. 54.

THURSDAY, January 27, 1898.

The 506th regular meeting of the SOCIETY OF ARTS was held this day at 8 P.M., Professor Swain in the chair.

The records of the previous meeting were read and approved. Mr. B. C. Batcheller, of Philadelphia, was introduced and read a paper on "A New System of Pneumatic Dispatch Tubes." He said that the history of pneumatic transit began practically in London in 1853, when the Electric and International Telegraph Company constructed the first tube for dispatching their telegrams. This tube was one and a half inches in diameter, and two hundred and twenty-five yards in length. Later longer tubes were constructed in London, Paris, Berlin, and Vienna.

The history of large pneumatic tubes for the transportation of mail began in Philadelphia in the winter of 1892–93. The system used in Philadelphia was described in detail, and also the methods employed in manufacturing the tubes. The straight pipes are of cast-iron similar to water pipes, and bored to a uniform diameter. The bends are made of brass tubing. Transmitters are provided with a time-lock to prevent collisions. Open receivers are used at far ends of the tubes, while in a loop a closed receiver is employed. This is provided with a circular plate to prevent the escape of air. The construction of stations, switches, electric block system, and carriers were described. The method of laying the pipes was discussed and then the Boston system, which was opened on the 17th of December, was discussed in detail. The speaker then turned to the theory of pneumatic transit, and exhibited curves showing the decrease of pressure with an increase of velocity as the carrier moves along the tube. In conclusion he gave an account of the ingenious method employed to determine the position of an obstruction in the tube. The calculation is made from observations of the time between a pistol shot and its echo. The paper was illustrated by the lantern, and specimens of carriers were exhibited. During the discussion which followed Mr. Batcheller said that the speed increases with the diameter of the tube.

It was voted to thank the speaker for the great pains he had taken to explain and illustrate the subject, and the Society then adjourned.

THURSDAY, February 10, 1898.

The 507th meeting of the SOCIETY OF ARTS was held at the Institute this day at 8 P.M., Mr. N. M. Lowe in the chair.

The record of the previous meeting was read and approved.

Messrs. Edward D. Densmore, of Boston, Will J. Sando, of Boston, and Rufus F. Herrick, of Rockaway, New York, were duly elected Associate Members of the Society.

Dr. Leonard Waldo, of Bridgeport, Connecticut, was introduced and read a paper on the " History and Present Development of the American Bicycle." He said that the first great advance in the manufacture of bicycles came with the introduction of machines of precision for the manufacture of interchangeable parts, and by the invention of pneumatic tires it became possible to use hard bearings with rolling balls. He spoke of the great care used in selecting the materials, all of which are forged and drawn cold, and he described in detail the various tests employed to determine the strength of materials and the efficiency of the machine.

Three difficulties were met with in the attempt to construct a chainless wheel with bevelled gear : (1) To find material for the tubular shaft that would resist torsion ; (2) to invent machines that would cut tooth gears rapidly ; (3) to harden without deformation. These have all been overcome. The paper was illustrated by lantern slides and a model of tooth gear. A discussion followed the reading of the paper. It was voted to thank Dr. Waldo for his interesting paper, and the Society adjourned.

THURSDAY, February 24, 1898.

The 508th meeting of the SOCIETY OF ARTS was held at the Institute at 8 P.M., Mr. Desmond FitzGerald in the chair.

The record of the previous meeting was read and approved.

Mr. Fred B. Forbes, of Somerville, was duly elected an Associate Member of the Society. The following papers were presented by title : "Best Resistance for a Sensitive Galvanometer," by F. A. Laws ; "Investigation of the Theory of the Solubility Effect in the Case of Tri-ionic Salts," by A. A. Noyes and E. H. Woodworth.

Mr. Henry ·B. Wood, Engineer of the State Survey, was then introduced and read a paper on "State, Town, and City Boundaries."

He gave the history of the attempts to determine the boundaries of Massachusetts, and described the present condition of the boundary lines. This was followed by a detailed account of the recent work of the Topographical Survey in determining the boundary lines between this State and New York and Rhode Island. In the work on the New York boundary the heleotrope was employed to obtain long lines of sight, one of these lines being forty-one miles in length. It was possible by this means to determine intermediate points to within five-eighths of an inch.

Mention was made also of the delineation of town boundaries; 2,500 corners have been located, and the work is less than half finished. In conclusion it was shown how the triangulation of the State is related to the determination of boundaries. The paper was illustrated by a large number of lantern views. A discussion followed, after which the Society adjourned.

THURSDAY, March 10, 1898.

The 509th meeting of the SOCIETY OF ARTS was called to order this day at 8 P.M., Mr. Blodgett in the chair.

The record of the previous meeting was read and approved.

Mr. A. T. Hopkins then read a paper, prepared by Mrs. Ellen H. Richards and himself, on "Certain Sanitary Aspects of Jamaica." He began with a general description of the country, and then gave an account of a journey which he made around the island for the purpose of collecting samples from the water supplies. The results of analyses made by Mrs. Richards were shown on a chart. It was found that distance from the sea, the presence of mountains, and the amount of rainfall and evaporation had marked influence upon the normal distribution of chlorine. The present methods of obtaining water and of disposing of sewage were described, and also a sewerage plant installed by Mr. Hopkins at Port Antonio. He concluded with an account of the recent epidemic of yellow fever, which he had traced to its source in two sailors recently from a Cuban port, who became sick in Kingston soon after their arrival. From them the disease spread after the usual manner of infectious diseases. The paper was illustrated.

During the discussion which followed Professor Sedgwick spoke of the importance of construction of a normal chlorine chart as a basis for the detection of injurious substances in water supplies, and complimented Mr. Hopkins upon his success in tracing the yellow fever epidemic.

After a vote of thanks to the speaker, the Society adjourned.

THURSDAY, March 24, 1898.

The 510th meeting of the SOCIETY OF ARTS was held in Room 22 of the Walker Building at 8 P.M., with Mr. Blodgett in the chair.

The record of the previous meeting was read and approved.

The Chairman introduced Mr. E. M. Smiles, of New York, who exhibited an apparatus for the production of X-rays.

A short discussion followed the exhibition, after which the Society adjourned.

THE CONSTITUTION OF STEEL CONSIDERED AS AN ALLOY OF IRON AND CARBON.

By ALBERT SAUVEUR.

Read April 28, 1898.

It has been conclusively established that in unhardened steel, at least, the totality of the carbon is combined with a portion of the iron forming the carbide Fe_3C, which is then distributed throughout the balance of the iron.[1] Steel may therefore be considered as a mixture or an alloy of iron and the carbide Fe_3C, and I shall endeavor to show that if we look upon steel in this light, the formation of its structure follows very closely the laws which govern the formation of the structure of a certain class of binary metallic alloys, namely, that group of alloys whose component metals form neither definite compounds nor isomorphous mixtures.

Great advance has been made in recent years in our knowledge of the true constitution of metallic alloys.   Much activity was created in this field of research through the work of special committees organized in England by the Institution of Mechanical Engineers, and in France by the Société d'Encouragement pour l'Industrie Nationale.   Among those investigators whose work has been most fruitful, the names of Dr. Guthrie, Roberts-Austen, H. Le Chatelier, Heycock and Neville, Charpy, Gautier and Behrens stand preeminent.

The study of the structure of alloys and of industrial metals has even called into existence a new department of science, for the microscope has revealed to us that all alloys and all industrial metals, which always contain a certain amount of impurities, are made up of constituents which may be regarded as minerals, for they possess all the characteristics of true minerals ; and as the study of rocks created the science of petrography, so from the study of the constitution of metals

---

[1] I purposely ignore here the small amount of graphitic carbon often found in high carbon steel after annealing.

and alloys was developed a new department of metallurgy, called metallography.

One of the most brilliant achievements of these recent investigations is to be found in the discovery that the structure of metallic alloys is controlled by the same laws which had been known for some time, to regulate the constitution of mixtures of melted salts and of frozen saline solutions. In other words, the greatest analogy exists between the structure of solid saline solutions and that of metallic alloys. The latter must be considered as true solutions. The fact that they are solid at the ordinary temperature, while saline solutions are liquid, has alone prevented the identity of the laws which govern the formation of the structure of both classes of substances, when they assume the solid state, from being discovered at a much earlier date. With a view of offering a rationale for the constitution of steel I shall briefly recall the working of these laws, although time permits me to do so only in their broadest outline.

Saline Solutions. — We all know that by dissolving common salt, sodium chloride, in water, we lower the freezing point of the water. By increasing the amount of salt the freezing point of the resulting mixture is, at first, correspondingly lowered, until it contains a certain percentage of salt. The lowest possible freezing point of a solution of sodium chloride in water is then reached, and further addition of salt will gradually raise the freezing point of the brine.

Dr. Guthrie found that the mixture which has the lowest possible freezing point contains about 23.50 per cent. of NaCl; and as the hydrate containing 10 molecules of water would require 24.50 per cent. of salt, Dr. Guthrie inferred that the solution of lowest freezing point was an hydrate of the formula $NaCl + 10\ H_2O$. He proposed for it and for all similar mixtures, *i. e.*, for all saline solutions of lowest freezing points, the name of *cryohydrate*, by which he meant to imply that they can only exist in the solid state, at a low temperature.

Figure 1 shows the curve of solubility, or the freezing curve, which is evidently the same thing, of a solution of sodium chloride in water. The abscisses represent the composition, the ordinates the temperatures at which the various mixtures freeze. The curve is made up of two branches which meet at a point E corresponding to a temperature of — 22° C, and to a solution containing 23.50 per cent. of NaCl, *i. e.*, to the composition and freezing point of the cryohydrate.

Supposing the mixture placed in a proper cooling medium, let us see what happens when a solution poorer in salt than the cryohydrate reaches its freezing point. It is at that instant saturated with water and further cooling will cause the *formation of ice.* If a thermometer be placed in the solution it will then indicate a retardation in the rate of cooling which, of course, denotes an evolution of latent heat and marks the beginning of solidification. If the corresponding temperature be plotted on the diagram it will give one point of the branch $M E$. The formation of pure ice causes the remaining liquid to become richer in salt, and as its freezing point is thereby correspondingly lowered, the deposition of ice does not take place at a constant temperature, but proceeds as the temperature is further lowered. The portion remaining liquid meanwhile becomes richer and richer in salt, until at a temperature of — 22° C it reaches the composition of the cryohydrate. The remaining liquid then solidifies *as a whole and at a constant temperature;* the thermometer placed in the mixture remains stationary until the whole mass has solidified. The heat which is here evolved is the latent heat of solidification of the cryohydrate, and if the corresponding temperature be plotted on the diagram it will give one point of the horizontal line. If the application of cold be continued after the whole mass is solid, the fall of temperature resumes again its normal rate.

A frozen saline solution, then, containing less salt than the cryohydrate, will be made up of crystals of ice surrounded by the frozen cryohydrate. If the solution contains more salt than the proportion found in the cryohydrate, when a certain temperature is reached, the thermometer indicates an evolution of heat which marks the beginning of solidification and corresponds to the formation of *crystals of salt* (sometimes hydrated), for the solution is at that temperature saturated with salt, and further cooling must cause the deposition of salt. The crystals of salt increase in quantity as the temperature continues to fall. The remaining liquid meanwhile becomes correspondingly poorer in salt, approaching more and more the composition of the cryohydrate which it reaches at a temperature of — 22° C. At that instant the portion remaining liquid solidifies as a whole and at a constant temperature, as indicated by the thermometer, which at this stage remains stationary until the entire mass has solidified. The plotting of the corresponding temperature gives another point of the horizontal line. The frozen mass is then made up of crystals of salt

Freezing of Crysthydrate

| 80 | 70 | | 50 | 40 | 30 | 20 | 10 | 0 |
| 20 | 30 | 40 | 50 | 60 | 70 | 90 | 90 | 100 |

COMPOSITION

VE OF SOLUBILITY OF AQUEOUS SOLUTIONS OF NaCl.

(See pages 79–80.)

surrounded by the solid cryohydrate. If the solution has a composition identical to that of the cryohydrate, it will not freeze until it has reached a temperature of — 22° C, when it will solidify as a whole and at a constant temperature.

The composition of the cryohydrate, it is seen, is independent of the composition of the solution, but the proportions of cryohydrate and of ice or of cryohydrate and of salt found in the frozen solution, depend, of course, upon the amount of salt in the solution. The branches *M E* and *N E* correspond, therefore, to the *beginning* of the evolutions of heat which accompany the beginning of solidification of ice or of salt. Hence they also represent the solubility of salt in water at various temperatures. The horizontal line corresponds to the evolutions of heat corresponding to the solidification of the cryohydrate, and indicates that such solidification takes place at a constant temperature, which is the same whatever the composition of the original solution.

While saline solutions, therefore, containing various proportions of salt *begin* to freeze at temperatures which depend upon their degree of concentration, they all *finish* freezing at the same temperature, namely, at — 22° C, the freezing point of the cryohydrate.

Seeing that one branch of the curve represents the formation of crystals of ice in brines poorer in salt than the cryohydrate, while the other branch corresponds to the formation of crystals of salt in solutions richer in salt, the conclusion is almost irresistible that their meeting point must correspond to a *simultaneous deposition of ice and salt*, and that the frozen cryohydrate, therefore, must be a mechanical mixture of ice and salt, probably in an extremely minute state of division, which would outwardly give it the appearance of a definite compound ; and indeed, Mr. Offer has shown that the cryohydrates do not form distinct and transparent crystals, but opaque masses, and that alcohol dissolves the ice and leaves a crystalline network of solid salt ; also that their heat of dissolution is equal to the sum of the heats of dissolution of the ice and of the salt ; finally, that the specific gravity is equal to the mean of those of both constituents, which facts argue strongly in favor of the cryohydrates being merely mechanical mixtures of ice and salt.

Mr. Ponsot, moreover, using colored salts, has ascertained that the cryohydrates were actually made up by the juxtaposition of crystals of pure ice and of salt, the salt being sometimes hydrated. We shall presently find some further evidences that the cryohydrates are merely mechanical mixtures of this description.

What has been said regarding brines applies to all aqueous saline solutions; they all give rise upon cooling to the formation of cryo-hydrates, *i. e.*, of solutions having a definite composition, and which freeze at a constant temperature which is also the lowest possible in each series.

If we now pass to mixtures of melted salts which form neither definite compounds nor isomorphous mixtures, we find that upon solidi-fying, the formation of their structure is regulated by exactly the same laws. They, too, form a mixture of definite composition and lowest melting point which solidifies at a constant temperature. Whichever salt is present in excess solidifies first until the composition of the molten mass has reached that of the mixture of lowest melting point. In these cases the water is simply replaced by another salt; and if we consider water as fused ice, we need not make any distinction between these two classes of substances, namely, between aqueous saline solu-tions and mixtures of melted salts.

If I have dwelt at such length upon the constitution of frozen saline solutions, it is because it will help us to account for the struc-ture of metallic alloys.

Metallic Alloys. — The present theory of the constitution of alloys so brilliantly worked out classifies all such mixtures into three classes:

I. Alloys which give neither definite compounds nor isomorphous mixtures.

II. Alloys which form definite compounds.

III. Alloys which form isomorphous mixtures.

For our purpose we need only consider the first group, namely, those alloys which form neither definite compounds nor isomorphous mixtures — and to make the matter clearer, let us take an individual instance, that of silver and copper alloys.

Figure 2 shows what is known as the curve of fusibility of silver and copper alloys. The abscisses represent the composition of the alloys; the ordinates the temperatures at which they solidify. Atten-tion need not be called to the striking resemblance between this curve and that of the saline solution just examined. The former was called a curve of solubility, but solubility and fusibility represent here, obvi-ously, the same phenomenon. Both are freezing curves. Fusibility applies to mixtures solid at the ordinary temperature; solubility to those liquid at that temperature.

Solidification of Eutectic Alloy

| 80 | 70 | 60 | 50 | 40 | 30 | 20 | 10 | 0 |
| 70 | 30 | 40 | 50 | 60 | 70 | 80 | 90 | 100 |

COMPOSITION

URVE OF FUSIBILITY OF SILVER AND COPPER ALLOYS.

(See pages 82-83.)

Like the curve of solubility of saline solutions, the curve of fusibility of metallic alloys is obtained by cooling from the molten state mixtures containing various proportions of the two component metals, and by carefully ascertaining, most conveniently by means of a Le Chatelier pyrometer, the temperature at which the cooling is momentarily arrested, or its rate retarded, which indicates an evolution of latent heat and marks the beginning of the solidification of the alloys.   As in the case of saline solutions all alloys of silver and copper, with one exception, exhibit a second evolution of heat occurring always at the same temperature and corresponding, therefore, to an horizontal line in the diagram.

The curve is composed of two branches starting respectively from the melting points of pure silver and of pure copper, and meeting at a temperature of 770° C, and for a composition corresponding to 28 per cent. of copper and 72 per cent. of silver.   The mixture of that composition has the lowest possible melting or freezing point of all silver and copper alloys.   It is called the *eutectic alloy*, a name proposed for it and for all similar alloys, by Dr. Guthrie.   It is seen at once that the greatest analogy exists between the eutectic mixtures of metallic alloys and the cryohydrates of saline solutions.

There are two cases to be considered.   The composition of the alloy may correspond exactly to that of the eutectic alloy or it may differ from it.   In the first case, as the mixture cools from a high temperature, it remains liquid until it reaches the solidification point of the eutectic alloy, 770° C, when it solidifies as a whole, and at a *constant temperature* as indicated by the pyrometer, which at this stage remains stationary until the whole mass has solidified.   Such a mixture behaves exactly like a saline solution having a composition identical to that of the cryohydrate ; the solid mass is made up entirely of the eutectic alloy.   See Plate I,[1] Figure 1.

If the composition of the alloy differs from that of the eutectic mixture, *i. e.*, if one of the constituents is present in excess with regard to that composition, the silver for instance, then when the cooling mass reaches a certain temperature the silver in excess begins to solidify exactly as did the water in the case of a saline solution.   This deposition is accompanied by an evolution of heat indicated by the

---

[1] The figures of this plate, with the exceptions of Figures 4 and 6, are reproductions from photo-micrographs taken by Mr. Osmond.   The Metallographist, Vol. I, No. 1, January, 1898.

pyrometer, and which, when the corresponding temperature is plotted, gives one point of the branch $M E$. The separation of pure silver continues until the portion remaining liquid, and which becomes all the while poorer in silver, has reached the composition of the eutectic alloy, 72% Ag + 28% Cu. At that instant the silver is saturated with copper, the copper is saturated with silver, and both metals solidify together at a constant temperature.

Alloys containing a larger percentage of silver than the eutectic mixture will, therefore, be made up of crystalline particles of silver in a matrix of the eutectic alloy. Plate I, Figure 2, exhibits the structure of an alloy containing 85 per cent. of silver. The large black areas represent the pure silver; the composite constituent forming the network is the eutectic alloy. The preparation was heated to a purple color, which accounts for the dark appearance of the silver.

If it be the copper which is present in excess with regard to the composition of the eutectic alloy, it is pure copper which begins to solidify when a certain temperature is reached. The solidification of copper continues until the portion remaining liquid has reached a composition identical to that of the eutectic alloy; it then solidifies as a whole and at a constant temperature, which when plotted gives one point of the horizontal line.

Alloys of silver and copper containing a larger proportion of copper, therefore, than the eutectic mixture, will be made up of crystalline particles of copper surrounded by the eutectic alloy. This is well shown in Figure 3, Plate I, which represents the structure of an alloy containing 65 per cent. of copper, *i. e.*, a large excess of that metal. The large white areas represent the pure copper; the constituent made up of grains or plates alternately light and dark is the eutectic alloy.

What, then, is the nature of these eutectic alloys? Seeing that they always had a definite composition, whatever the composition of the original alloy, and also on account of the closeness of their fractures, which often have a conchoidal appearance, it was thought for many years that they were definite compounds, although their compositions seldom correspond to exact simple ratio of the atomic weights of the components.

On the other hand, seeing that one branch of the curve of fusibility corresponds to the solidification of pure copper, the other to the solidification of pure silver, it would seem highly probable that their

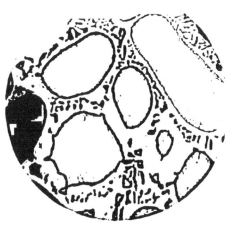

FIG. 1. — ALLOY OF SILVER AND COPPER.
Cu, 28 per cent.; Ag, 72 per cent.
Magnified 600 diameters.

FIG. 2. — ALLOY OF SILVER AND COPPER.
Cu, 15 per cent.; Ag, 85 per cent.
Magnified 600 diameters.

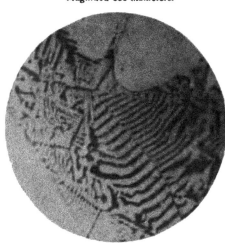

FIG. 3. — ALLOY OF SILVER AND COPPER.
Cu, 65 per cent.; Ag, 35 per cent.
Magnified 600 diameters.

FIG. 4. — ALLOY OF SILVER AND ANTIMONY.
Ag, 66 per cent.; Sb, 34 per cent.
Magnified 500 diameters.

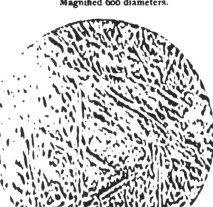

meeting point must correspond to the simultaneous solidification of silver and copper, and that the eutectic alloy is merely a mechanical mixture of the two constituents. The microscope has shown that such, indeed, is the case. All eutectic alloys are made up of extremely minute crystals or plates of the two components in close juxtaposition. This is well illustrated in Figure 1, Plate I, which shows the structure of the eutectic alloy of silver and copper under a magnification of 600 diameters. Figure 4, Plate I, reproduced from a photomicrograph of Mr. Charpy, presents another beautiful instance of the characteristic structure of eutectic mixtures. The component metals are here silver and antimony, and the alloy contains an excess of silver which is represented in the photograph by light areas.[1]

In the fact that eutectic alloys are mechanical mixtures, we find an additional important evidence that such must also be the constitution of cryohydrates, whose formation is so similar to that of eutectic alloys.

On account of the minuteness of their constituents, eutectic mixtures often require very high power for their resolution, and they frequently present under the microscope a beautiful play of interference colors, strongly suggestive of mother-of-pearl.

Alloys of Iron and Fe_3C. — With the laws which control the formation of the structure of alloys fresh in our minds, we shall have no difficulty in accounting for the structure of steel as revealed to us by the microscope. In Figure 3 the abscisses represent the composition of the metal in terms of iron and the carbide Fe_3C, the corresponding percentages of carbon being indicated between brackets. The ordinates represent the temperatures at which some evolutions of heat occur during the undisturbed cooling of the metal.

By allowing samples of steel containing various amounts of carbon to cool from a high temperature, and carefully observing the rate of cooling by means of a Le Chatelier pyrometer, when a certain temperature is reached, which varies with the carbon content, a sudden retardation occurs which denotes, of course, some evolution of heat. By plotting the corresponding temperatures, the curve *M E N* is obtained. The diagram also indicates that on further cooling a second evolution of heat occurs, this time at nearly the same temperature, whatever the degree of carburization of the metal. Graphically,

[1] This sample was treated with sulphuretted hydrogen, which blackens the silver while it has no action upon the antimony. The reproduction, however, was prepared from a negative print, so that the silver here appears white instead of dark.

therefore, it is represented by a nearly horizontal line.[1] This second evolution of heat is generally very marked. It is sometimes so intense as to produce an actual rise in the sensible temperature of the steel, a "recalescence" of the cooling metal. It is why the phenomenon is known by the name of recalescence, and the temperature at which it occurs called the temperature or the point of recalescence. It is a critical temperature of vital importance; in passing through it the structure of the steel is entirely changed, and nearly all its properties, chemical, physical, and mechanical, markedly altered.

The appearance of the diagram obtained in this way recalls at once the curve of solubility of saline solutions and that of fusibility of metallic alloys. Here again we have two branches of curves, meeting in this case at a temperature of about 670° C., and for a composition corresponding to 12 per cent. of Fe_3C (or 0.8 per cent. carbon) in the steel. We have also a nearly horizontal line passing by the point of intersection. The analogy between the three curves (Figures 1, 2, and 3) is, indeed, so striking that it will readily suggest, I believe, the inference that the branch $M E$ (Figure 3) corresponds to the separation or segregation (we cannot use the word solidification here, for at this temperature the whole mass is solid) of the iron, and the branch $N E$ to the segregation of the carbide Fe_3C, while the point E marks

---

[1] The temperatures at which retardations occur during the cooling of steel do not, of course, when plotted, give absolutely straight lines; experimental errors alone would preclude such possibility. Moreover, slight amounts of impurities have a notable influence upon the position of the retardation, and it is quite impossible to obtain various grades of steel absolutely free from impurities or containing exactly the same amount. It is sufficient for our reasoning that the curve should be made up, as it is, of two branches which meet in the region indicated. The two branches appear not to deviate very much from straight lines, and might possibly coincide exactly with them if experimental errors and other disturbing factors could be eliminated. It should also be stated that these evolutions of heat do not begin and end at the same temperature, but, on the contrary, cover a noticeable range — sometimes as much as 50° C. or more — so that the first and second evolutions run together long before a carbon content of 0.8 per cent. has been reached, with the result that steel containing over 0.50 per cent. carbon appears to have only one retardation extending over a considerable range of temperature: the end of the first retardation merges into the beginning of the second

Finally, when the metal contains less than about 0.25 per cent. carbon, a third retardation is detected, located between the first evolution and the point of recalescence. It has been purposely left out of this diagram because it would have rendered the resemblance which the latter bears to the curve of solubility or of fusibility just examined, less marked, and to little purpose, for the existence of a third evolution of heat confined to these narrow limits does not sensibly affect the strength of the deduction drawn from the striking analogy of the curves.

95		90	88	85		80		75		70
5		10	12	15		20		25		30
(.33)		(.67)	(.80)	(1.00)		(1.33)		(1.67)		(2.00)

COMPOSITION

INDICATING EVOLUTIONS OF HEAT (ACCOMPANIED BY STRUCTURAL
NGES) OCCURRING DURING THE COOLING OF STEEL.

(See pages 85–86.)

PLATE II.

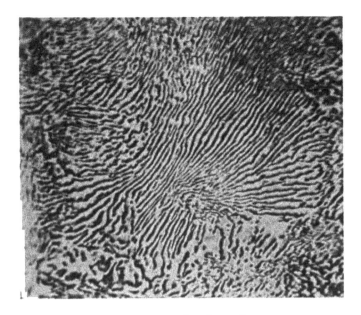

FIG. 1. — STEEL — 1 PER CENT. CARBON.

Magnified 1,000 diameters.

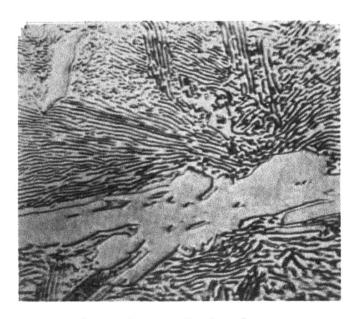

FIG. 2. — STEEL — 1.5 PER CENT. CARBON.

Magnified 1,000 diameters.

the formation of a eutectic alloy of iron and Fe_3C, *i. e.*, a mechanical mixture whose structure should be made up of small crystalline plates or grains alternately of iron and Fe_3C, and resulting from the simultaneous segregation of the two constituents. If we are right, steel containing less than 12 per cent. of Fe_3C (0.8 per cent. carbon) should be made up of crystalline grains of pure iron (at least of carbonless iron) surrounded by the eutectic alloy of Fe and Fe_3C; steel more highly carbonized should be formed of grains of the carbide Fe_3C surrounded by the eutectic alloy, while if the metal contains exactly 0.8 per cent. carbon it should be entirely made up of the eutectic alloy.

Such conclusions are confirmed in every particular by the microscopical examination of the structure of various grades of steel. Plate II, Figures 1, 2, and Plate III, Figure 1, show the reproduction of some beautiful photomicrographs taken by Mr. Osmond under a magnification of 1,000 diameters.[1] Figure 1 exhibits the structure of the eutectic alloy of Fe and Fe_3C. It presents all the structural characteristics of eutectic mixtures, being made up of thin plates alternately of Fe and Fe_3C. These plates seldom exceed $\frac{1}{40,000}$ of an inch in thickness; the plates of iron are somewhat darkened (or rather deprived of their metallic luster) by the' polishing and etching, while the plates of Fe_3C are left white and brilliant, and stand slightly in relief. The arrangement is, therefore, similar to that of the reflection gratings of physicists, and explains the pearly appearance of this remarkable constituent.

The steel of Figure 2, Plate II, contains 1.50 per cent. of carbon, therefore an excess of Fe_3C with regard to the composition of the eutectic alloy. The white portion standing in relief represents areas of Fe_3C. It is surrounded by the eutectic alloy. Figure 1, Plate III, shows the structure of a steel containing less carbon, and therefore more iron, than the eutectic mixture. The light background indicates the excess of iron. The accuracy of the deduction drawn from the appearance of the curve of Figure 3 is further illustrated by the lowest row of drawings of Plate IV, which show the structure of a series of slowly cooled steels of ascending carbon content. They were drawn directly from the microscope under an original magnification of 250 reduced in the reproduction to 72. The shaded constituent represents the eutectic alloy; the pure iron has been left white, which is its appearance under the microscope; while the carbide Fe_3C is here rep-

---

[1] The Metallographist, Vol. I, No. 1, January, 1898.

resented by black areas to distinguish it more readily from the iron, although in reality it has a brilliant, metallic appearance.

Mineralogical names have been given to these constituents : Pure iron has been called *ferrite*. The carbide Fe_3C, *cementite*, because abundant in cement steel, while the name of *pearlyte* has been given to the eutectic alloy of Fe and Fe_3C, *i. e.*, of ferrite and cementite, because of its pearly appearance.[1] The composition of pearlyte, like that of any eutectic alloy, remains the same, whatever the composition of the steel, *i. e.*, whatever its carbon content. It always contains in the neighborhood of o.8 per cent. of carbon.[2] The relative proportions of ferrite and pearlyte, however, or of cementite and pearlyte in the steel, vary, of course, according to the degree of carbonization.

Very low carbon steels, then, are made up of a matrix of iron or ferrite, with here and there a particle of pearlyte (Plate IV, Figure 4) ; the fine black lines indicate the junction lines between the grains of ferrite. As the carbon increases the amount of pearlyte increases proportionally (Figures 7 and 10), until with o.8 per cent. of carbon the metal is made up entirely of pearlyte (Figure 12). It is then said to be saturated. With further increase of carbon, cementite makes its appearance, increasing in quantity with the carbon content, causing, of course, a corresponding decrease in the amount of pearlyte (Figures 14 and 16).

The structural composition of any carbon steel may readily be calculated from its carbon content. If the steel contains less than o.8 per cent. of carbon it is made up of ferrite and pearlyte. Let x be the percentage of ferrite, y that of pearlyte ; we have

$$(1) \quad x + y = 100 ;$$

and since the totality of the carbon is found in the pearlyte and forms o.8 per cent. of its composition, we have the second equation :

$$(2) \quad \frac{0.8}{100} y = C,$$

in which C represents the known amount of carbon in the steel.

---

[1] These very appropriate names were suggested by Professor Henry M. Howe, and have been quite universally adopted.

[2] Professor Arnold experimenting with exceptionally pure carbon steels, especially prepared, finds that pearlyte contains nearly o.9 per cent. carbon. In the case of commercial steel, however, the carbon content of pearlyte is nearer o.8 per cent. Impurities, and it would seem manganese especially, have a tendency to lower the percentage of carbon in pearlyte.

PLATE III.

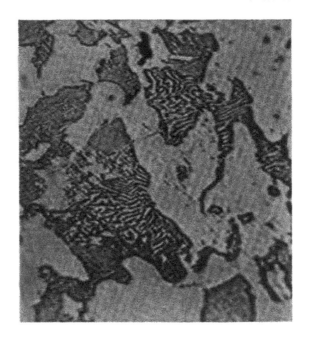

FIG. 1. — STEEL.—0·45 PER CENT. CARBON.
Magnified 1,000 diameters.

FIG. 2. — STEEL — 1.5 PER CENT. CARBON.
Quenched at 1050° C. in ice water.
Magnified 1,000 diameters.

If the metal contains more than 0.8 per cent. carbon, it is composed of pearlyte and cementite. If z be the percentage of cementite, we have

$$(1) \quad y + z = 100.$$

Pearlyte contains 0.8 per cent. C, and cementite 6.67 per cent. C (which is the proportion required by the formula Fe_3C), and since the sum of the pearlyte carbon and of the cementite carbon must be equal to the total carbon in the steel, we have the second equation:

$$(2) \quad \frac{0.8}{100} y + \frac{6.67}{100} z = C.$$

While the similarity between the formation of the structure of alloys or of frozen saline solutions and that of carbon steel is indeed striking, there is one feature in which the two phenomena differ, which is momentous. In the case of alloys or of saline solutions the curves of fusibility or of solubility, which indicate the temperature at which the constituent present in excess begins to segregate from the mass, also represent the beginning of the *solidification* of that constituent. The heat evolved is a latent heat of solidification. It indicates a change of internal energy *accompanied by a change of state.*

In the case of steel, on the contrary, the whole metallic mass is already in the solid state when segregation of the constituents takes place. The beat evolved here indicates a change of internal energy which is *not accompanied by a change of state.* A change of this kind is of course very suggestive, if not conclusive, of an allotropic transformation, and in it the believers in the allotropy of iron find their strongest argument.

Such considerations lead us to the notion of solid solutions. The existence of solid solutions can no longer be reasonably contested. We all know with what readiness carbon diffuses through solid steel. The phenomenon is in every way similar to the diffusion of salt in water. Many instances of solid solutions — diffusions' of one metal in another, etc. — have been described recently by Professor Roberts-Austen and others.

The constitution of steel before the segregation of one of the components has begun, *i. e.*, above the curve $M E N$ (Figure 3), remains to be ascertained. This region evidently corresponds to the liquid state in metallic alloys and in saline solutions, from which it might naturally be inferred that the carbide Fe_3C is here uniformly diffused or dissolved through the iron.

It is hardly possible to examine the structure of the metal at a red heat, but by cooling the steel from one of these high temperatures very rapidly — by immersing it in cold water or in some other cooling mixture — the changes, structural and others, which take place during the retardations, do not occur, at least not in their entirety (being denied the necessary time), and we retain in the cold metal the conditions which existed at a high temperature. Such treatment constitutes the process of *hardening*, and we must ascertain the character of the constitution of *hardened* steel.

Calling the microscope to our assistance, we find that if the metal be quenched before the first retardation, unless it be very slightly or very highly carbonized, it is made up of a single constituent represented by dotted areas in the figures of Plate IV, and which has been called *martensite*. Figures 5, 8, and 11 illustrate the structure of steels containing respectively 0.21, 0.35, and 0.80 per cent. of carbon, and quenched above the curve $M E N$ (Figure 3)   Figure 5, Plate I, shows the structure of martensite under a magnification of 1,000 diameters. It is the reproduction of a photo-micrograph by Mr. Osmond.[1] If the metal contains very little carbon, a small amount of iron (ferrite) is found together with the martensite, even after quenching from a very high temperature. (See Figure 1, Plate IV, and Figure 6, Plate I.) In highly carbonized steel Mr. Osmond finds, upon sudden cooling from a high temperature, a new constituent which he has named *austenite*.[2] Figure 2, Plate III, shows the structure of a steel containing 1.50 per cent. carbon and quenched in ice water from a very high temperature (1,050° C.). The light background represents the austenite ; the dark needles are made up of martensite.

If the metal be quenched between the first and the second evolutions of heat, it is found to be made up of a mixture of martensite and ferrite, or of martensite and cementite ; the former when the metal contains less than 0.8 per cent. carbon, the latter when more highly carbonized. This is well illustrated by Figures 3, 6, 9, 13, and 15, of Plate IV. In steels quenched below the recalescence temperature no

---

[1] The Metallographist, Vol. I, No. 1, January, 1898.
[2] Idem.

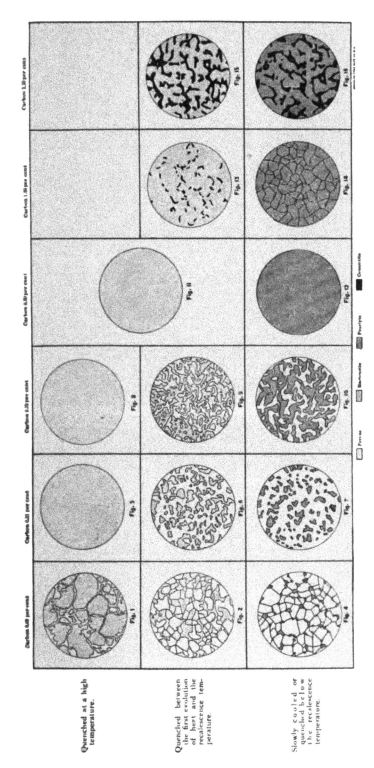

PLATE IV. — MICROSTRUCTURE OF STEEL.

Magnified 72 diameters   (Reduced from original magnification of 250.)

martensite is to be found.   Their microstructural compositions do not differ from those of the slowly cooled metals (Plate IV, lowest row), and we know that steel suddenly cooled from a temperature below that of recalescence is not materially hardened, if at all.

We all know that within certain carbon limits the most striking difference between the properties of steel suddenly cooled from a high temperature, and those of the same metal slowly cooled, is to be found in the enormously greater mineralogical hardness of the former.   This increase of hardness it evidently owes to the presence of martensite. The statement, however, does not in any way help us to solve the problem of hardening, which has been the subject of so much investigation and so much controversy, for the cause of the hardness of martensite remains to be determined.

Martensite may be compared to the liquid portion of saline solutions and of metallic alloys.   Above the curve MEN (Figures 1 and 2) the former are entirely liquid, while steel is made up entirely of martensite (with the exceptions noted above).   Between the first and the second evolutions of heat we have in the one case crystals of one of the constituents suspended in the portion remaining liquid; in the other, crystalline particles of ferrite or of cementite imbedded in a matrix of martensite; while below the second retardation no portion of the saline solution or of the alloy remains liquid, and no martensite remains in the steel.

It is certainly reasonable to suppose that martensite is an homogeneous solid solution of carbon or of the carbide Fe_3C in iron.   It can hardly be anything else.   The contention that it is a definite compound of iron and carbon (to which Professor Arnold would give the formula $Fe_{24}C$) must be abandoned, for it is evident that its composition, like that of the portion remaining liquid during the cooling of metallic alloys, varies both with the composition of the steel and with the temperature.

The fact, however, even if it were conclusively established, that martensite is a solution of iron and carbon or of iron and Fe_3C, could hardly account for its extreme hardness, although certain metallurgists, the "carbonists" as they have been called, see in it a satisfactory explanation of the hardening of steel.   The theory of the "allotropists," who consider martensite as a solution of carbon or of Fe_3C in an allotropic condition of the iron, itself very hard, is more plausible; it is supported by much more cogent evidences, and is steadily gaining ground.

While unhardened steel is generally made up of two of the three constituents, ferrite, cementite, and pearlyte, it may be readily conceived, however, that even if no other impurities were present, the physical properties of the steel, which give to this metal such an unique place in the arts, do not depend exclusively upon the relative proportions of these constituents — in other words, upon the carbon content; but that they depend also upon the distribution, mode of occurrence, size, and shape of the individual grains or crystalline particles, and these features are regulated by the treatment, thermal and physical, to which the metal is subjected.

The structure of steel is extremely sensitive to slight changes of treatment, and an alteration of the structure, however slight, always implies a corresponding alteration of physical properties.

The microscope gives us a means of studying those structural changes which are so closely related to the properties of the metal, and thus opens up possibilities in the art as well as in the science of metallurgy, whose value could hardly be overestimated.

THE RELIABILITY OF THE DISSOCIATION-VALUES DETERMINED BY ELECTRICAL CONDUCTIVITY MEASUREMENTS.

By ARTHUR A. NOYES.

1. *Van Laar's Correction of the Heat of Solution Formula.*

There has recently appeared[1] an article by van Laar in which he claims with the greatest positiveness that the dissociation-values calculated from the electrical conductivity are incorrect. In order to sustain this view, he brings forward certain "proofs" and offers an hypothesis concerning the cause of the supposed unreliability of the conductivity method. The matter is of so great importance and the view of van Laar is, in my opinion, so completely unjustifiable, that I feel compelled to reply to his arguments, and to present the existing proofs of the contrary idea.

As an evidence of his view, van Laar attempted to show that the heats of solution calculated thermodynamically from the changes in solubility caused by variations of temperature agreed much better with the experimental values when the theoretical (Ostwald) dilution-law was used in the calculation, than when the empirical (Rudolphi-van't Hoff) law derived from conductivity values was employed. Unfortunately, however, van Laar has not, it seems to me, derived correctly the relation between the change in solubility and heat of solution in the case of dissociated substances. It is apparent that his equation in the derivation of which the empirical dilution-law was assumed is incorrect, since this furnishes an evidently false result in the limiting case of complete dissociation. The equation to which I refer is as follows :

$$\frac{L}{RT^2} = \frac{3}{3-a}\frac{d \log s}{dT} \qquad (1),$$

in which L expresses the heat of solution; R, the gas constant; T,

---

[1] Ztschr. phys. Chem., 21, 79.

the absolute temperature, a the degree of dissociation, and s the solubility. When $a = 1$, this equation becomes:

$$\frac{L}{RT^2} = \frac{3}{2} \frac{d \log s}{dT} \qquad (2).$$

For complete dissociation, however, this equation must evidently be quite independent of any dilution-law whatever; for no change in the degree of dissociation takes place in this case. Also, since in the case of a wholly dissociated di-ionic salt the equilibrium constant K, which occurs in the general van't Hoff equation:

$$\frac{d \log K}{dT} = \frac{Q}{RT^2},$$

is equal to s^2, the product of the concentrations of the ions, it follows that:

$$\frac{L}{RT^2} = \frac{2 \, d \log s}{dT}, \qquad (3),$$

an equation which has already[1] been employed and confirmed. Even the equation derived by van Laar under the assumption of the theoretical dilution-law,

$$\frac{L}{RT^2} = \frac{2}{2-a} \frac{d \log s}{dT} \qquad (4),$$

becomes simplified to equation (3), when $a = 1$. His two equations (2) and (4) therefore lead to different results in the case where the influence of the dilution-law must disappear. Moreover, even the last equation of van Laar cannot be correct in general, although it fulfils the conditions of the two limiting cases; for it contains no term which represents the change of the dissociation with the temperature. This is, however, obviously necessary; for the heat of solution of a partially dissociated substance may be regarded as the sum of two quantities, one of which depends on the heat of solution of the undissociated substance, and the other on its heat of dissociation. This latter amount of heat is, however, determined thermodynamically by the temperature-change of the dissociation-constant, while the former is independent of it. Accordingly the sum, the actual heat of solution, must be a function of the temperature-change of the dissociation.

---

[1] Nernst, Theoretische Chemie, 516.

The original equation of van't Hoff [1]

$$\frac{L}{RT^2} = \frac{i \, d \log i \, s}{d T} = (1 + a) \frac{d \log s}{d T} + \frac{d a}{d T} \quad (5)$$

satisfies this· theoretical requirement.   That of van Laar is an attempted improvement on this.   According to him the assumption of van't Hoff [2] and Rudolphi [3] that : $L = W + a \, Q$, where L is the actual (total) heat of solution ; W, the heat of solution of the undissociated substance, and Q, its heat of dissociation, is not correct ; but since in the solution of the quantity ds there takes place, besides its own change in dissociation, a change in the degree of dissociation of the total quantity s of the substance present, by an amount $\frac{da}{ds} \, ds$, it follows that $L \, ds = W \, ds + \left(a + s \frac{da}{ds}\right) Q \, ds$, in which the quantity $\frac{da}{ds}$ is to be obtained from the dilution-law.   For the·heat of solution, L, of a gram molecule, it follows, therefore, according to van Laar, that : $L = W + \left(a + s \frac{da}{ds}\right) Q$.

This "correction" seems to me, however, to be wholly erroneous, and to arise from a misunderstanding of the significance of the heat of solution.   The molecular heat of solution is, in a thermodynamical sense, the heat which is absorbed by the dissolving of one molecule of the substance in a solution of *unchangeable* concentration ; namely, that corresponding .to the equilibrium.   Even differential changes of concentration must not occur if the thermodynamic relations are to hold strictly; and if they actually occur, as van Laar assumes, the heat of solution then obtained will differ from the thermodynamical heat by an amount corresponding to the influence of those changes in concentration.   That is to say, the dissolving must take place with a simultaneous increase in the volume of the solution by the addition of the solvent, in such a way that the concentration remains constant, just as in the reversible evaporation of a liquid the volume of the vapor must increase continuously.

By these considerations it seems to me that the falsity of van Laar's attempted correction of van't Hoff's equation is made evident. The matter is, however, so important, as well from a general point of

[1] Ztschr. phys. Chem., **17**, 147 ; $i = (1 + a)$ for a di-ionic electrolyte.
[2] Ztschr. phys. Chem., **17**, 547.   [3] Ztschr. phys. Chem., **17**, 299.

view as from the standpoint of this discussion, that it seems proper to present a direct derivation of the van't Hoff equation by means of a cyclical process, and especially so, since van't Hoff himself, according to the assertion of van Laar,[1] has acknowledged the validity of the latter's proposed correction.

2. *Derivation of the Relation Between the Heat of Solution of Dissociated Substances and the Change in their Solubility with the Temperature.*

In order to derive this relation we will consider a cyclical process to be carried out by means of an osmotic machine. The machine consists, in its original condition, of a cylinder provided with a semipermeable piston, in the bottom of which, below the piston, is one molecule of the solid substance, while above is that amount of the pure solvent in which the solid substance is soluble. The solid substance is moistened with an infinitely small amount of the saturated solution, and the osmotic pressure on the piston is just balanced by a corresponding weight. The machine is placed in a very large heat reservoir at $T°$, and the following cyclical process consisting of five parts is carried out :

Part I. By an indefinitely small decrease of the weight on the piston the latter is allowed to rise until the solid substance is dissolved. Let the concentration of the resulting saturated solution be expressed by s.

Part II. The piston is fastened and the machine placed in a very large heat reservoir at $(T + dT)°$.

Part III. The piston is set free after loading it with a weight equal to the osmotic pressure ; the weight is then increased by an indefinitely small amount, and the piston allowed to sink until the concentration of the solution has risen from s to that of the solution saturated at $T + dT$, that is, $s + ds$.

Part IV. The piston is now allowed to sink still further, until the molecule of the substance has separated out of the solution.

Part V. Finally the machine is placed in the heat reservoir at $T°$.

The cyclical process is now completed and the original condition restored. The Second Law of Energetics gives now the relation which must exist between the work dA performed on the system and

---

[1] Ztschr. phys. Chem., 17, 547.

the amount of heat Q which is transferred from one temperature to the other. This relation is: $dA = Q \dfrac{dT}{T}$. (6).

We will next determine the amount of work performed in the above-described process. The work done in each part consists in a change of volume taking place at a constant pressure, and is therefore equal to the product of these two quantities. According to the *Avogadro-van't Hoff* principle, the osmotic pressure is equal to nRT, where n denotes the number of molecules present. As above, the concentration of the solution saturated at $T°$ is s, and at $T + dT°$ is $s + ds$. Further, let the number of molecules which is present in the solution of one molecule of the solid substance at $T°$ be i, and at $T + dT°$ be $i + di$. Then the following values of the change in volume ($\triangle v$), of the pressure (P), and of the work performed (A), in the separate parts of the process, are:

Part I. $(\triangle v)_\mathrm{I} = \dfrac{1}{s}$; $P_\mathrm{I} = i s R T$; $A_\mathrm{I} = - i R T$.

Part II. $(\triangle v)_\mathrm{II} = 0$; $A_\mathrm{II} = 0$.

Part III. $(\triangle v)_\mathrm{III} = - \left(\dfrac{1}{s} - \dfrac{1}{s + ds} \right)$; $P_\mathrm{III} = i s R (T + dT)$;
$A_\mathrm{III} = i R T\, d \log s$.

Part IV. $(\triangle v)_\mathrm{IV} = - \dfrac{1}{s + ds}$; $P_\mathrm{IV} = (i + di)(s + ds) R (T + dT)$;
$A_\mathrm{IV} = i R T + i R dT + R T di$.

Part V. $(\triangle v)_\mathrm{V} = 0$; $A_\mathrm{V} = 0$.

The total work done in the process is therefore:

$$dA = i R\, (T\, d \log i s + dT).$$

On the other hand the amount of heat, Q, absorbed at the temperature T is evidently the sum of the heat of solution L in the ordinary sense (without work being done) and the external work; that is to say, $Q = L + i RT$. If these values of dA and Q are substituted in equation (6), the expression of the Second Law, we obtain $\dfrac{L}{RT^2} = \dfrac{i d \log i s}{dT}$, which is the original van't Hoff equation. Attention may be especially called to the fact that in this cyclical process no opportunity is offered for the introduction of van Laar's correction. These equations are evidently independent of any dilution-law whatever, and of the number of ions into which the electrolyte is dissociated. If the electrolyte breaks up to an extent a into n ions, $i = 1 + (n-1) a$, whence follows :

$$\frac{L}{RT^2} = [1 + (n-1) a] \frac{d \log s}{dT} + (n-1) \frac{da}{dT}.$$

. Now if the electrolyte follows the dilution-law : $K = \dfrac{(a s)^\nu}{(1-a)s}.$

where ν may have any value, then we obtain by differentiation :

$$\frac{da}{dT} = \frac{a(1-a)}{\nu-(\nu-1)a} \frac{d \log K}{dT} - \frac{(\nu-1)a(1-a)}{\nu-(\nu-1)a} \frac{d \log s}{dT}.$$

By substitution of this value in equation (6) we obtain :

$$\frac{L}{RT^2} = \frac{\nu + (n-\nu)a}{\nu-(\nu-1)a} \frac{d \log s}{dT} + \frac{(n-1)a(1-a)}{\nu-(\nu-1)a} \frac{d \log K}{dT} \quad (7).$$

This equation is rigidly exact and general ; in it we presuppose only that the electrolyte follows some exponential dilution-law of the assumed form.

In the two limiting cases where $a = 1$, and $a = 0$, the equation becomes simplified to :

$$\frac{L}{RT^2} = \frac{n\, d \log s}{dT}, \text{ and } \frac{L}{RT^2} = \frac{d \log s}{dT},$$

as the theory requires. In general, however, in the calculation of the heat of solution, a knowledge not only of the temperature-change of the solubility, but also that of the dissociation-constant, is necessary. The latter change has in the case of most electrolytes been shown to be so small that for practical purposes the last member may ordinarily be neglected without a large error, a simplification which is permissible especially in the case of very weakly and very strongly dissociated electrolytes, for the coefficient of $d \log K : dT$ disappears when a approaches zero or unity.

We will now consider two special cases : first, where the electrolyte is di-ionic and follows the theoretical (Ostwald) dilution-law, where it is, for example, a weak acid or base ; and second, where the electrolyte is di-ionic and follows the empirical (van't Hoff) dilution-law, where it is, for example, a salt. In both cases $n = 2$; in the first case $\nu = 2$, and in the second $\nu = \frac{1}{2}$. By substituting these values in equation (7) and omitting the last member we obtain:

$$\frac{L}{RT^2} = \frac{2}{2-a} \frac{d \log s}{dT} \quad \text{when } \nu = 2 \quad (8).$$

and $\quad \dfrac{L}{RT^2} = \dfrac{3+a}{3-a} \dfrac{d \log s}{dT} \quad \text{when } \nu = \frac{1}{2} \quad$ (9).

The first formula is identical with that obtained by van Laar; the second is, however, essentially different from his, which reads:

$$\frac{L}{RT^2} = \frac{3}{3-a} \frac{d \log s}{dT} \quad (10).$$

3. *Comparison of the Calculated and the Observed Heats of Solution.*

In the following tables the results of the calculation of the heats of solution according to the three formulæ (8), (9), and (10), given above, are placed together beside the experimentally found values for the four substances used by van Laar. Only those values in the next to the last column are calculated by me;[1] the rest are those of van Laar.

HEATS OF SOLUTION.

	Ostwald's dilution-law assumed. Calculated by (8).	Van't Hoff's dilution-law assumed.		Experimentally found.
		Calculated by (10).	Calculated by (9).	
Silver acetate	4369	3688	4562	4613
Silver propionate . . .	3789	3148	3928	3980
Silver isobutyrate . . .	2715	2289	2836	2860
o-Nitrobenzoic acid . .	7167	6766	7449	7083

From the fact that the values in the first column of numbers, which were obtained with the help of the Ostwald law, agree much more closely with the experimental values than do those of the second

---

[1] These values were obtained from those of van Laar in the preceding column by multiplication by $\dfrac{3+a}{3}$ The values of a used are those derived by van Laar from conductivity measurements, namely: for silver acetate, 0.713; for silver propionate, 0.744; for silver isobutyrate, 0.718; and for *o*-nitrobenzoic acid, 0.303.

column, which were calculated by means of van't Hoff's dilution-law and the equation of van't Hoff as corrected by van Laar, van Laar concluded that salts in reality follow Ostwald's dilution-law, and not van't Hoff's, as is to be inferred from the conductivity; in other words, the conductivity gives, according to him, incorrect values of the dissociation. Further support of this latter assumption he finds in the fact that in the case of the three salts, even the values calculated with the help of Ostwald's law differ from the experimental ones by several per cent. But I have proved above, as I believe, that the equation of van Laar, in the deduction of which van't Hoff's dilution-law was presupposed, is incorrect, and that the true expression is equation (9). If one is to decide between the two dilution-laws, the values of the first and *third* columns of numbers must, therefore, be compared with those of the fourth. This comparison shows now just the opposite result: in the case of the three salts, the values deduced from van't Hoff's law agree almost completely (within 0.8 to˙ 1.3 per cent.) with the actual ones, while those from Ostwald's law deviate by about 5 per cent. Moreover, it is especially worthy of note that the nitrobenzoic acid, which like all weak acids follows Ostwald's law, possesses a heat of solution which is in accord with the assumption of the validity of this law, but which is not consistent with van't Hoff's law. The results are therefore in perfect agreement with the usual assumptions based on conductivity in regard to the dissociation of the substances in question, and not only is the argument of van Laar refuted, but at the same time a new proof is furnished of the validity of van't Hoff's dilution-law in the case of salts, and of the reliability of the dissociation-values determined from conductivity measurements.

4. *Electrical Conductivity and Inversion Velocity in Mixtures of Water and Alcohol.*

Van Laar finds a further proof of his view in the fact that the rate of the inversion of sugar by hydrochloric acid, in mixtures of alcohol and water, is essentially different from that in pure water, although according to Cohen[1] the degree of dissociation of the acid, as determined by its conductivity in the two solvents, is the same. Cohen[2] also considers this fact as leading to the conclusion that "in solutions in mixtures of water and alcohol the electrical conductivity is not an entirely correct measure of the degree of dissociation."

---

[1] Ztschr. phys. Chem., 25, 41.   [2] Ibid, p. 44.

It is remarkable that this conclusion has been drawn by two separate investigators; for it appears to me obviously unjustified. Even if the concentrations of the hydrogen ions are the same, it is by no means to be expected that the rates of inversion in the different solvents would be equal. For, aside from the fact that the catalytic activity of the hydrogen ions is not necessarily constant, it is clear that the concentration of one of the reacting substances, the water, undergoes a change; and it would be indeed remarkable if this change had no influence on the rate of inversion. It is not improbable that the ions of the water play an important part in the hydrolysis of the sugar, and it has been shown by Löwenherz[1] that the addition of alcohol greatly reduces the dissociation of water. It is therefore entirely inadmissible to ascribe to the inversion results such a significance as van Laar and Cohen do.

5. *Van Laar's " Explanation of the Deviations from Ostwald's Law."*

It seems superfluous to discuss the hypothesis of van Laar on the cause of the incorrectness of the dissociation-values derived from the conductivity, since, from what has been said above, no reason exists to doubt their correctness. A remark in regard to it may be, however, not without interest. If I correctly understand his hypothesis, its essential feature is that the Joule heat-effect, which is continuously produced by the current during the measurement of the conductivity, causes a localized increase of temperature around the ions, as a result of which the dissociation is temporarily changed. Against this hypothesis, however, there exists the following fatal objection: if the dissociation and conductivity are affected by the heat produced by the current, then the observed values of these would depend on the strength of current employed, which is known not to be the case.

6. *Existing Determinations of the Dissociation by Independent Methods.*

In his article, van Laar pays no attention to the previously published experiments which bear upon the reliability of the conductivity method of determining dissociation; yet certain investigations have been carried out, the purpose of which was to answer this very question. In the first place may be considered the recent investigations of

---

[1] Ztschr. phys. Chem., 20, 294.

Jones,[1] Loomis,[2] Abegg,[3] and Raoult,[4] on the lowering of the freezing-point of water by certain salts. Of these the two salts, potassium chloride and sodium chloride, have been most carefully studied ; for the lowerings of freezing-point caused by them have been determined by all four of the mentioned investigators.

In order to show now the extent of the agreement of the values with one another and with the results of the conductivity method, I have brought together in the following table the calculated dissociation-values for the concentrations 0.01 to 0.1 normal. In the calculations from the freezing-point-lowerings 1.85° was assumed as the molecular lowering of water.[5] The conductivity determinations of Kohlrausch were made use of in calculating the dissociation-values by this method.

DISSOCIATION–VALUES.

Normal concentrations.	POTASSIUM CHLORIDE.						SODIUM CHLORIDE.					
	Jones.	Loomis.	Abegg.	Raoult.	Mean.	Kohlrausch.	Jones.	Loomis.	Abegg.	Raoult.	Mean.	Kohlrausch.
0.01	95	95	96	..	95.3	94.4	95	99	105	..	99.7	93.7
0.03	92	90	90	91	90.8	91.1	92	93	92	97	93.5	89.6
0.05	90	89	88	88	88.8	89.1	90	91	92	93	91.5	87.3
0.10	87	86	..	85	86.0	86.2	88	88	88	89	88.2	84.2

It is seen that, in the case of potassium chloride, the four separate values by the freezing-point method agree well with one another ; and the averages of them are almost identical with the dissociation-values calculated from Kohlrausch's conductivity measurements. In the case of sodium chloride, however, the agreement between the freezing-point results of the separate observers is not so complete, and their averages are about 4 per cent. higher than the values to which

[1] Ztschr. phys. Chem., 11, 110, 529; 12, 623.
[2] Wied. Ann., 51, 500; 57, 495.
[3] Ztschr. phys. Chem., 20, 207.
[4] Comptes rendus, 124, 885; 125, 751.
[5] Compare Abegg., Wied. Ann., 64, 499.

the conductivity leads. Whether this latter result arises from a constant error in the freezing-point determinations or from a theoretical inaccuracy in one of the two methods can not, of course, be determined at present. In view of this fact it must be admitted, to be sure, that these freezing-point measurements do not contribute much towards answering the question regarding the reliability of the conductivity method. Nevertheless a certain significance is to be attributed to the close agreement in the case of the potassium chloride, the best investigated salt.

In another way, however, more decisive evidence of the reliability of that method has been furnished. From the phenomenon of solubility effect it is possible to determine dissociation-values, and experiments of Noyes and Abbot[1] have shown that the values so determined in the case of certain thallium salts are in accordance with those derived from the conductivity. For details reference is made to the original article; but the final results may be presented here. The first two columns of figures in the following table show the dissociation-values which were calculated from independent solubility experiments with two pairs of salts. The last column contains the corresponding dissociation-values, which were derived from the conductivity.

Salt.	From solubility experiments. I.	From solubility experiments. II.	From the conductivity.
Tl Cl	86.5	86.5	86.6
Tl SCN	86.7	86.6	85.6
Tl Br O$_3$	89.9	91.1	89.0

These entirely distinct methods, freezing-point-lowering and solubility-effect, lead therefore to nearly the same dissociation-values. That the electrical conductivity furnishes essentially correct dissociation-values in the case of di-ionic salts in moderately dilute solution is therefore probable. At any rate, more weighty reasons than those brought forward by van Laar must be discovered, before this method should be discredited.

---

[1] Ztschr. phys. Chem., 16, 136.

EXPERIMENTAL STUDY OF TRANSFORMER DIA-
GRAMS.

By HARRY E. CLIFFORD, NATHAN HAYWARD, AND ROBERT ANDERSON.

THE method of graphically representing the various currents and electromotive forces concerned in two mutually related circuits has been most beautifully and extensively developed by Bedell.[1] For two such circuits, in one of which, the primary, a constant periodic current is caused to flow, the loci of primary impressed voltage and secondary current and electromotive force, when the secondary resistance is varied, are shown in Figure 1. To pass to the case in which the electromotive force impressed on the primary circuit is constant, requires a mere alteration in length of every line in the diagram, in a ratio which is readily determined for a given primary E. M. F. The character of the necessary

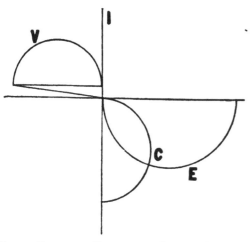

FIG. 1 — TRANSFORMER DIAGRAM WITH CONSTANT CURRENT IN THE PRIMARY.

V = Locus of vectors representing primary impressed E.M.F. at different loads.

E = Locus of vectors representing voltage across the secondary at different loads.

C = Locus of vectors representing secondary current at different loads.

I = Vector representing primary current.

All phases are measured from I, the clock-wise direction being taken as positive for lag. All vectors are in the extreme left-hand position at no load; moving in the clock-wise direction as the load comes on.

changes in the original diagram will be at once appreciated from a consideration of Figure 2. The effect of magnetic leakage or external inductance in the secondary circuit is to cause a displacement of

[1] Bedell. Principles of the Transformer.

the locus of the primary E. M. F. from the locus of the primary current. This effect is well illustrated in Figure 3.

It seems somewhat strange that the relations indicated by these various transformer diagrams should have received no especial attention from the standpoint of experimental verification — the only tests, so far as we are aware, being those carried on by Bedell,[1] in which there is, as it seems to us, one serious error.   Throughout his work it

is a s s u m e d, without measurement, that the secondary E. M. F. is 90° behind the primary current — an assumption which is hardly justifiable in the light of our results.

As our work demanded a transformer in which the magnetic reluctance a n d magnetic leakage could be changed through wide limits, we chose a modified Ferranti type, in which the coils could be placed in any relative position, the core removed or used with either open or closed magnetic circuit.   To accomplish t h i s the primary was wound on

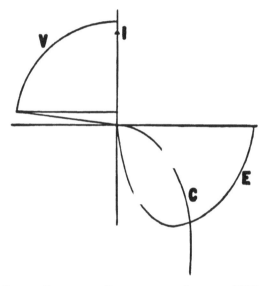

Fig. 3. — Transformer Diagram, with Constant E.M.F. Impressed on Primary.

V = Locus of vectors representing primary E.M.F. at different loads.
E = Locus of vectors representing voltage across secondary at different loads.
C = Locus of vectors representing current in secondary at different loads.
I = Direction of vector representing primary current; the length of vector being a variable.

a bobbin, into which the core would just fit: the secondary slipped over the primary.   The core was built up of seventy-three mild steel stampings 20" x 2.5" of the kind used by the General Electric Company in the modern "H" transformer, and when placed in position might be left either unbent, as in a hedgehog transformer, or bent into a more or less compact magnetic circuit.   We were thus able to vary

[1] Proceedings of International Electrical Congress held at Chicago, 1893, p. 234.

the magnetic reluctance from that of an air transformer to that of a commercial Ferranti transformer, and at each step vary the leakage by changing the relative position of the coils. We stepped up by a ratio of 4 to 1, using a bank of lamps as our load, which we assumed to be non-inductive.

The primary current was measured by the ordinary type of Siemen's dynamometer of a range from 0.5 to 20 amperes. The dynamometer was calibrated by means of Weston ammeters. The voltages, both primary and secondary, were determined with an electrostatic voltmeter having a range from 30 to 150 volts. The range of the voltmeter necessitated the use of a non-inductive drop in the secondary circuit, but the use of this enabled the instrument to be read in the most precise portion of the scale. Calibration of these instruments was made with a Weston voltmeter. The phase angles were determined by the ordinary contact method,[1] measuring the angles between the zero points. When the wave was distorted by hysteresis, the lag angle was obtained from the power factor. The frequency was determined by comparison with a Konig standard fork, the dynamo being found to vary comparatively little throughout the entire time of the experiments, and to give very closely a sine wave of electromotive force.

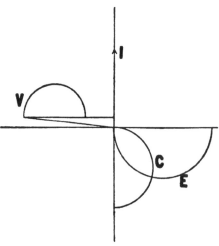

FIG. 3.—TRANSFORMER DIAGRAM, WITH CONSTANT CURRENT IN PRIMARY, AND EITHER EXTERNAL SELF-INDUCTANCE OR MAGNETIC LEAKAGE IN SECONDARY.

V = Locus of vectors representing primary E.M.F.
   (The semi-circle has been moved away from vector I.)
E = Locus of vectors representing secondary E.M.F.
C = Locus of vectors representing secondary current.
I = Vector representing primary current.

CONSTANT CURRENT AIR TRANSFORMER.

The first series of tests was carried on without the use of the iron core, the secondary coil being placed directly over the primary. The

[1] Electrical World, Vol. xviii, p. 141, August 29, 1891.

inductances of the coils were determined by the Wheatstone bridge method of comparison with a known variable inductance.

Owing to the fact that in this run the values were not within the ranges of our instruments, the primary current and the primary and secondary electromotive forces were determined from the plot taken by the contact method, the ordinates being measured at twelve points, and the square root of the mean square found. A summary of the results of this test is given in the following table:

TABLE I.

	Effective Values.				Phase with Respect to P. C. of	
Loads.	P. E.	S. E.	P. C.	S. C.	P. E.	S. E.
No. . .	9.17	32.4	14.2	....	$-77.4°$	$+96.0°$
1st . .	9.11	31.4	14.2	.342	$-71.1°$	$+107.4°$
2d . .	8.32	22.9	14.2	.832	$-68.4°$	$+132.6°$
3d . .	6.90	8.38	14.2	1.20	$-67.8°$	$+161.4°$
4th . .	6.52	3.29	14.2	1.22	$-67.2°$	$+170.4°$
5th . .	6.44	1.17	14.2	1.22	$-70.5°$	$+174.9°$

It is to be observed that the current is kept constant throughout the test. A no-load run could not be made, as the contact method required us to use a drop across our secondary. This, however, was allowed for in diagram. A correction was also made for the primary drop.

The effect of magnetic leakage is clearly shown in the following diagram, Figure 4, in which the plotted points represent the experimental results, the continuous lines indicating the theoretical polar diagram. Considering the fact that a precision closer than 2 per cent. could hardly be hoped for in the measurements themselves, it seems reasonable to conclude that the air transformer follows the theoretical constant current diagram with great exactness.

CONSTANT ELECTROMOTIVE FORCE IN PRIMARY.

A series of tests was next conducted with the coils arranged as at first, but with the iron core introduced in the hedgehog form, and the

primary E. M. F. kept sensibly constant at 49 volts.   A summary
follows in Table II :

TABLE II.

LOADS.	P. E.	S. E.	P. C.	S. C.	P. E.	S. E.
	EFFECTIVE VALUES.				PHASE WITH RESPECT TO P. C. OF	
No. . .	48.9	179.	8.17	....	−85.5°	+ 96°.
1st . .	48.9	174.	9.50	1.02	−66.6°	+ 123°.
2d . .	49.0	169.	11.42	1.76	−52.5°	+ 137°.
3d . .	49.5	166.	13.37	2.34	−48.3°	+ 147°.
4th . .	49.0	158.	15.68	3.00	−45.0°	+ 153°.
5th . .	49.0	152.	17.75	3.57	−43.6°	+ 157°.

The graphical representation of this series is shown in Figure 5.
Although no comparison can be made with the theoretical diagram,
the form of the curves is at once seen to be similar to that which
would be expected from the theory itself, and which is illustrated in
general by the diagram, Figure 2.   A second series was next taken
with the core precisely the same as in the preceding tests, but the
coils slipped apart instead of being one directly outside the other.

TABLE III.

LOADS.	P. E.	S E.	P. C.	S. C.	P E	S. E.
	EFFECTIVE VALUES.				PHASE BETWEEN P. C AND	
No. . .	45.1	147.	7.77	....	−83.5°	97°.
1st . .	45.1	140.	8.73	.79	−68.9°	122°.
2d . .	45.3	132.	10.18	1.35	−62.2°	139°.
3d . .	45.0	126.	11.45	1.72	` −60 5°	146°.
4th . .	44.9	100.	14.47	2.58	−61.2°	159°.

The graphical representation of this series, Figure 6, is of course
similar to that shown in Figure 5, and is of interest as evidencing very
strongly the effect of increased magnetic leakage.   With this type of

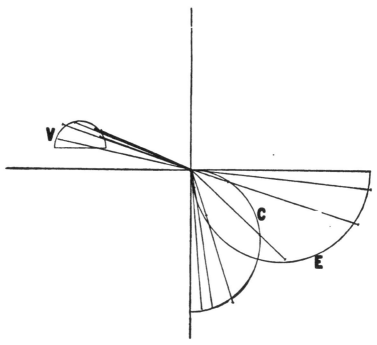

FIG. 4. — AIR TRANSFORMER

V = Locus of primary E.M.F.   E = Locus of secondary E.M.F.   C = Locus of secondary current.

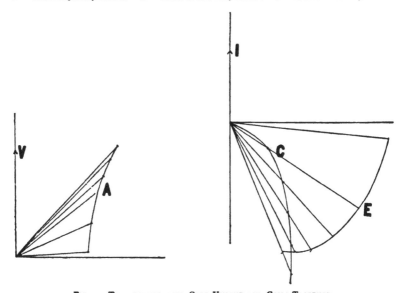

FIG. 5. — TRANSFORMER WITH CORE UNBENT AND COILS TOGETHER.

C = Locus of the current in secondary: the phase measured from primary current "I."
E = Locus of the E.M F in secondary; the phase measured from primary current "I."
A = Locus of the current in primary, the phase measured from primary E M.F. " V."

(These are direct plots of the observed values, the clock-wise direction being taken as the direction of lag.)

transformer the current wave was not sensibly distorted from its sine form. From this fact we should expect no distortion of the polar diagram. In order to see if this is the case, we have reduced by direct ratio Figures 5 and 6 to constant current diagrams, and superimposed one on the other in Figure 6 (a). These two diagrams, like that in Figure 4, seem to show most conclusively that the theory sensibly holds for a transformer of this type. The effect of magnetic leakage is here shown most satisfactorily. The larger semicircles represent the values with the coils together, the smaller those with coils apart.

CLOSED MAGNETIC CIRCUIT.

The core was next bent to form a closed magnetic circuit, giving a loop of such size that the coils might be arranged to show the effect of leakage to a greater or less extent. In this series of tests hysteresis comes into play to a marked degree, causing a distortion of the primary current wave, and producing a change in the phase relations of primary and secondary electromotive force. The lag angle in the

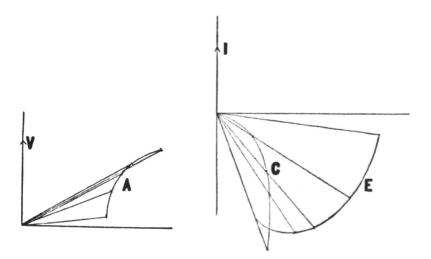

FIG. 6. — TRANSFORMER WITH CORE UNBENT AND COILS APART.

C = Locus of the current in secondary; the phase measured from primary current " I."
E = Locus of the E.M.F. in secondary; the phase measured from primary current " I."
A = Locus of the current in primary; the phase measured from the E M F. of primary " V."

(These are direct plots of the observed values; the clock-wise direction being taken as the direction of lag.)

Fig. 6 (a).

(Figures 5 and 6 reduced to constant current)

The larger circles are obtained from 5, the smaller from 6. The circles are drawn as the ones best representing the points obtained from 5 and 6.

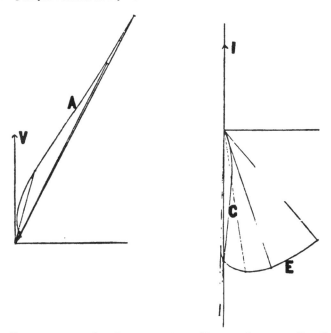

FIG. 7.—TRANSFORMER WITH CORE BENT INTO A CLOSED MAGNETIC CIRCUIT AND COILS TOGETHER.

A = Locus of the primary current ; the phase measured from primary voltage " V."
C = Locus of the secondary current . the phase measured from primary current " I "
E = Locus of the secondary E M F : the phase measured from primary current " I "

(These are direct plots of the observed values , the clock wise direction being taken as the direction of lag) primary circuit was consequently determined from the power factor deduced from the watt curve. A summary of these results is given in Table IV, and the graphical result is shown in Figure 7.

TABLE IV.

Loads.	Effective Values.				Phase between P. C. and	
	P. E.	S. E.	P. C.	S. C.	P. E.	S. E.
No. . .	46.8	192.	.41	0	— 41°.	138°.
1st . .	47.2	192.	.56	.096	— 21°.	159°.
2d . .	47.1	187.	4.92	1.10	— 16°.	171°.
3d . .	46.8	170.	13.33	3.08	— 28°. —	180°.
4th . .	46.8	152.	17.00	4.07	— 28.° +	181°.

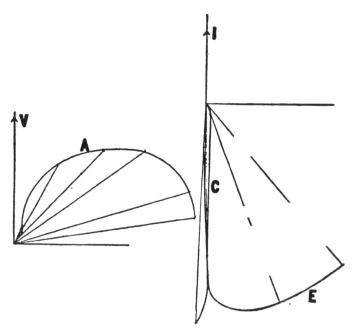

Fig. 8.—Transformer with Closed Magnetic Coils at Right Angles.

C = Locus of the secondary current; the phase measured from primary current " I."
E = Locus of the potential across secondary; the phase measured from primary current " I."
A = Locus of the primary current; the phase measured from primary voltage " V."

(These curves are direct plots of the observed values; the clock-wise direction taken as direction of lag.)

It will be observed that although the theoretical diagram is followed to a certain extent, yet the distortion is very marked. It will be noticed that the lag of secondary E. M. F. with reference to the primary current is no longer 90° on open secondary, but 138°, which effect is undoubtedly to be ascribed to hysteresis. The distortion of the current wave is also much more marked when no load is present than at full load.

Keeping the magnetic circuit constant in character, the coils were next moved so as to occupy planes at right angles to each other, thus increasing the leakage. Table V gives a summary of the results when the coils were in this position.

TABLE V.

| | Effective Values. | | | | Phase between P. C. and | |
Loads.	P. E.	S. E.	P. C.	S. C.	P. E.	S. E.
No. . .	46.6	192.	.407	0	− 42°.	139°.
1st . .	46.6	188.	.88	.094	− 25°.	158°.
2d . .	46.1	163.	4.05	.93	− 29°.	179°.
3d . .	46.3	150.	5.75	1.34	− 44°.	178°.
4th . .	46.7	133.	7.20	1.68	− 55°.	178°.
5th . .	45.5	38.	8.25	1.93	− 74°.	182°.
6th . .	45.5	....	8.25	Short circuit.	− 82°.	182°.

It is very interesting to observe in the graphical representation of the results of this test, Figure 8, the peculiar form for the primary current curve, differing as it does from the form obtained in any of the previous tests, and indicating the nature of variation of the primary current in those transformers whose secondaries may be safely short-circuited.

As a final test it was decided to bring the transformer into as nearly a practical form as possible. Over four pounds (29 per cent.) of iron were removed from the core, which was then made in the closed circuit form, the coils being placed one directly over the other, the arrangement being a good Ferranti transformer. The tests were conducted exactly as in the two immediately preceding cases. Summary and graphical representations are shown in Table VI and Figure 9.

TABLE VI.

	EFFECTIVE VALUES.				PHASE BETWEEN P. C. AND	
LOADS.	P. E.	S. E.	P. C.	S. C.	P. E.	S. E.
No. . .	46.7	....	.396	....	— 55°.	+ 125°.
1st . .	47.0	202.	.652	.115	— 19°.	+ 162°.
2d . .	46.8	185.	4.55	1.079	— 13°.	+ 174°.
3d . .	46.8	174.	.10.05	2.34	— 22°.	+ 174°.
4th . .	46.8	149.	16.8	4.02	— 31°.	+ 176.°

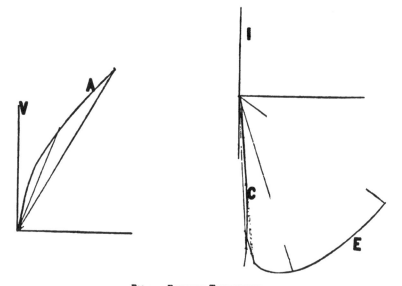

FIG. 9. — FERRANTI TRANSFORMER.

C = Locus of secondary current; the phase measured from "I."
E = Locus of voltage across secondary; the phase measured from "I."
A = Locus of primary current; the phase measured from primary E.M.F. "V."

(These curves are direct plots of observed values; the clock-wise direction being taken as direction of lag.)

It is of some interest to observe the diminished distortion in this particular diagram, due, without doubt, to the lessened effect of hysteresis, owing to the removal of a portion of the iron forming the magnetic circuit. It will also be observed that the lag of secondary E. M. F. with respect to primary current has been reduced from 138° to 125°, as one would expect.

If we now compare the last three figures with Figure 2, we see how much they are distorted from the theoretical, while as we have seen in the first three cases the distortion, if any, is negligible. This distortion, therefore, appears first when hysteresis begins to have its effect in altering the sine wave of primary current, and what is more, the distortion increases as the effect of hysteresis increases. We therefore feel safe in assuming that this distortion is due to hysteresis.

If we compare Figures 7, 8, and 9 with 2, we see that the secondary diagram is not only revolved in the clockwise direction, but is shut up, *i. e.*, the no-load vector is revolved further than the full-load. This is what we should expect, as at no-load the hysteresis effect, as shown by distortion of the primary current curve, is much greater than at full-load.

ROGERS LABORATORY OF PHYSICS.
December, 1897.

ON THE BEST RESISTANCE FOR A SENSITIVE GAL-VANOMETER.[1]

By FRANK A. LAWS.

IT is well known that in the practice of most methods of electrical testing, the precision attainable depends on the proper adjustment of the galvanometer resistance to the work in hand, and that in planning new work a solution for the best galvanometer resistance is always made. It is our purpose to derive a solution under conditions other than those usually imposed.

To render the discussion more complete, a statement of the usual solution of the problem will be given. It is as follows: If a simple circuit, consisting of a battery, a resistance and a galvanometer, be arranged, the deflection of the instrument will be a maximum when the galvanometer resistance is equal to that of the remainder of the circuit. In obtaining this solution two assumptions are made:

First. That the coils have fixed dimensions.

Second. That the ratio of the diameter of the covered to that of the bare wire is the same for all sizes.

That this second assumption is incorrect is shown by the fact that the ratio for a No. 20 B. & S. gauge double silk-covered wire of the American Electrical Works is 1.12, while for a No. 40 wire it is 2.77.

In the course of the discussion we shall use the following symbols:

G, the galvanometer constant, field at center of coil per unit current.

I_G, galvanometer current.

D, deflection of instrument.

V, volume of the coil.

b, linear constant of the coil, $b = k (V)^{\frac{1}{3}}$.

R_G, the galvanometer resistance.

R, the resistance of the circuit external to the galvanometer.

C, the diameter of the covered wire.

[1] Reprinted from the Physical Review, Vol. V, No. 5 (1897) pp. 300-305.

B, the diameter of the bare wire.

$y = C/B$.

w, the resistance per unit volume of the wire.

n, the number of turns per unit area.

E, E. M. F. of battery.

k, constants.

We shall assume an indefinitely short needle, and for the present retain the first assumption.

The usual demonstration is based on the equation,

$$G = k \sqrt{R_G}.$$

This equation requires correction. It is easily shown that

$$G = 4\pi n b \ (\phi. s),$$

where $(\phi. s)$ is a function the shape of the coil. As the same bobbin is to be used in all cases, we have

$$G = k_1 n = \frac{k_1}{C^2} = k_2 \frac{\sqrt{w}}{y} = k_3 \sqrt{\frac{R_G}{y^2}}.$$

To apply this to the typical circuit we have

$$I_G = \frac{E}{R + R_G}, \quad D = k_4 \ I_G \ G = k_5 \frac{E \sqrt{\dfrac{R_G}{y^2}}}{R + R_G}.$$

Differentiating and solving for a maximum, we have

$$R_G = \frac{R}{\dfrac{2}{y^2 \dfrac{d\left(\dfrac{R_G}{y^2}\right)}{dR_G}} - 1}.$$

If y be constant, this reduces to

$$R_G = R,$$

the relation usually given.

This general formula was first given by Professor Silas W. Holman. In order to employ it the properties of the wire must be found by winding experimental coils.

The assumption of fixed dimensions will now be discarded, and attention given to the galvanometer constants of a family of similar instruments. For coils of rectangular cross section we have, referring to the diagram:

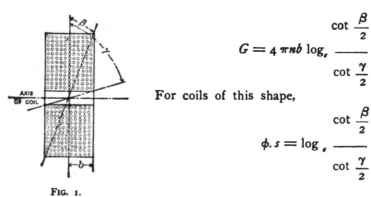

FIG. 1.

$$G = 4\,\pi n b \, \log_e \frac{\cot \dfrac{\beta}{2}}{\cot \dfrac{\gamma}{2}}$$

For coils of this shape,

$$\phi. s = \log_e \frac{\cot \dfrac{\beta}{2}}{\cot \dfrac{\gamma}{2}}$$

For instruments of the same family, β and γ constant, we have:

$$G = k_6 n b = k_6 \frac{b}{yy B^2}$$

$$b = k_7 \left(\frac{R}{w}\right)^{\frac{1}{3}},$$

$$y = \frac{k_9}{B^2 w^{\frac{1}{3}}},$$

$$\therefore G = k_{10} \left(\frac{w^{\frac{1}{3}}}{y}\right) R_G^{\frac{1}{3}}.$$

This shows that the sensitiveness of similar galvanometers wound to various resistances with the same size of wire, is proportional to the cube root of the galvanometer resistance. Also for a family of instruments of the same resistance the best size of wire will be that for which $\dfrac{w^{\frac{1}{4}}}{y}$ is a maximum. By winding experimental coils we obtain data for determining this function, a table of values of which is given below.

WIRE OF AMERICAN ELECTRICAL WORKS.

B. and S. Gauge No.	Single, Silk covered.	Double, Silk covered.	Ratio.
	$\dfrac{w^{\frac{1}{4}}}{y}$	$\dfrac{w^{\frac{1}{4}}}{y}$	
20	.94	.87	1.08
26	1.40	1.12	1.25
30	1.81	1.27	1.42
34	2.21	1.34	1.65
36	2.19	1.33	1.65
38	2.10	1 28	1.64
40	....	1.15	....

The results are plotted in Fig. 2.

The best size for both single and double-covered wire is seen to be No. 34. This result, of course, applies only to the particular make of wire examined.

We see that there may be two instruments of a family having the same resistance which will give equal deflections with the same current.

We are now in position to find the best galvanometer resistance. We have

$$D = k_{10} \left(\frac{w^{\frac{1}{4}}}{y} \right) \frac{R_G^{\frac{1}{2}}}{R + R_G}.$$

Considering for the moment that all the instruments are wound with the same size of wire, we have a maximum value of D when

$$R_G = R/2.$$

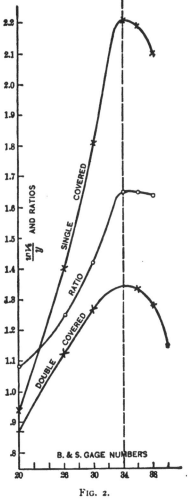

FIG. 2.

RELATIVE GALVANOMETER CONSTANTS FOR SIMILAR COILS WOUND TO THE SAME RESISTANCE WITH VARIOUS SIZES OF SILK-COVERED WIRE.

Therefore the best galvanometer of the family is one which is wound with a No. 34 wire to a resistance equal to one-half that of the remainder of the circuit.

We will now consider two special examples :

The coils of two-spool reflecting galvanometers very commonly have a volume of about eight cubic inches. Suppose the resistance external to the instrument to be 10ω and that double silk-covered wire is to be used, then by the ordinary rule the ideal galvanometer resistance is 10ω. In accordance with our first deduction No. 20 wire will give the best winding. R_G will then be 6ω.

$$D = k\,I_G\,G = k_1\,I_G\,nb\,;$$

$$D_{20} = k_1\,\frac{E}{16.}\;775.\;(8.)^{\frac{1}{3}}.$$

For the instrument wound with No. 34 wire we have $R_G = 5.\omega$, corresponding to a volume of coil of .026 cubic inches.

$$D_{34} = k_1\;\frac{E}{15}\,.\;7780.\;(.026)^{\frac{1}{3}}\,;$$

$$\frac{D_{34}}{D_{26}} = 1.6\;\text{appox.}$$

The deflection of an instrument similar to the first, but designed in accordance with our suggestion, will be 60 per cent. greater than that of the instrument wound in accordance with what is called the solution for the best galvanometer resistance.

The coils of eight cubic inches would have a diameter of about 2.6″ and an axial breadth of about 1.5″. The approximate dimensions of the second coil would be diameter .4″, breadth .25″. The weight of wire in the first case would be 30. ounces, in the second, .032 ounces.

The results we have just obtained are in emphasis of the plea of Professor Boys for small instruments. In this connection we may observe that if we have two similar galvanometers of equal resistance, the first having a constant G_1, and wound with wire for which the ratio of the diameter covered to bare wire is y_1, while these quantities for the second instrument are G_2 and y_2, and if the linear dimensions of the first are N times those of the second, then,

$$\frac{G_2}{G_1} = \sqrt{N}\left(\frac{y_1}{y_2}\right).$$

The factor $\frac{y_1}{y_2}$ shows that there is a limit below which it is unprofitable to reduce the size of the instrument, this being the size for which a No. 34 wire is appropriate.

For galvanometers originally wound with wire finer than No. 34, the more sensitive instrument would be the larger.

We will now consider the case of a Wheatstone bridge where the resistances of the arms are small and nearly equal. Let them be of about one ohm each. Such an arrangement would be used in standardizing one-ohm coils, or in bolometric work.

The ordinary solution gives one ohm as the best galvanometer resistance. If the coil has a volume of 8 cubic inches, the winding should be of No. 16 wire making $R_G = .88\omega$. We will suppose the E. M. F. of the battery to be 1 volt, its resistance 1 ohm, and that the bridge is out of balance by $\frac{1}{1000}$ ohm ; then with due regard to carrying capacity we have

$$I_G = \tfrac{1}{15000}. \text{ approx.}$$
$$D_{16} = k_1 \, 325. \; (8)^{\frac{1}{3}} \, \tfrac{1}{15000}.$$

If wound as here indicated, we should have

$$R_G = .5\omega \quad I_G = \tfrac{1}{12000}.$$
$$D_{34} = k_1 \, 7780. \; (.0026)^{\frac{1}{3}} \, \tfrac{1}{12000}.$$
$$\frac{D_{34}}{D_{16}} = 2.04.$$

The deflection may be doubled by properly designing the galvanometer. This result is for double-covered wire. If single-covered be used, we should multiply by the ratio factor 1.65, giving

$$\frac{D_{34}}{D_{16}} = 3.36,$$

showing that by the use of a proper galvanometer wound with single-covered wire the sensitiveness of the bridge may be increased about 200 per cent.

The gain indicated in the examples is much greater with very low than with high resistance instruments, the former as usually wound departing more widely from the conditions here imposed.

Of course the numerical results indicate chiefly the order of magnitude of the gain, for it would be impossible to reduce the size of the instruments as much as is here indicated.   Also we have not considered the magnetic system, or the fact that the space at the center of the coil must be large enough for free movement of the needle, and that turns in very close proximity to a finite needle produce less than the effect indicated by the strength of field at the center of the coil.

Our conclusions are, that if an existing arrangement of apparatus be taken, we can increase its sensitiveness by replacing the galvanometer by a similar instrument wound with wire of a particular size to a resistance equal to one-half of that of the remainder of the circuit. This best size of wire is that for which $\dfrac{w^4}{y}$ is a maximum.

For low resistance instruments the reduction in size renders magnetic shielding possible without undue clumsiness.

ROGERS LABORATORY OF PHYSICS,
MASSACHUSETTS INSTITUTE OF TECHNOLOGY,
July, 1897.

THE EXACT ESTIMATION OF TOTAL CARBOHYDRATES IN ACID HYDROLYZED STARCH PRODUCTS.[1]

By GEORGE W. ROLFE and W. A. FAXON.

Received June 11, 1897.

THE determination of the exact amount of carbohydrates present in solutions of commercial glucose has always been conjectural, since the evaporation method, the only available means of estimation, has always caused in the residue an indeterminate amount of decomposition, usually attributed to oxidation or the destructive effect of high temperature. Since, however, in acid hydrolyzed starch products, it seems certain that the component carbohydrates preserve their individuality throughout the combinations which they may make with each other, at least so far as to have constant optical and chemical properties, as well as a constant influence on the specific gravity per unit weight of each present in solution, it is quite possible to determine the *proportional amount* of each constituent by the use of one arbitrary specific factor throughout the calculation. This has been explained in detail in several publications of O'Sullivan, Brown, Heron, Morris, and others. It is also well known that the factor used is that representing the increase in specific gravity caused by 1 gram of cane sugar in 100 cc. of solution.

It follows that if the specific gravity influences of the isolated carbohydrates were known it would be quite possible to predict the specific gravity factor of any hydrolyzed starch product. A year ago an attempt was made by one of us to predict these factors of the hydrolyzed starch products for all rotation between the limits of possible specific rotatory powers for the factor 386. A provisional curve was plotted from the following formula:

$$\Sigma = m\Sigma_m + g\Sigma_g + d\Sigma_d.$$

Σ being the specific gravity factor of the hydrolyzed starch product identified by its specific rotatory power; m, g, and d, the respective

[1] Reprinted from *Journal American Chemical Society*, Vol. XIX, No. 9.

percentages of maltose, dextrose, and dextrin present; and Σ_m, Σ_g, Σ_{d}, their corresponding specific gravity factors.

We considered this curve a provisional one, as we were obliged to approximate the factor for dextrin from the imperfect data of O'Sullivan, Salomon, and others, taking 0.00400 as the most probable value. The factors for dextrose and maltose were those of Salomon [1] for 10 per cent. solutions, 0.00381 and 0.00390, respectively. We found that the plotted values of the calculated factors, where the specific rotatory powers were expressed as abscissæ, formed a *straight line joining the plots corresponding to the factors taken for pure dextrin and dextrose* at the corresponding rotation of 195° and 53.5°.

Having defined approximate theoretical values we have sought to confirm them by actual determinations of the total solids present in a number of representative solutions prepared as previously described.[2]

Much ingenious apparatus [3] has been devised for overcoming the decomposition in drying already referred to. The principle of one class is the introduction of a presumably inert atmosphere, such as coal gas or hydrogen. Another class uses a vacuum, which mitigates at the same time oxidation and influence of high temperature.

Apparently, experience places so little confidence in the advantages of any of these multifarious appliances, at least in the case of sugars, that the ordinary method of drying at about 100° till the continual loss does not exceed a given rate per hour, is that usually recommended. We therefore at first dried 10 cc. each of solutions of hydrolyzed starch products on paper rolls at 105°–110° C.,[4] the rolls being placed in weighing beakers, and dried to practically constant weight in an oven kept at the given temperature. The results obtained were discordant, and in general lower than those of our provisional curve.

A second series of determinations were made by drying paper rolls, prepared in a similar way in a vacuum pan, the temperature being about 40°. In breaking the vacuum, air was passed through sulphuric acid and made to enter at the bottom of the pan to avoid the descent of moist vapors. Drying by this method was exceedingly tedious, and the values obtained are much lower than given by our

---

[1] J. prakt. Chem., 2, 28.
[2] This Journal, 18, 871–872.
[3] Wiley's Principles and Practice of Agricultural Analysis, Vol. III; see also recent bulletin by the U. S. Department of Agriculture.
[4] Rolfe and Defren : this Journal, 18, 872.

assumed curve, besides being discordant. They indicate the need of more heat in drying.

At this point in our work we received the important paper by ' Brown, Morris, and Millar,[1] giving the details of an elaborate investigation of the primary carbohydrates and of the products of diastase hydrolysis. In the latter the actual values were successfully predicted by a method similar to ours, but by a somewhat more exact formula, as it took into consideration variations in the factors due to those of concentration. These investigators used the drying apparatus designed by Lobray de Bryn and Von Laent[2] for the drying of maltose. This apparatus had worked so successfully for diastase converted solutions that we at once adopted a modification of it for our next essay. Several pieces of apparatus of the type described below (Figure 1) were

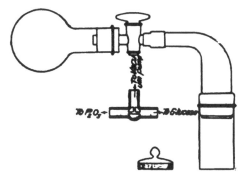

FIG. 1. — APPARATUS FOR DRYING SUGARS AND GLUCOSE SYRUPS.

improvised from the stock of material available. Our apparatus is arranged so that the greater bulk of the water can be first removed without coming in contact with the phosphorus pentoxide. This avoids previous evaporation on a water-bath. The tared weighing beaker containing the solution to be evaporated is slipped on the end of the adapter, an air-tight joint being made with "bill-tie" tubing. By the three-way stop-cock communication can be had with the vacuum pump, and with a 250 cc. flask containing phosphorus pentoxide, or either can be shut off.

The method of drying is as follows : The cock is opened to the pump only, and the air exhausted to 680–690 mm. The beaker is

[1] J. Chem. Soc., January, 1897.
[2] Rec. Trav. Chim., 1894, 13, 218.

lowered into an oil-bath and heated to about 100° till most of the water is evaporated. Communication is then made with the pentoxide flask, and the mass dried to constant weight at about 120°. For convenience, the vapors from the various pieces of apparatus were passed through a four-liter vacuum pan. This large receiver produced an almost instantaneous exhaustion when a stop-cock was opened. Complete drying usually required from eighteen to twenty-four hours. The slowness of our drying apparatus was possibly due to the contracted opening through the stop-cock, which may have prevented free circulation of vapors over the pentoxide. This drawback was the result of the necessary improvising of our apparatus from stock at hand.

The results obtained are tabulated in the following table and also

TABLE.

Number.	Time of drying. (Hours.)	Vol. of solution. (cc.)	$d_{15.5}$	$a_{D36.}$	Weight of residue.	Factor obtained.	Factor calculated.
1	18	10	1.02835	164.2	0.7104	0.003992	0.003980
2	18	10	1.03134	160.0	0.7833	0.003999	0.003974
3	24	10	1.03657	144.8	0.9223	0.003965	0.003954
4	18	10	1.03364	141.8	0.8509	0.003954	0.003950
5	24	10	1 04212	138 7	1 0644	0.003957	0.003946
6	18	10	1.03650	136.1	0.9222	0.003953	0.003942
8	18	10	1.03701	128.0	0.9376	0.003943	0.003931
9	24	10	1.03701	128.0	0.9417	0.003930	0.003931
11	18	10	1.03841	115.7	0.9781	0.003927	0.003918
12	18	10	1.03704	111.7	0.9454	0.003918	0.003911
13	18	10	1.03704	111.7	0.9465	0.003914	0.003911
15	24	10	1.03843	104.7	0.9827	0.003910	0.003901
16	18	10	1.03545	90.5	0 9127	0.003885	0.003883
17	18	10	1.03594	81.3	0.9282	0.003883	0.003871
18	18	10	1 03549	73.4	0.9162	0.003870	0.003860
19	18	10	1.03881	70.5	1.0081	0.003850	0.003856

(The omitted numbers are those of samples which were lost by breakage, or scorching due to loss of vacuum.)

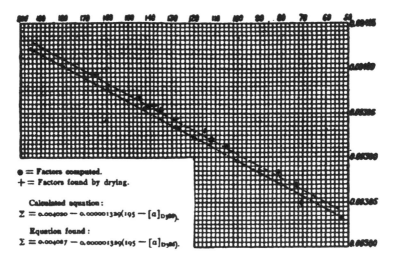

FIG. 2. — SPECIFIC GRAVITY FACTORS FOR APPROXIMATELY 10 PER CENT. SOLUTIONS OF HYDROLYZED STARCH PRODUCTS.

plotted in Figure 2. As will be seen, they form a straight line which is very slightly higher than the line calculated by using the revised carbohydrate values of Brown, Morris, and Millar. The variation is very slight, as the plot is drawn on a scale by which one division represents about 0.06 per cent. of the values obtained. The two equations obtained are more nearly concordant when we take into consideration that the average of the specific gravities of the samples is about 1.0360. This would raise the calculated value 0.000003 on the scale, or a little more than one-half of a division. We, therefore, have adopted as the most probable values what happens to be the mean. Our conclusion, then, is that within the concentrations expressed by the specific gravity factors, 1.035 and 1.045, we can calculate the absolute specific gravity influence of any acid hydrolyzed starch solution by the equation $\Sigma = 0.004023 - 0.000001329$ (195 $- [a]_D$), when the specific rotatory power (obtained by the factor 0.00386) is known. These values within the limits of concentration given are correct to less than $\frac{1}{10}$ per cent. of their values. For commercial glucoses, the factor 0.00393, taken as a constant, is sufficiently exact for most determinations.

While this equation will now enable us to determine the exact amount of carbohydrate in solution when the specific gravity has been previously corrected for the influence of other dissolved material, the simpler computation based on the factor 0.00386 will doubtless continue in use as more convenient for those calculations where proportion of carbohydrates is alone desired.

We have also under investigation the action of heat on commercial glucoses when samples are boiled down to candies, as well as the study of certain disturbing influences on the determination of cupric reducing powers of glucoses. The results are not yet complete enough for publication.

TECHNOLOGY QUARTERLY

AND

PROCEEDINGS OF THE SOCIETY OF ARTS.

VOL. XI.   SEPTEMBER, 1898.   No. 3.

PROCEEDINGS OF THE SOCIETY OF ARTS.

THIRTY-SIXTH YEAR, 1897-98.

THURSDAY, April 14, 1898.

THE 511th regular meeting of the SOCIETY OF ARTS was held at the Institute this day at 8 P.M., Professor Swain in the chair.

The record of the previous meeting was read and approved. A paper on "The Exact Estimation of the Total Carbohydrates in Acid Hydrolyzed Starch Products," by G. W. Rolfe and W. A. Faxon, was presented by title, also a paper on "The Reliability of the Dissociation Values Determined by Conductivity Measurements," by Arthur A. Noyes.

The Chairman then introduced Mr. S. Albert Reed, of New York, who read a paper on "Fireproof Construction." The various methods of rendering buildings resistant to fire were described, and then an account was given of tests made by the Underwriters' Association of New York. These sbowed that iron columns would give way at a dull red heat. The speaker discussed the effects of several large fires, and spoke of the danger of extreme height. The paper was illustrated by the lantern. A discussion followed, after which the Chairman thanked Mr. Reed in the name of the Society, and the meeting adjourned.

THURSDAY, April 28, 1898.

The 512th regular meeting of the SOCIETY OF ARTS was held at the Institute at 8 P.M., Mr. Blodgett in the chair.

The record of the previous meeting was read and approved. The following persons were duly elected Associate Members of the Society: Messrs. Charles H. Fish, of Dover, New Hampshire, A. S. Garfield, of Paris, France, and Walter E. Spear, of Lawrence.

The Chairman then introduced Mr. Albert Sauveur, of Boston, who read a paper[1] on the "Constitution of Steel Considered as an Alloy of Iron and Carbon." Steel was defined as an alloy of iron and the carbide of iron, Fe_3C. Its structure follows the law of that class of binary compounds which do not form definite compounds nor isomorphous mixtures. The curve of freezing points of a saline solution of varying density was compared with the curve of melting points of an alloy of silver and copper of varying proportions, and this in turn was compared with a similar curve of steel containing various amounts of carbon. The three curves were found to be essentially similar, and the eutectic alloy of silver and copper, which is the one having the lowest melting point, and which melts at a constant temperature, was shown to have properties corresponding with those of the cryohydrate of the saline solution on one hand, and with those of pearlite in steel on the other. Pearlite, like the eutectic alloy, is composed of very fine plates of the two substances, alternating in close juxtaposition, and where steel is cooled slowly it forms the matrix containing larger masses of iron (Ferrite = Fe) or of the carbide (Cementite = Fe_3C, containing 6.67% C). When steel is cooled suddenly, however, the matrix is Martensite, the composition of which varies with the proportion of carbon in the steel, and the temperature at which it was quenched.

A discussion followed, after which the Chairman thanked Mr. Sauveur for his very interesting paper, and the Society adjourned.

THURSDAY, May 12, 1898.

The 36th annual meeting (513th regular meeting) of the SOCIETY OF ARTS was held at the Institute on this day at 8 P.M., with Mr Blodgett in the chair.

---

[1] Published in full in the TECHNOLOGY QUARTERLY, June, 1898, Vol. XI, No. 2, pp. 78–92.

The record of the previous meeting was read and approved.

Messrs. C. F. Baker, of Boston, and James G. Melluish, of Bloomington, Illinois, were duly elected Associate Members of the Society.

The annual report of the Executive Committee was read and placed on file.

ANNUAL REPORT OF THE EXECUTIVE COMMITTEE

Presented at the Thirty-sixth Annual Meeting of the Society of Arts, May 12, 1898.

The first meeting of the SOCIETY OF ARTS during the present year was held on October 28, the meeting of October 14 having been omitted to allow members to attend the Memorial Meeting in Music Hall, in honor of President Walker. In all, twelve meetings have been held, and the following papers have been read : " Recent Work in Heat Measurement at the Institute," by Mr. Charles L. Norton; "Cable and Underground Electric Roads," by Mr. Louis J. Hirt; "Contributions to our Knowledge of the Micro-organisms and Sterilizing Processes in the Canning Industries. II. The Souring of Sweet Corn," by Messrs. S. C. Prescott and W. Lyman Underwood; "The Concentration of Ores, with Special Reference to Recent Investigations in this Subject," by Professor R. H. Richards; " A New System of Pneumatic Dispatch Tubes," by Mr. B. C. Batcheller; "The History and Present Development of the American Bicycle," by Dr. Leonard Waldo; "On Town, State, and City Boundaries," by Mr. Henry B. Wood; "Certain Sanitary Aspects of Jamaica," by Mrs. Ellen H. Richards and Mr. Arthur T. Hopkins; "On Fireproof Construction," by Mr. S. Albert Reed; "On the Constitution of Steel Considered as an Alloy of Iron and Carbon," by Mr. Albert Sauveur; and on " The Economic Relation of Deep Inland Waterways to the State of New York," by Mr. George W. Rafter.

The membership of the SOCIETY OF ARTS continues to increase, although there have not been added so many new members as during the previous year. At the close of the year 1896-7 the number of life members was 59; five have died since then, leaving the present number 54. The number of associate members a year ago was 311; of these one has died and nine have resigned. These losses are offset by the election during the year of 22 associate members, making the total number 323.

Early in the autumn the Executive Committee lost one of its members by the death of Mr. Thomas Doane, who was distinguished as an engineer and as the founder of the Doane College in Nebraska. He had been a life member of this Society for many years. The other life members who have died during the year are John M. Forbes, Joseph S. Fay, John Lowell, and Theodore Lyman. We have lost one associate member by death, Mr. C. H. Parker, of Cambridge.

A notable event in the history of the SOCIETY OF ARTS was the election of Professor James M. Crafts to be President of the Institute. He was chosen by the Corporation, October 20, and was introduced to the Society as its President at the meeting of October 28. An amendment to the By-Laws of the SOCIETY OF ARTS was adopted by the Society December 9. This amendment provides that three members shall constitute a quorum of the Executive Committee for the transaction of business.

The Board of Publication, which has charge of the TECHNOLOGY QUARTERLY AND PROCEEDINGS OF THE SOCIETY OF ARTS, was reappointed, with the exception of Professor Henry M. Howe, who has removed to New York. Professor Charles R. Cross was chosen in his place. The following members, therefore, constitute the Board of Publication as now organized: Professor W. T. Sedgwick, Chairman, Professor Charles R. Cross, Professor Dwight Porter, A. Lawrence Rotch, Esq., and the Editor, Dr. R. P. Bigelow. From 1880 to 1891 the SOCIETY OF ARTS published an annual volume of " Proceedings." The TECHNOL-

OGY QUARTERLY was founded in 1887 by the students, and was published by a board of editors chosen from the Junior and Senior Classes, William S. Hadaway, Jr., '87, being the first editor-in-chief. Later Mr. James P. Munroe became sole editor, and in 1892 the SOCIETY OF ARTS assumed control of the journal and united to it the "Proceedings." It thus became the official organ of the Institute. By a vote of the Society the Executive Committee was authorized to appoint annually a Board of Publication to exercise a general control and supervision over the QUARTERLY. The active management, however, falls upon the editor, who is appointed by the Executive Committee upon the nomination of the Board of Publication.

During the present year the TECHNOLOGY QUARTERLY has published not only the minutes of proceedings of the Society, and a number of papers at its meetings, but it has also contained a number of original articles which have been presented to the Society only by title. The series of *Results of Tests Made in the Engineering Laboratories* has been continued, as well as the monthly *Review of American Chemical Research.*

Respectfully submitted,

For the Executive Committee,

GEORGE W. BLODGETT, *Chairman.*

The following officers were elected for the year 1898–99 : Mr. Arthur T. Hopkins, *Secretary,* and Messrs. George W. Blodgett, Desmond FitzGerald, Edmund H. Hewins, Frank W. Hodgdon, and Charles T. Main, *Members of the Executive Committee.*

Mr. N. M. Lowe offered the following resolution, which was adopted :

The SOCIETY OF ARTS desires to acknowledge its appreciation of the untiring and efficient labors of its Secretary, Dr. Robert P. Bigelow, and to express its regret that he finds it impossible longer to serve the Society in that capacity. Therefore *Resolved,* that the foregoing minute be adopted by the Society and spread upon the records.

The following papers were presented by title : "Experimental Study of Transformer Diagrams," by H. E. Clifford, N. Hayward, and R. Anderson ; "On the Telescope-Mirror-Scale Method, Adjustments and Tests," by S. W. Holman.

The Chairman then introduced Mr. George W. Rafter, of Rochester, New York, who read a paper on "The Economic Relation of Deep Inland Waterways to the State of New York." The early history, he said, of the New York state canals is an exceedingly fascinating study, especially when we consider the extraordinary changes which have taken place since the canals were opened in 1825. The great interest in the Erie Canal is due to the position of the State, which is in the exact line along which the produce of the West must move to the East. The Mohawk Valley furnishes the great opportunity. The history of early attempts to construct waterways by private corporations was sketched, the history of the canals as at present constructed was described, and then an account was given of the projects for making a deep waterway from the Great Lakes to the sea. Two main routes were described, one by way of Oswego and the Mohawk Valley, and the other through the St. Lawrence River and Lake Champlain. Which is the better route is not yet determined, but the one through the St. Lawrence has the advantage of being a continuous down grade, and requiring less than a hundred miles of artificial channel.

Thirty years ago the canals carried annually from six to eight million tons of freight. In 1896 and 1897 they carried only about three and one-half millions, of which three million tons was through freight, while the New York Central Railroad carried about twenty-two million tons. These figures show that the canal is of little value to the state, and the opinion was expressed that the canal should be turned over to the national government. New York was described as the greatest water power state in the Union, being capable of developing one and a half million horse power, and it would gain by abandoning its canals and developing its water power. The speaker closed by a description of a trip across the state along the Erie Canal, illustrated by lantern slides.

The Chairman thanked Mr. Rafter for his interesting paper, and the Society then adjourned.

ROBERT PAYNE BIGELOW, *Secretary.*

THE TELESCOPE–MIRROR–SCALE METHOD; ADJUSTMENTS AND TESTS.[1]

By SILAS W. HOLMAN.

Received May, 1898.

In the telescope-mirror-scale method for measuring small angles, an illusory estimate of the accuracy attained may easily arise through inattention to the requisite adjustments, tests, and corrections. Yet an adequate and well-ordered presentation of them is not to be found. Partial statements are given in text-books on manipulation, but Czermak[2] alone has given a discussion approaching completeness. Even this omits some essential points, and moreover is not arranged in a manner to lend itself readily to the practice of the method.

The following presentation of the subject is designed for the observer who would put the method into direct service for obtaining measurements of a specified or of a determinate accuracy. It therefore not only discusses the several sources of error, but describes the various adjustments and tests in the sequence in which they should ordinarily be made, and gives a numerical measure of the closeness with which each must be carried out to secure a specified precision in the result. Some remarks on the selection of instruments are appended.

The general plan adopted in treating each adjustment or test is as follows: To deduce a general, though usually approximate, expression for the error attending its omission, or preferably for the correction therefor. To deduce therefrom a numerical measure of the closeness with which the adjustment must be made or the test fulfilled, in order to insure a designated precision in the use of the method, so far

---

[1] Copyrighted, 1898, by S. W. Holman. Published separately by John Wiley & Sons, 53 East 10th Street, New York. Cloth, 75 cents.

[2] CZERMAK. "Reduction Tables for Readings by the Gauss-Poggendorff Mirror Method." Besides a discussion of many of the errors of the method, this book gives extended tables of corrections and reductions, of much service in long series of observations. The text is in German, French, and English, in three parallel columns.

as that source of error is concerned ; and to make certain comments based thereon.

To render the problem definite, the tangent of the angle of deflection of the mirror is assumed to be the desired result of an observation. The deductions are easily adaptable to other less frequently employed functions. The expressions for the errors or corrections are brought into the form of fractional errors or corrections ; that is, they are expressed as a fraction of the value of the tangent of the deflection. The correction is of course equal to the error, but has the opposite algebraic sign, a positive error requiring a negative correction, and vice versa. It is more convenient in the present work to deal directly with corrections than with errors. The numerical solutions are computed on the basis of a desired precision of one part in one thousand, or one-tenth of one per cent., in the resulting value of the tangent. The precision discussion based on methods elsewhere stated[1] is sufficiently obvious. As there are some fifteen sources of error, the average effect to be assigned to each consistently with the prescribed limit of 0.001 in the result is $0.001 \div \sqrt{15} = 0.00\ 030$ nearly enough, as several of the errors are usually rendered insignificant. The solutions are also made for a scale of millimeters at a distance of one meter from the mirror, and for a maximum deflection of 500 mm. It will be found by inspection of the results, especially under I, III, XII, XIII, XIV, XV, that 0.1 per cent. is about the limit of accuracy attainable under these conditions.

In the employment of the apparatus absolute values of the tangent of the deflection are sometimes sought, in which case the measurements may be called *primary*. More often, however, only relative values are required, or rather we are concerned as to the accuracy of only relative values of the tangents. The measurements may then be called *secondary*. An example of the secondary use is where the telescope and scale are employed with a reflecting galvanometer to measure currents, and where the "constant" of the apparatus is found by sending a known current and reading the deflection. In such cases any constant fractional errors in the telescopic method enter into the "constant," as well as into subsequent observations, but with opposite signs, so that they are eliminated from the results. It is therefore needful to discuss the errors with respect to both primary and second-

---

[1] "Precision of Measurements." John Wiley & Sons, New York.

ary use of the method, and relief is thus found possible in the latter
from some of the exactions of' the former.

When reversals are taken, that is, when the mirror is deflected so
that a reading can be made first on one and then on the opposite half
of the scale, certain sources of error are reduced by averaging the two
deflections. This is usually precluded in practice, however, by such
conditions as continual change in the reading, avoidance of delay, etc.
Both cases must therefore be discussed.

The desired tangent or other function of the angle of deflection
is computed from the observed scale-reading d and scale-distance r as
stated under XVI. We will assume that it is found, with due allowance for the approximation involved (cf. XVI), from the expression

$$\tan \varphi = \frac{1}{2} \cdot \frac{d}{r} \left(1 + u + v + w + \ldots \right)$$

where u, v, w, etc., are the fractional corrections to the observed
tangent to allow for the various sources of error to be pointed out.
As d enters as a direct factor in this expression for the tangent, the
fractional corrections may be applied directly to d. In fact it is more
convenient in deducing the formulæ to find an expression for either
δd, the numerical correction to d, or for $\delta d / d$, the fractional correction
to d. Thus δd is such a quantity, expressed in scale divisions, that
when added to the observed scale-reading d it will give the correct
scale reading, as far as the designated source of error is concerned.
And $\delta d / d = u$ or v, etc., is this correction expressed as a fraction of
the observed reading d. The algebraic sign of any correction may be
either determinate or indeterminate. Also it may be noted that as r
enters as a factor in the denominator, the corrections u, v, etc., may
be applied to or deduced for r instead of d, if desired, but with the
difference that the algebraic sign would be reversed. Any corrections
which may prove to have a constant value may be included once for
all in the numerical constant $1 / 2r$. Constant fractional corrections
disappear when merely relative deflections are used, or when the constant is determined by calibration, as above stated. The assumption
is made that none of the errors with which we have to deal are in
excess of 1 or 2 per cent. Greater ones must be reduced by instru-
mental rearrangement before they can be determined with sufficient
closeness.

The algebraic expressions which will be deduced for u, v, w, etc., are useful in two ways: First, they enable us to compute the correction to be applied to any observed reading; second, they are used in the precision discussion, which shows how closely each correction must be worked out to secure a prescribed accuracy in the value of tan, φ or how close an adjustment is needed in order that the correction may be omitted without introducing more than its due share of error into the result. These will be called *negligible corrections.* Any error of this magnitude produces an effect not more than is admissible on the result, and the effect of one as small as one-third of this amount will be inappreciable.

The results deduced for the telescope-mirror-scale method are in general directly applicable to the mirror method where a beam of light replaces the telescope.

THE PROCEDURE.

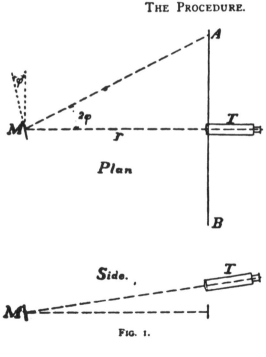

Preliminary.— Focus the cross-wires in the eye-piece. Set up the apparatus nearly in proper position as in Figure 1, adjusting telescope, mirror and scale so that the scale is visible. The following description will assume the disposition of the apparatus to be the most usual and simple one, shown in Figure 1: The scale A B is horizontal with its middle point just below or above the optic axis of the telescope, which is perpendicular to the scale. The mirror (plane) faces the telescope and scale at a convenient. distance. The horizontal distance r between scale and mirror should then

Plan

Side.

FIG. 1.

be roughly measured, say to 1 per cent., for use in the approximate computations in the several tests, etc. The selection of this distance is determined largely by convenience, and usually lies between one and three meters. It must be sufficient to insure to the smallest scale-readings to be observed, a sufficient number of divisions to be read with the requisite fractional precision. This means greater distances and higher magnifying power for very small angles. Labor in computation may be saved by making this distance some simple round number, *e.g.*, 1,000 or 2,000 scale divisions, but the work of this adjustment will usually outweigh that of computation. The distance must, however, be maintained accurately constant. The telescope may be at a distance different from that of the scale, if more convenient, but further removal reduces the magnification. As far as the principle of the method is concerned, the telescope may be at any position whatever relatively to the scale, consistent with having the line from the middle of the scale to the mirror nearly horizontal, as.pointed out in test X. If there is annoyance from duplicate reflections from the glass front of the mirror house, the glass may be slightly tilted forward.

The following tests and adjustments are then to be made in the order in which they are given, a repetition of any of them being subsequently made if disarrangement may have occurred : —

Adjustments and Tests.

A list is here given of the various adjustments and tests, together with a brief statement for convenient reference of the closeness necessary to correspond with an accuracy of one-tenth of one per cent. in the resulting value of tan φ, assuming a straight millimeter scale at a distance of 1 meter, with deflections not exceeding 500 mm. and not less than 100 mm., readings being taken to tenths of a millimeter by the eye.

I. Cross–Wire Focus. — By parallax test there must be no apparent motion of wires, or less than $\frac{1}{20}$ division. Telescope must also be perfectly focussed. Primary and secondary measurements the same.

II. Optic Axis of Telescope Radial. — P. M. Cross-wires centered within 1 mm., or better, to 0.2 mm., if possible, when tested on scale at distance of 1 m. Optic axis must pass within 1 mm. of axis of suspension.

III. SCALE GRADUATION. — The average error in the distance of any ruling from the central ruling must be less than 0.03 mm. P. M. and S. M. the same.

IV. COVER–GLASS THICKNESS. — P. M. Glass negligible when less than 1 mm. in thickness. S. M. Glass always negligible. Glass must permit good definition.

V. COVER–GLASS SURFACES PARALLEL. — P. M. To be negligible, the maximum displacement must be less than 0.3 mm. on rotating the glass 180° in its own plane. If more is found, the glass must be kept in the position of minimum displacement. S. M. The same.

VI. COVER–GLASS CURVATURE. — P. M. To be negligible, the focal length of the cover-glass must exceed 70 meters. This will be the case when an object at a distance of 10 meters from the telescope, and sharply in focus with the cover-glass interposed, does not require to be moved through more than about 1 meter towards or from the telescope to maintain the focus when the glass is withdrawn. S. M. The error is here always negligible.

VII. MIRROR THICKNESS. — P. M. Negligible thickness is 0.5 mm. S. M. Always negligible.

VIII. MIRROR ECCENTRICITY. — P. M. Negligible up to an eccentricity of 10 mm. S. M. Also negligible.

IX. MIRROR CURVATURE. — P. M. Correction depends on amount of eccentricity. It vanishes with no eccentricity. For specific case see later. S. M. Curvature has no sensible effect.

X. MIRROR VERTICAL AND VERTICAL ANGLE *TMO* SMALL. — P. M. Telescope and scale must not be separated vertically by more than 60 mm. S. M. Negligible, and distance need not be very small.

XI. SCALE HORIZONTAL. — P. M. Error insignificant when end of scale is not more than 1 or 2 mm. out of horizontal through middle. Negligible up to 12 mm. S. M. Negligible for several centimeters of tipping.

XII. SCALE PERPENDICULAR TO *OM* AND PLANE. — P. M. Without reversals, the difference in distance of the ends of the scale (the 500 mm. marks) from the suspension fiber must be less than 0.6 mm. With reversals, 25 mm. is close enough. Scale must be plane within these limits. S. M. About the same as in primary. Hence advantage in reversals where practicable.

XIII. NULL POINT. — P. M. The scale once fixed in place can-not be disturbed without impairing XII. Hence accidental change of null reading must be remedied by turning the mirror, not by mov-ing the scale. With reversals, the null reading may be as great as 17 mm. without correction. Without reversals, the null reading must be less than 0.6 mm., even if allowed for. S. M. Same as in primary.

XIV. DISTANCE OF SCALE FROM MIRROR. — P. M. This must be measured and *kept constant* within 0 3 mm. This demands more than the usual attention, and renders some special device important. See XIV later. S. M. The same.

XV. ESTIMATION OF TENTHS OF DIVISION. — Nearest tenth is to be read in both P. M. and S. M.

XVI. TO COMPUTE THE DESIRED FUNCTION OF φ from the ob-served deflection d. Details are given under XVI later. Five places of significant figures should be used in d, etc.[1]

I. CROSS–WIRE FOCUS.

The cross-wire intersection must be brought accurately into the focus of the eye-piece on each occasion of use of the apparatus, and by each observer for himself. Inattention to this point may easily give rise to an error as great as half a scale division. The focussing should be done by the parallax method, as follows : Focus the wires as sharply as possible by moving the eye-lens. Then focus the telescope very carefully on the scale or on some object showing some sharply marked point of reference. Move the head to and fro sidewise, so that the pupil of the eye shall travel from one side to the other of the aperture of the eye-piece. If the wires are not in focus they will appear to move over the scale. If so, refocus them, and then refocus the tele-scope on the scale. Continue until no apparent motion of the wires is perceptible. Good focussing is promoted by looking away from the telescope frequently ; also in some cases by fixing the attention of the other eye on a printed page held at the distance of most distinct vision beside the telescope. The accurate focussing of the telescope is not less important than that of the wires. The error from imperfect focus will be indeterminate in sign and magnitude.

PRIMARY MEASUREMENTS. — The negligible correction will then be $\pm \delta d$, whose magnitude must be such that $\delta d / d = 0.00\,030$ for

---

[1] Computation Rules and Logarithms. The Macmillan Co., New York.

the smallest value of d to be used. Now deflections of less than 100 divisions are not employed in exact work for reasons shown in XV. Hence substituting $d = 100$ mm. we have $\pm \delta d = 0.03$ mm. The extreme motion of the eye in the parallax test will obviously produce a displacement of the maximum error, and more than double the average error. Hence the focus will be good enough when the maximum displacement by the parallax test is less than about 0.06 divisions, or about $\frac{1}{30}$ of the smallest scale division. It is clear, then, that the utmost attention to this detail is necessary.

SECONDARY MEASUREMENTS. — This source of error affects primary and secondary measurements equally.

II. OPTIC AXIS OF TELESCOPE RADIAL.

This adjustment requires two operations : First, centering the cross-wire intersection, that is, bringing it into the optic axis ; second, directing the optic axis towards the axis of suspension, *e.g.*, by focussing the wire intersection upon the suspending fiber.

The second operation is obvious enough, but must be repeated from time to time by way of precaution. It is not easily executed when the telescope draw-tube does not permit of focussing on objects as near as the mirror.

The well-known method of centering the wires is briefly as follows : Lay the telescope in a pair of V grooves — temporary wooden ones will suffice. Focus on the scale, or on a few rulings roughly equal to the scale divisions, placed at a distance about equal to TM. Rotate the telescope in the grooves. The wires are centered when the intersection shows no motion on the scale when thus rotated. If adjusting screws are provided, the diaphragm carrying the wires should be centered until the motion is less than 0.1 or 0.2 mm., as demonstrated later. If, as is commonly the case, there is no means of adjustment of the wires, they must be remounted if the extreme apparent displacement on turning through 180° is more than about 1 mm., for work of the accuracy here assigned.

DEMONSTRATION. — We will take the simplest case, where the optic axis TF of the telescope is at right angles to the scale AB. If adjustments XII and XIII had been made the relative positions would be slightly different, but the result would be essentially the same as far as this source of error is concerned. Suppose the telescope so located that TF meets the mirror at c' instead of c, the latter being in the vertical axis of suspension.   Let cc' be denoted by z.   Then we seek an expression for the fractional correction $\delta d / d$ for the error which z will produce.

If c were at c', as it should be, and were deflected through an angle φ, the observed reading OA would be the true reading d desired.   With c at c', the observed reading becomes OA', and $A'A = \delta d$.   As $c' = r$, we have $\delta d : c' c'' = d : r$.   But $c' c'' = z \tan \varphi = z (d / 2r)$ approx. Hence $\delta d / d = zd / 2r^2$ approx., for a deflection on the side considered.   For a reverse deflection the correction would obviously be of the same amount, but the opposite sign.

Imperfect centering of the wires, besides misdirecting the optic

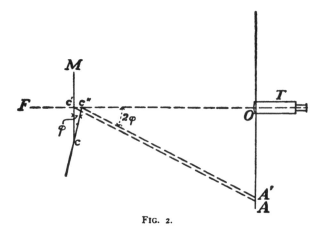

FIG. 2.

axis, introduces a slight error through the smaller magnification of the outer scale divisions, but this may be easily shown to be small compared with the foregoing.

PRIMARY MEASUREMENTS. — *Without reversals.*   As the negligible correction above assigned is $\delta d /d = 0.00\,030$, the negligible value of z for the worst case, where $r = 1,000$ mm. and $d = 500$ mm., is to be found from $z = (\delta d / d) / (d / 2r^2) = 0.00\,030 \times 2 \times (1,000)^2 \div 500 = 1.2$ mm.   Thus the optic axis must pass within about $\pm$ 1.2 mm. of the axis of suspension, which demands no inconsiderable care.   As the adjustment requires two operations, both must be made closer than this limit, or one must be rendered imperceptible.   The latter, by careful centering of the wires, is usually easier of accomplishment.

If this is done so that the displacement in the above test is less than $\frac{1}{4}$ of *s*, *i.e.*, 0.4 mm. on a scale at the distance *TM*, or better to 0.1 or 0.2 mm., this part of the error will be negligible.

With reversal of deflections, this source of error vanishes when *s* does not exceed 2 or 3 mm.

SECONDARY MEASUREMENTS. — Substantially the same degree of care is needed as in primary work.

III. SCALE GRADUATION.

The graduation must be uniform to such an extent that the actual correction δd to any observed deflection *d* shall be small enough to make $\delta d / d$ less than the assigned limit. The only thorough way to test this is to measure out from the middle point of the scale to each successive smallest division by some comparator or dividing engine, thus determining every value of $\delta d / d$, and then to apply the corrections thus found unless they are negligible. If an error not exceeding $\frac{1}{20}$ mm. is admissible on a mm. scale, the test may be made by laying a standard (Brown and Sharpe) mm. scale against the unknown one, and inspecting the coincidences of the rulings. The first 50 or 100 divisions either way from the middle are not employed in work of even moderate accuracy, because the value of *d* is then so small that the unavoidable fractional error from the eye estimation and other sources is excessive.

PRIMARY MEASUREMENTS. — The requisite accuracy of graduation for the above assigned limit would be attained when for the smallest value likely to be used for *d* in such work, viz., $d = 100$ mm.,

$$\frac{\delta d}{d} < 0.00030 \quad \therefore \quad \delta d = 0.030 \text{ mm.}$$

That is, stated briefly, the average error in the distance of any scale division from 0 should not exceed 0.030 of a division. This limit can hardly be attained in paper scales without great care.

SECONDARY MEASUREMENTS. — Any systematic error of the scale which makes it too long or too short by a constant fractional part, *i.e.*, for which $\delta d / d$ is constant, will be without effect, since it will cause the value of the "constant" found in calibration to be too small or too large by an equal fraction, so that the error in the constant and in deflection readings will exactly offset each other. The *inequalities* of graduation must be, however, less than 0.030 division.

IV. Cover Glass Thickness.

If the glass has plane parallel faces and is of a thickness t, it will displace the ray parallel to itself at each passage. As the simplest case, suppose the glass to be so placed that a ray OM from the center of the scale to the mirror is normal to the glass. Then the ray will experience no displacement in passing from O to M, but would be displaced from MA to gA' in passing from M to the scale. So that with the telescope at O a reading $OA' = d$ would be observed instead of the true reading OA. The correction δd_1

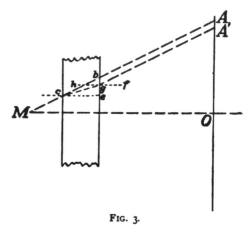

Fig. 3.

to d_1 is therefore $AA' = gb$. Now $gb = be - ge$, $be = t \tan 2\varphi$, $ge = t \tan gce$. By law of refraction, n being the index of refraction,

$$\frac{\sin hgc}{\sin fgA'} = \frac{1}{n} = \frac{\tan gce}{\tan 2\varphi} \text{ approx., as } 2\varphi \text{ is small.}$$

$$\therefore \delta d_1 = t \left(\tan 2\varphi - \frac{1}{n} \tan 2\varphi\right) \text{ approx.}$$

$$= t \left(1 - \frac{1}{n}\right) \tan 2\varphi \text{ approx.}$$

For ordinary glass $n = \dfrac{3}{2}$

$$\therefore \delta d_1 = \frac{t}{3} \cdot \frac{d_1}{r} ; \therefore \frac{\delta d_1}{d_1} = \frac{t}{3r}$$

Inspection of the diagram shows that the sign of the correction is positive on either side of O. Hence $\dfrac{\delta d}{d} = \dfrac{1}{3} \cdot \dfrac{t}{r}$. The reflecting surface of the mirror is supposed to contain the axis of suspension in all cases unless otherwise specified, so that $OM = r$.

If, however, the glass is not perpendicular to OM, the error becomes unsymmetrical, and dependent upon the horizontal angle between OM and the mirror. It is best, therefore, that the angle should be kept very nearly 90°, as may easily be done. With a very thick cover-glass this point would require special investigation.

The effect of the thickness of the cover-glass may be readily shown from the above formula to be equivalent to the reduction of the scale distance r by one-third of the thickness of the cover-glass.

PRIMARY MEASUREMENTS. — The requisite thinness of the cover-glass, in order that the omission of the correction may be admissible, may be found from

$$\frac{\delta d}{d} = \frac{t}{3r} < 0.00\ 030 \ \therefore \ t = 3.1000 \cdot 0.00\ 030 = 0.9 \text{ mm.}$$

The correction must therefore be applied when, as is almost always the case, the cover-glass is more than about 1 mm. in thickness.

SECONDARY MEASUREMENTS. — Since the fractional correction $\delta d / d$ is constant, the correction disappears in this work as in other similar cases.

V. COVER–GLASS SURFACES PARALLEL.

Defect with regard to this is common. Its effect depends so much upon the various possible angular positions of the surfaces relatively to OM that it is not readily reducible to a general expression, but an easy and sufficient test can be developed. Give the mirror a deflection to about the end of the scale. By clamping, or in some effectual way, hold the mirror so that this reading shall remain fixed; focus sharply; then using due care not to disturb the apparatus, place the cover-glass flat against the objective of the telescope, or better against some diaphragm arranged to hold it not far from the objective; read closely. Now rotate the cover-glass through 180° in its own plane, pressing it against the objective or diaphragm, and read again. A change of reading due to twice the refraction by the prismatic or wedge shape of the glass will be found. Turn the glass in its own plane into various (marked) angular positions and thus locate the diameter along which the displacement is greatest. This will locate the direction of the edge of the wedge, that is, the intersection of the two plane surfaces, since this must be vertical when that diameter is horizontal.

PRIMARY MEASUREMENTS. — It is more prudent not to use a glass which shows a maximum displacement S exceeding 0.3 mm. at a greatest deflection of $d = 500$ mm. ; for the negligible correction will be

$$\frac{\delta d}{d} = \frac{1}{2} \cdot \frac{S}{500} < 0.00\,030 \quad \therefore\ S = 0.3 \text{ mm.}$$

If this limit is not easily reached, the wedge axis may be placed horizontal, which will sensibly eliminate the error. But this requires that during all use of the apparatus the cover-glass be continually inspected to see that it is in the proper position.

In making this test the cover-glass may often, to good advantage, be rotated in position instead of being removed and placed in front of the objective. In that case special care must be taken not to disturb the mirror when the glass is being rotated through 180°. The maximum displacement observed in this method of test in either case is in excess of the worst error which the wedge would cause in the observations, unless it was reversed in position from time to time — which must be guarded against if the glass is poor.

SECONDARY MEASUREMENTS. — No relaxation from the foregoing requirement is admissible.

VI. COVER–GLASS CURVATURE.

Minor irregularities in the surface of the cover-glass of the mirror house produce merely a blurring of the image, such as is seen in looking through ordinary window glass with a telescope. The cover-glass must be sensibly free from this. If either surface of the cover-glass is systematically curved, the glass will act as a lens. The focus of points seen obliquely through the glass is then changed in both distance and direction. The change in distance is either unnoticed or is corrected by the focussing of the telescope. The change of direction causes an error in d. The glass may be equivalent to a spherical lens or to a cylindrical lens. The effect of the change in direction may be studied by the central ray of any beam ; we are concerned here with the curvature as revealed by a horizontal section only. For simplicity, suppose the glass to be equivalent to a spherical lens $D\,G$ of focal length f with its axis coincident with $O\,M$. If the lens is convex, the true reading A will be shifted to A' toward O; if concave, in the opposite direction.

The correction required and the method of test may be developed as follows : The lettering of the diagram (plan) being as before, let

$O A = d =$ true deflection reading,

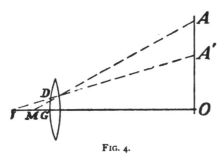

$O A' = d_1 =$ observed deflection reading,

$M G = e = p_1 =$ distance of cover-glass G from mirror (roughly),

J and $M =$ conjugate foci of lens G when $M G \lessgtr f$, the focal length of G,

$J G = p_2,\ A M, A' J =$ straight lines.

FIG. 4.

As we have treated separately the thickness of the glass, $A\,M\,O = a$, $A' J O = \beta$.

$$\frac{\tan a}{\tan \beta} = \frac{p_2}{c} = \frac{d}{r} \cdot \frac{p_2 + r - c}{d_1}$$

$$\therefore \frac{d_1}{d} = \frac{c}{r} + \frac{c}{p_2} - \frac{c^2}{p_2 r}$$

But by the law of lenses, inserting for convenience the negative sign because $p_1 < f$,

$$\frac{1}{p_1} - \frac{1}{p_2} = \frac{1}{f} \text{ or } \frac{1}{c} - \frac{1}{p_2} = \frac{1}{f}.$$

Hence multiplying by c and transposing $\dfrac{c}{p_2} = 1 - \dfrac{c}{f}$,

$$\therefore \frac{d_1}{d} = \frac{c}{r} + 1 - \frac{c}{f} - \frac{c}{r} + \frac{c^2}{rf} = 1 - \frac{c}{f}\left(1 + \frac{c}{r}\right).$$

The signs will obviously be the same for deflections on the opposite side of O. The correction is therefore

$$\frac{\delta d}{d} = \frac{d - d_1}{d} \text{ approx.} = 1 - \frac{d_1}{d} = \frac{c}{f}\left(1 + \frac{c}{r}\right).$$

But in all ordinary cases $c < \dfrac{1}{10}\, r$, so that for the present purpose $\dfrac{c}{r}$ is negligible compared with 1, and $\dfrac{\delta d}{d} = \dfrac{c}{f}$.

This evidently applies to deflections in either direction, provided that the axis of the lens coincides with OM; otherwise the correction would be unsymmetrical and would contain another term. The effect of this want of symmetry, however, would be detected under test V for wedge shape of cover-glass, and therefore need not be here discussed.

If the cover-glass proves to be cylindrical in the test given later, the worst effect of the cylindrical surface will be produced when its axis is vertical. The correction is then the same as for the spherical surface.

PRIMARY MEASUREMENTS. — The negligible correction may be found from $\dfrac{\delta d}{d} = \dfrac{e}{f} < = 0.00\,030.$

Assuming a value of $e = 20$ mm., not often exceeded, this yields $f = 20/0.00\,030 = 7.10^4$ mm. $= 70$ m. Hence the cover-glass, whether equivalent to a spherical or a cylindrical lens, must have a focal length exceeding 70 m. How the actual value of f is best determined will presently be shown.

SECONDARY MEASUREMENTS. — The fractional correction to the deflection is constant. Hence in secondary work the error introduced in the "constant" by neglect of the correction exactly offsets the errors entering into the subsequently observed deflections by the same cause, so that the systematic curvature has no effect.

MEASUREMENT OF f. — Direct the telescope (any good telescope other than that of the apparatus will answer if more convenient) upon any well defined object, such as a printed page, held at a distance of several meters (10 or more if practicable). Interpose the cover-glass, placing it directly in front of the telescope. Focus sharply ; remove the cover-glass. If the glass is equivalent to a lens with spherical surfaces, the object will cease to be in focus. Without changing the focus of the telescope, bring the object towards it (concave) or move it away (convex) until the focus is again sharp. Let a denote the distance from glass to object in the first position and b in the second. Then $\dfrac{1}{b} - \dfrac{1}{a} = \dfrac{1}{f}$, as the two positions are conjugate foci. In actual measurements it is, of course, better to measure $(a - b)$ directly with some care, and then a roughly (or b), thence computing b and f. To detect unequal curvature of different parts of the glass, it is well to measure f for each of its four quarters successively, covering its remaining surface.

It may sometimes be better merely to test whether f is sufficiently great without actually measuring it. Thus if the limit is $f > 70$ m., then if the object be set up with $a = 10$ m., for example,

$$\frac{1}{b} = \frac{1}{70} + \frac{1}{10} = \frac{80}{700} \quad \therefore \; b = 8.8 \text{ m.}$$

or $a - b = 1.2$ m.

If, therefore, b were distant from a by as much as 1 m., the glass would be good enough. Cover-glasses should never be found so poor as this on good instruments, but this cannot be relied upon, as makers frequently send out very unfit glasses. The glass should be tested by focussing through it separately on vertical, horizontal, and oblique lines. Difference of focus shows a cylindrical form which may or may not be superposed upon the spherical. The test should be applied to that value of f which makes $a - b$ the greatest.

VII. MIRROR THICKNESS.

The mirror usually consists of a thin plane-parallel piece of glass "silvered" on its rear surface. If in any case the mirror were very thick it would require special investigation as to corrections for want of parallelism or planeness of faces; but with thin mirrors ordinarily used, and the quality which is insured by the requirement of good definition, these points may be safely neglected.

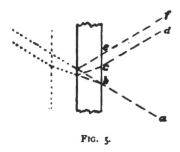

Inspection of the figure will show that a ray $a\,b$ passing obliquely into the front surface of the mirror will pass out along $c\,d$, parallel to $e\,f$, the path it would have travelled had there been no glass in front of the reflecting surface.

FIG. 5.

Also, the emergent ray $c\,d$ is displaced from $e\,f$ by the same amount it would have been had it passed through a plane-parallel glass of twice the thickness of the mirror. Since by the process of adjustment the line $O\,M$ is sensibly normal to the mirror, the correction will be symmetrical with respect to O. If, then, t' represents the mirror thickness, the correction to be applied to it will obviously be $\dfrac{\delta d}{d} = \dfrac{2}{3} \cdot \dfrac{t'}{r}$.

PRIMARY MEASUREMENTS. — Requisite thinness of mirror to be negligible $= 0.5$ mm.

SECONDARY MEASUREMENTS. — The fractional correction being constant may be wholly neglected.

VIII. MIRROR ECCENTRICITY.

If the vertical axis of suspension does not lie in the reflecting surface of the mirror, the eccentricity will cause an error in the scale reading. As the simplest case, let $c\,M$ represent the reflecting surface (plane) in its normal position, $O\,M$ being perpendicular to it; and let Y be the axis of rotation. When the mirror turns through an angle q to the position M', the reflected ray going to the telescope at O will be $A'\,N$ instead

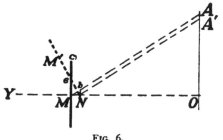

FIG. 6.

of $A\,N$ which it would have been had Y passed through M. The correction to the observed reading $d' = O\,A'$ is therefore $A\,A'$; and for this an expression must be found. From the figure,

$$\delta d_1 = A\,A' = b\,N = M\,N \cdot \tan 2q,$$

$$M\,N = c\,M \cdot \tan q = \frac{1}{2}\,c\,M \cdot \tan q = \frac{1}{2}\,Y\,M \cdot \tan^2 q,$$

$$\tan q = \frac{1}{2}\,\tan 2q \ (\text{approx.}) = \frac{1}{2}\cdot\frac{d_1}{r}. \text{ Also let } Y\,M = X,$$

$$\therefore\ \delta d_1 = \frac{X}{8}\cdot\frac{d^3}{r^3}.$$

This correction will obviously be the same in sign and amount for equal deflections in either direction. Hence

$$\frac{\delta d}{d} = \frac{1}{8}\cdot\frac{X}{r}\cdot\frac{d^2}{r^2}.$$

PRIMARY MEASUREMENTS. — The requisite smallness of X may be found from

$$\frac{\delta d}{d} = \frac{1}{8}\cdot\frac{X}{r}\cdot\frac{d^2}{r^2} < 0.00\,030,$$

$$\therefore\ X = \frac{8\cdot 1000\cdot 1000^2}{500^2}\cdot 0.00\,030 = 10 \text{ mm.}$$

It is rare that the arrangement of the apparatus is such that the eccentricity exceeds a few millimeters, so that this correction is ordinarily quite negligible. If the normal position of the mirror is oblique instead of perpendicular to OM, the correction becomes unsymmetrical. and would require further investigation for special cases if the eccentricity were large.

If the apparatus is such that the correction is not small and must be applied, the above form may be insufficient, owing to the inaccuracy produced by the approximation $\tan \varphi = \dfrac{1}{2} \tan 2\varphi$ employed in deducing it. It is, however, easy to modify the expression so that it may have any desired accuracy, by inserting in place of the approximation the series expressing $\tan \varphi$ in terms of d/r given later.

SECONDARY MEASUREMENTS. — As the fractional correction to d has a constant value it vanishes for all values of d as in former cases.

IX. MIRROR CURVATURE.

This produces no error if the axis of suspension of the mirror is tangent to the spherical reflecting surface. If the curvature of the mirror is irregular, the definition will be impaired; but if the telescope collects the light from the whole mirror, no sensible irregularity of reading will be caused.

In the diagram, which is exaggerated for clearness, let Y be the vertical axis of suspension, MR the mirror undeflected (assumed concave), and OY the line from Y to the center of the scale, and to which the undeflected mirror will be normal. Suppose the mirror to be turned through an angle φ about Y, then its new position will be $M'R'$, M having moved to

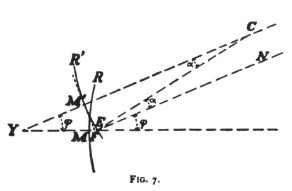

FIG. 7.

M' through the angle $MYM' = \varphi$. $YM'C$ is now normal to the mirror, and let $M'C$ be the radius ρ of curvature of the mirror. A ray along OY would now meet the mirror at E, whereas, if the mirror had been plane, the reflection would have been from F; and if there had been no eccentricity, from M, since Y would then have passed through M. The error due to the forward motion from F to E would be capable of correction by the same expression as that for MF, but it may be shown to be so small as to be negligible except when ρ is very short, in which case, however, the mirror is not safe for use in accurate work even with the application of the correction.   It is therefore at first necessary to deal only with the error which comes from the increased angle between the normal and OE.   Owing to the curvature of the mirror, the normal to the mirror at E, instead of lying parallel to YC and therefore making an angle φ with OE, makes with it a greater angle $CEO = \varphi + a$, where a is of course equal to the angle between the tangents at M and M'.

We desire an expression for the fractional correction $\delta d / d$.   But

$$\frac{\delta d}{d} = \frac{\tan 2\varphi_1 - \tan 2\varphi}{\tan 2\varphi_1} = \frac{\tan 2\varphi_1 - \tan 2\varphi}{\tan 2\varphi} \text{ approx.,}$$

$$= \frac{\tan \varphi_1 - \tan \varphi}{\tan \varphi} \text{ approx.}$$

Now $\varphi_1 = \varphi + a$, and a is always very small;

$$\therefore \tan \varphi_1 = \frac{\tan \varphi + \tan a}{1 - \tan \varphi \tan a} = \tan \varphi + \tan a \text{ approx.}$$

$$\therefore \frac{\delta d_1}{d_1} = \frac{\tan a}{\tan \varphi}.$$

But as $M' \therefore C$ and EN are parallel, $M' CE = a$; and as E and F are nearly coincident and FM' is perpendicular to YC,

$$\frac{\tan a}{\tan \varphi} = \frac{X}{\rho},$$

$$\therefore \frac{\delta d_1}{d_1} = \frac{X}{\rho} \text{ approx.}$$

For a deflection in the opposite direction the correction would be the same, so that the general expression for the correction is

$$\frac{\delta d}{d} = \frac{X}{\rho}.$$

For a concave mirror ρ is $+$, for a convex mirror $-$.

PRIMARY MEASUREMENTS. — The limiting value of ρ may be found from

$$\delta d_1 / d_1 \lesseqgtr 0.00030 = X / \rho.$$

With $X =$ zero, that is, no eccentricity, the correction vanishes. For $X = 1$ mm., $\rho = 3300$ mm. $= 3.3$ m. For $X = 10$ mm., $\rho = 33$ m. and so on. An accidental curvature of less than 3 m. is not unlikely, and an eccentricity of 1 mm. is not uncommon and by no means always avoidable. It is therefore necessary to measure ρ roughly as shown below, but it is unlikely that the correction will be large when a presumably plane mirror is used, and a little consideration of the character of the error will show that a mirror requiring a large correction cannot be employed.

SECONDARY MEASUREMENTS. — As the fractional correction is constant, it may be entirely omitted.

TO MEASURE ρ. — Focus the telescope sharply on the reflection of the middle of the scale from M as in using the apparatus. Measure MO within a few mm. Turn the telescope slightly so as to be able to look beyond the mirror, and place a printed page or other suitable object in the line of sight at a distance about equal to MO. Without changing the former focus of the telescope, move the object towards and from the telescope until a point P is found at which the focus is again sharp. Measure MP within a few mm. Then the radius of curvature of the mirror is

$$\rho = 2 \cdot \frac{MO}{1 - \dfrac{MO}{MP}}$$

in which the numerical values of both MO and MP are considered positive. If $MP > MO$, clearly ρ is positive and the mirror is concave. If $MP < MO$, ρ is negative and the mirror convex.

If it is merely necessary to know whether ρ exceeds a specified limit, this expression may be transformed into

$$MP = \frac{MO}{1 - \dfrac{2\,MO}{\rho}}.$$

It is then merely necessary to calculate the value of MP and to see that the object is in focus at a distance less than this in the above test.

X. MIRROR VERTICAL AND VERTICAL ANGLE TMO SMALL.

Let $MH =$ horizontal line through M,

$MN =$ normal to mirror,

$TMO =$ line of sight when $\varphi = 0$.

The reflecting surface of the mirror is here assumed to coincide with the axis of suspension, so that $MH = r$.

If the mirror is vertical and the telescope and scale are both at H, no correction will be required. The mirror may be made vertical or nearly so, but T must usually be either above or below the scale, hence the angle TMO cannot be zero. Figure 8 represents the case when neither M nor TMO are zero, and where $NO < NT$. For this case the fractional correction will be shown to be

$$\frac{\delta d}{d} = \frac{NH \cdot NO}{r^2}.$$

This obviously approaches zero as NH approaches zero, *i. e.*, as M becomes more and more nearly vertical. It will disappear if the mirror is exactly vertical whatever the angle TMO (within the limits for which the formula holds). But as it is of course impossible to render M exactly vertical, it is necessary to inquire how nearly so it can probably be rendered, and what would be the corresponding limit of TMO. With considerable care the mirror may be made so nearly vertical that N will fall within 10 mm. of H.

Inserting then $NH = 10$, $r = 1000$, and $\delta d / d \lessgtr 0.00030$ and solving for NO gives $NO = 0.00030 \cdot 1000^2 \div 10 = 30$ mm.

PRIMARY MEASUREMENTS. — Thus as NO is approximately equal to NT, the telescope and scale must not be farther apart than 60 mm. Many forms of telescope and scale are so faulty in design that this closeness of approach is impossible. Most forms provide for motion over a much greater range. No such motion is necessary, and none is desirable.

SECONDARY MEASUREMENTS. — Since the correction is a constant one so long as T, N, and O remain fixed, no correction is necessary in secondary work, and no special care to have NH and NT small. Since, however, the correction formula is only approximate and applies only to small values of NO, the adjustment should be made somewhat nearly to the above limit. The values of NH and NT must not change during a series of measurements by as much as the above amounts.

DEMONSTRATION. — Let Figure 8 represent the apparatus in side view, M being the mirror, O the middle point of the scale, T the telescope, MN the normal to the mirror when deflected, and MH

a horizontal line through M. Let the dotted line through O be a vertical line, and assume T, N, and H to lie upon that line, so that all the lines of the figure lie in a vertical plane through $O M$. The mirror, which should be vertical, is inclined to its vertical axis of rotation by the small angle a. Thus the normal $M N$, which should coincide with $M H$ when undeflected, makes with it the angle $N M H = a$. The telescope and scale should coincide (T with O), but cannot. Hence the telescope is at any position T above or below the scale O, and $N M T = N M O = \beta$.

Let Figure 9 represent a vertical plane through the scale and viewed along $M H$. Then $S S$ will represent the scale, $H' H''$ a horizontal line through H and at right angles to $M H$, and T', N', O', and H' the points T, N, O, and H respectively. Suppose now the

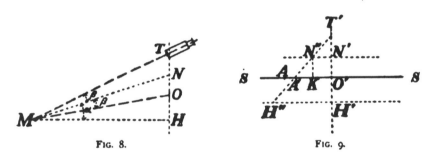

FIG. 8. FIG. 9.

mirror is turned about its vertical axis of suspension, assumed to pass through M, through a small horizontal angle φ. Thus the normal $M N$ will describe a conical surface with a vertical axis and having its vertex at M. This cone would intersect the vertical plane through the scale in an hyperbola with its vertex at M'. Let $N' N''$ be a horizontal straight line; then if $N' N''$ be the horizontal projection on this plane of the part of this hyperbola described when M turns through φ, we have,

$$\tan \varphi = N' N'' / M H = N' N'' / r.$$

It is also true that as φ is small, the hyperbola will sensibly coincide with $N' N''$, but we need not make this assumption. Since the reflected ray will lie in a plane containing T and $M N''$, the observed scale reading when M is deflected through φ will be at A', the intersection of a straight line $T N''$ (prolonged) with the scale. Since it is read by a horizontal scale having vertical rulings, the distance $O' A'$

along the scale will be the horizontal projection upon the scale of the parabolic path of a ray from T reflected by M upon the plane of the scale. This short portion of the parabola of the ray will sensibly coincide with the horizontal straight line $S\,S$.

Let A represent the point (unknown) of the scale at which the true reading corresponding to the observed reading A' would fall; then

$$\delta d_1 = O'\,A - O'\,A'.$$

For values of φ so small that $\tan 2\varphi = 2\tan\varphi$ nearly enough,

$$\delta d_1 = 2\,r\tan\varphi - O'\,A' = 2\,N'\,N'' - O'\,A'.$$

But also for larger values of φ, within the usual limits, although the value of $\tan 2\varphi$ exceeds $\tan\varphi$ in a continually increasing ratio, the value of $O'\,A'$ increases from the same cause in sensibly the same ratio; so that the above expression for $\delta\,d_1$ holds nearly enough for all cases. From the diagram,

$$O'\,A' = O'\,K + K A' = N'\,N'' + N''\,K \cdot \tan O'\,T'\,A' = N'\,N''' +$$
$$N\,O \cdot \frac{N'\,N''}{N\,T}.$$

Whence as $2\,N'\,N'' = d_1$ nearly enough (as a factor),

$$\frac{\delta\,d_1}{d_1} = \frac{1}{2}\left(1 - \frac{N\,O}{N\,T}\right).$$

But,

$$\frac{N\,O}{N\,T} = \frac{M\,O}{M\,T} = \frac{r\,\cos\,(\alpha + \beta)}{r\,\cos\,(\alpha - \beta)} = \frac{\cos\,\alpha\,\cos\,\beta\,-\,\sin\,\alpha\,\sin\,\beta}{\cos\,\alpha\,\cos\,\beta\,+\,\sin\,\alpha\,\sin\,\beta}$$

$$= 1 - 2\sin\,\alpha\,\sin\,\beta \text{ approx. when } \alpha \text{ and } \beta \text{ are small;}$$

$$= 1 - 2\,\frac{N\,H \cdot N\,O}{r^2} \text{ approx.}$$

$$\therefore \frac{\delta\,d_1}{d_1} = \frac{N\,H \cdot N\,O}{r^2} \text{ approx.}$$

The correction has the same sign on either side of O, hence

$$\frac{\delta\,d}{d} = \frac{N\,H \cdot N\,O}{r^2} \text{ approx.}$$

In the last approximation $N\,T$ might have been introduced instead of $N\,O$. The choice is determined by the fact that O and N lie nearer to H than T and N, so that the approximation is closer in assuming $\sin \beta = N O / r$. If, therefore, in an actual case the telescope is much nearer to the horizontal than the scale, it will be slightly better to insert $N T$ instead of $N O$. The sign of the correction to r is — when $N O < N T$ and $+$ when $N O > N T$.

XI. SCALE HORIZONTAL.

To adjust the scale horizontal, focus T on the scale, with M swinging free. Note the height of the cross-hair intersection upon the divisions of the scale. Deflect M to the right and left. The intersection should remain at the same height. If it does not, raise one end of the scale.

If the scale be tipped upward or downward from the horizontal through a small angle γ, remaining in the same vertical plane, the reading on either side will be shortened by the versed sine of the angle γ; that is, the fractional correction to d will be

$$\frac{\delta d}{d} = \text{versin } \gamma = 1 - \cos \gamma.$$

PRIMARY MEASUREMENTS. — For the requisite closeness of adjustment,

$$\frac{\delta d}{d} = 1 - \cos \gamma \lessgtr 0.00\,030,$$

$\therefore \cos \gamma = 0.99\,970$, and $\gamma = 1.°4$.

In making the test the observed change in the height h of the cross-hair intersection in passing from the middle to the end of the scale, would be

$$\delta h = 500 \tan \gamma.$$

Hence the requisite closeness will be

$$500 \cdot \tan \gamma = 500 \cdot \tan 1.°4 = 12 \text{ mm.}$$

This adjustment can be made with perfect ease to 1 mm. or less, so that this error disappears.

SECONDARY MEASUREMENTS. — As $\dfrac{\delta d}{d}$ is constant so long as γ is constant, the amount of tipping of the scale within wide limits makes no difference so long as it remains always at the same angle.

XII. SCALE PERPENDICULAR TO $O\,M$ AND PLANE.

To render $M\,O\,A$ a right angle, lay off or select two points A and B near the ends of the scale, exactly equidistant from O.   Measure carefully the distances $M\,A$ and $M\,B$ from the suspension fiber of M. Adjust the scale until the two are equal.   When this and the preceding adjustment have been completed, the scale should be rigidly and permanently fixed in place, and means provided to enable the fiber to be always brought back to its present position.   A method of support for the scale quite separate from the support of the telescope, as described later, contributes to stability and convenience.   It may be noted that inequality of scale readings on reversal of M is not a proof of imperfection in this adjustment, nor equality a sufficient proof of correct adjustment.

The fractional correction will be shown to be

$$\frac{\delta d}{d} = \mp \sin \beta \tan 2\,\varphi - \frac{1}{2} \sin^2 \beta \text{ approx.},$$

where β is the *small* angle $A\,O\,A'$ by which the scale is out of adjustment.   The negative sign of the first term applies to deflections towards A, the positive towards B, used right-handedly.

PRIMARY MEASUREMENTS. — If deflections are taken on one side only, *i. e.*, without reversals of M, the requisite closeness in β will be attained when

$$\pm \sin \beta \tan 2\,\varphi = 0.00\,030$$

as $\frac{1}{2} \sin^2 \beta$ is negligible when the apparatus is closely adjusted.   For $r = 1000$ mm., $d = 500$ mm.,

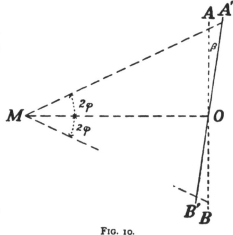

FIG. 10.

$$\sin \beta = 0.00\,030\,\frac{1000}{500} \text{ roughly} = 0.00\,060.$$

This corresponds to $\beta^\circ = 0.00\,060\,/\,0.0174$
$$= 0.°035 \text{ roughly} = 2'.$$

But it is more convenient to have the requisite closeness expressed in terms of $MA' - MB'$. Now, nearly enough when β is very small, $AA' = BB' = OA \cdot \tan\beta = OA \cdot \sin\beta = 0.00060 \cdot 500 = 0.30$ mm., and also $MA' - MB' = AA' + AB' = 2 \cdot OA \cdot \tan\beta = 0.60$ mm. Hence, the requisite closeness in $MA' - MB'$ *without reversals* is 0.6 mm.

If reversals of deflections are taken, giving d_1 on one side and d_2 on the other, the corrected deflection is

$$\frac{1}{2}\left[d_1 + d_2 - \sin\beta\,\tan 2\varphi - \frac{1}{2}\sin^2\beta + \sin\beta\,\tan 2\varphi - \frac{1}{2}\sin^2\beta\right]$$

$$= \frac{1}{2}(d_1 + d_2) - \frac{1}{2}\sin^2\beta.$$

Hence the requisite closeness will be attained when

$$\frac{1}{2}\sin^2\beta' = 0.00030, \quad \therefore \ \sin\beta' = 0.025.$$

This corresponds to $\beta'^\circ = 0.025 / 0.0174 = 1.°4$ roughly. Then $2 \cdot OA \cdot \tan\beta = 25$ mm. Hence the requisite closeness in $MA' - MB'$ *with reversals* is only 25 mm. This demonstrates a great advantage in reversals in primary work when practicable. For not only is the scale more readily adjusted, but a slight undetected deviation of it or a slight lateral displacement of the suspension is much less serious.

The scale must evidently be plane within the same limits within which it must be perpendicular to OM.

SECONDARY MEASUREMENTS. — In these nearly the same closeness is requisite in $\sin\beta$ as in primary work. If, however, deflections are restricted to the outer half of the scale, *i. e.*, 250 to 500 mm., a little less closeness will suffice, since the error at the deflection reading (whether with or without reversals) at which the calibration is made will enter into the value of the constant, and this will eliminate the error at the *same point* in subsequent readings. It will also reduce the average error about one-half if the calibration deflection is about 350 to 400 mm., but the difference in error of a small and of a large deflection will still remain the same.

DEMONSTRATION. — Let A and B be any points on the scale equidistant from O and on opposite sides. Then OA and OB will be the true scale readings with the scale properly adjusted, and will be equal. OA' and OB' will be the observed readings.

The fractional correction $\delta\, d_1 / d_1$ on the side towards A' will be

$$\frac{O\,A - O\,A'}{O\,A'} = \frac{O\,A}{O\,A'} - 1.$$

But $\dfrac{O\,A}{O\,A'} = \dfrac{\sin\,(90° - 2\,\varphi - \beta)}{\sin\,(90° + 2\,\varphi)} = \dfrac{\cos\,(2\,\varphi + \beta)}{\cos\,2\,\varphi}$

$= \dfrac{\cos\,2\,\varphi\,\cos\,\beta - \sin\,2\,\varphi\,\sin\,\beta}{\cos\,2\,\varphi} = \cos\,\beta - \sin\,\beta\,\tan\,2\,\varphi.$

Now $\cos\,\beta = (1 - \sin^2\,\beta)^{\frac{1}{2}}$

$\qquad\qquad = 1 - \dfrac{1}{2}\,\sin^2\,\beta$ approx., as β is very small.

$\therefore\ \dfrac{\delta\, d_1}{d_1} = \dfrac{O\,A}{O\,A'} - 1 = -\dfrac{1}{2}\,\sin^2\,\beta - \sin\,\beta\,\tan\,2\,\varphi.$

Similarly on the side toward B,

$$\frac{\delta\, d_2}{d_2} = \frac{O\,B}{O\,B'} - 1 = -\frac{1}{2}\,\sin^2\,\beta + \sin\,\beta\,\tan\,2\,\varphi.$$

Or, in general,

$$\frac{\delta\, d}{d} = \mp\,\sin\,\beta\,\tan\,2\,\varphi - \frac{1}{2}\,\sin^2\,\beta.$$

XIII. Null Point.

The " null point " or " zero reading," *i. e.*, the reading when the mirror is undeflected, must be the middle point O of the scale, adjustment XII having been made ; and $O\,M$ must lie in a vertical plane perpendicular to the scale. In other words, the line of sight $O\,M$, when M is undeflected, must lie in a vertical plane which is perpendicular to the scale at its middle point O. This does not require either that any normal to M should lie in or parallel to that plane, or that the axis of the telescope should lie in that plane. Or, as less precisely expressed, the plane of the mirror need not be parallel to the scale, nor the axis of the telescope be exactly above or below O. The telescope may be at any position off at one side if more convenient, but ease and accuracy of adjustment, as well as other considerations, lead to the customary location of the axis of the telescope more or less exactly above or below O. Provided that the axis of rotation (suspension) and the reflecting surface of M are sensibly coincident, and that

X, XI, and XII have been completed, this adjustment may be accurately made for any position of the telescope thus: Focus the telescope sharply on the suspension fiber (axis of rotation) of M, with M swinging free. Turn the telescope, or shift it laterally without disturbance of the scale, until the cross-hair intersection falls upon the fiber. Still without disturbing the scale, tip the telescope about a horizontal axis parallel to the scale until the cross-hairs are approximately central on the mirror, and change the focus until the scale is sharply defined. The reading will in general not be exactly at the middle point, but the telescope and scale must now be clamped rigidly in position, and all subsequent adjustment of the null reading to zero must be made *by the mirror*, that is, by changing the direction of the suspended system until the null reading is exactly the middle point O from which A' and B' are measured off in making adjustment XII. This will be effected according to the nature of the apparatus; *e.g.*, by changing the directive field, in a sensitive galvanometer; by turning the torsion head or the whole instrument if it is an electrodynamometer or electrometer, or by twisting the mirror upon its suspension rod, and so on.

If the axis of the suspension and the reflecting surface of M do not coincide, the above method of adjustment becomes inaccurate to an extent depending upon the eccentricity of the reflecting surface, and on the departure of the telescope from the vertical through O. If the eccentricity is but a few millimeters, its effect is wholly negligible if T is within a few millimeters of this vertical, as may be seen from adjustment VIII. In that case no attention to the adjustment of T and O beyond casual inspection is called for.

If when this stage of the adjustment is reached it is found that the cross-hair intersection does not fall at the right height upon the scale for good reading, the telescope may be tipped slightly as a remedy. But if when this is done the illumination or extent of field is not all that the apertures of telescope and mirror should give, then the adjustments must all be repeated, beginning by raising and lowering the telescope and scale together, and tipping the former more or less, focusing centrally on the mirror and then on the scale alternately.

During observations it is by no means always practicable to bring the null reading exactly to the middle point of the scale. This reading d_0, however, is taken before and after each deflection, or with sufficient frequency, and the observed deflection reading d' is corrected for d_0, or reversed readings are taken, giving the same result. The advantage of numbering the scale from one end continuously to the other instead of both ways from a middle point is apparent in this connection. For in the former case no attention to the sign of either d' or d_0 will be necessary, and corrected deflections, viz., $d' - d_0$ will be always — on the side towards the zero and + on the other side, the sign taking care of itself.

But even with the correction for the observed null reading applied, there remains an error from the fact that the ray $M d_0$, or rather the vertical plane through it, is not exactly at right angles to the scale. The question remains how closely this angle must approach 90° ; or in other words, what is the limit of displacement of the null reading when corrected for, either by observing d_0 or by reversing.

Suppose the scale to be properly adjusted at $A B$, and that the null reading becomes subsequently displaced (*e.g.*, by change of torsion in the suspension, of direction of field, etc.) from O to O''.

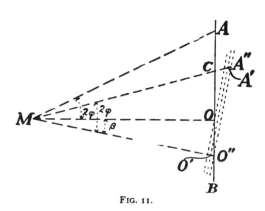

FIG. 11.

The observed reading towards A for a deflection q of M corrected for d_0 (= $O O''$) will be $O'' C$. The true reading would have been $O A$. It is desired then to find an expression for the fractional correction to reduce $O'' C$ to $O A$. If a horizontal scale were placed perpendicular to $M O''$ at a point O' such that $M O' = M O$, then the observed reading $O' A'$ on such a scale would be equal to the true reading $O A$. Imagine also a scale $O'' A''$ parallel to $O' A'$ but drawn through O''. Then

$$\frac{O' A'}{O'' A''} = \frac{O' M}{O'' M} = \frac{O M}{O'' M} = \cos \beta.$$

But as shown under XII (case of $O B / O B'$),

$$\frac{O'' A''}{O'' C} = \cos \beta + \sin \beta \tan 2 q$$

$$\therefore \frac{O'A'}{O''C} = \cos^2 \beta + \cos \beta \, \sin \beta \, \tan 2\,\varphi$$

$$\therefore \frac{OA}{O''C} = 1 - \sin^2 \beta + \sin \beta \, \tan 2\,\varphi, \text{ approx.,}$$

as with β very small, the factor $\cos \beta$ is so nearly unity as not sensibly to affect the last term ;

$$\therefore \frac{\delta d_1}{d_1} = \frac{OA - O''C}{O''C} = - \sin^2 \beta + \sin \beta \, \tan 2\,\varphi$$

and in general

$$\frac{\delta d}{d} = \mp \sin \beta \, \tan 2\,\varphi - \sin^2 \beta,$$

the $+$ and $-$ signs in the first term applying respectively to right and left-hand deflections q.

PRIMARY MEASUREMENTS. — *With Reversals of M.* Disregarding the observational sign of d, the mean deflection used would be $\frac{1}{2} (d_1 + d_2)$, and the correction would be

$$\frac{\delta d}{d} = \frac{1}{2} \, [\sin \beta \, \tan 2\,\varphi - \sin^2 \beta - \sin \beta \, \tan 2\,\varphi - \sin^2 \beta]$$

or $= - \sin^2 \beta$ for any value of q.

The requisite closeness of β would therefore be found from

$$\frac{\delta d}{d} = 0.00\,030 = \sin^2 \beta$$

$$\therefore \sin \beta = 0.017 \; ; \; \beta = 1.°0, \text{ or}$$

$$O''O = 1000 \tan \beta = 17 \text{ mm.}$$

Without Reversals, the correction would depend chiefly on q as β would be very small compared with any practicable value of q. Hence nearly enough,

$$\frac{\delta d}{d} = \mp \sin \beta \cdot \tan 2\,\varphi.$$

The requisite closeness of β for the worst case where $r = 1000$ mm., $d = 500$ mm., would then be found from

$$\frac{\delta d}{d} = 0.00\,030 = \pm \sin \beta \cdot \frac{500}{1000}$$

$$\therefore \sin \beta = 0.00\,060 \quad \beta = 0.°035 = 2', \text{ or}$$

$$OO'' = 1000 \tan \beta = 0.6 \text{ mm.}$$

Thus with reversals a rather rough adjustment of the null reading from time to time is sufficient; but with deflections on one side only, even with the null reading allowed for, this reading must never exceed about half a millimeter without the application of the above special correction $\delta d_1 / d_1$ or $\delta d_2 / d_2$. It is also essential to note that the adjustments of the null reading must be made by turning the mirror, *not* by shifting the scale.

SECONDARY MEASUREMENTS. — The requirements here are precisely the same as in primary work, and the frequent and often unavoidable practice of reading without reversals makes the above remarks of special importance. It is obviously much better to reverse where practicable.

XIV. DISTANCE OF SCALE FROM MIRROR.

This is invariably *the shortest horizontal distance from the axis of suspension to the vertical plane through the scale rulings.* The reflecting surface of the mirror should lie near to this axis when possible, but the measurement of r must be from the actual suspension horizontally to a vertical line through the middle point of the scale after XII has been performed. It is important to note, as the following pages will show, that most of the adjustments become more easy and the corrections smaller as the scale distance is increased. Therefore, as scale errors are easily reduced, it is better to make r as large (up to 3 or 4 m.) as is consistent with the magnifying power of the telescope, using a scale of not more than 1 m. in length. The telescope, scale, and instrument containing M are set in their proper positions and completely adjusted before r is measured. This is then carefully determined by a horizontal wooden rod, wire, or steel tape, using plumb lines if necessary at M or at S. Before being used in computations, r must be expressed in the same unit as d. The telescope should be so far from M that its draw tube will permit focussing on M.

PRIMARY MEASUREMENTS. — To find the requisite closeness in the measurement of r we have to note that r enters as a direct factor in the denominator of tan φ. Hence the percentage or fractional error in tan 2φ is proportional to that in r, but with the opposite sign. (See

"Computation Rules," Proposition I, page xii.) The limiting value of $\delta r / r$ will therefore be the same, neglecting sign as $\delta d / d$. Therefore

$$\delta r / r = 0.00\,030, \quad \delta r = 0.3 \text{ mm.}$$

Hence the requisite closeness in r is about 0.3 mm.; that is, r must be measured and must be constant to about 0.3 mm. See remark in next paragraph.

SECONDARY MEASUREMENTS. — In secondary work r is not measured, but must remain constant, and suitable means must be provided to see that it is so, within the same closeness of 0.3 mm. This requires much more attention than is usually given to this point, especially when the mirror hangs on a long suspension so that change of level of the instrument may easily change r by 1 or 2 mm.

A special device is almost a necessity in careful work, to facilitate the test for constancy in the distance OM. In many instruments, especially those having a long suspending fiber, a slight difference of level alone produces considerable displacements of the mirror towards or from the scale or laterally, — all equally objectionable. The direct application of XII and XIII for the elimination of these displacements on each day of use of the apparatus would be very laborious. This fact together with inadvertence frequently leads to the assumption that the distance, once adjusted, remains sufficiently constant. The foregoing figures as to the requisite closeness show the danger in such neglect. Devices will readily suggest themselves. The following will sometimes answer: Place the points of the leveling screws in positions from which they are not easily displaceable. Level the instrument until the suspension swings as it should. Then make upon the instrument, near the lower end of the suspending fiber, say on opposite sides of the suspension tube, reference marks so located that the fiber lies in the line of sight between them. Similarly locate another line of sight, nearly at right angles to the first and also passing through the fiber. For convenience, one of these lines should be parallel to a pair of the leveling screws, and also to the scale. At each time of use, adjust the screws to bring the fiber into both of these lines of sight. Measure OM once for all, and at the same time measure between two points chosen for convenience, one on the base of the instrument, the other on the scale. It is then necessary from day to day merely to bring the fiber into the reference lines by turning the leveling screws, and then to measure the distance between the chosen points, for which purpose a rod cut to the right length is convenient.

XV. Estimation of Tenths in Reading.

Fractions of a division must be read, and this can be done only by estimation by the eye. With a little practice, so much facility in the estimation of tenths of a division is attainable that the error need never exceed one-twentieth of a division, with an average error of half this amount, or 0.025 division.

Primary Measurements. — The limit of attainable accuracy being fixed at $\pm\ d = 0.025$ division, and the desired fractional accuracy being $\pm\ \delta d\ /\ d = 0.00030$, the minimum admissible value of d will be $d = 0.025\ /\ 0.00030 = 83$, or in round numbers $d = 100$ divisions, whatever the size of the division. Smaller deflections than 100 must not be used in the most careful work, and preferably not less than, say, 200 divisions, whatever the scale distance.

Secondary Measurements. — Same as in primary.

XVI. To Find q, Tan q, Sin q, etc., from d and r.

In the application of the telescope and scale method it is desired to find from the observed values of d and r the corresponding values of q, tan q, sin q, etc., according to circumstances.

For the most accurate work, especially in primary measurements, the best way is to compute tan $2\,q = (d\ /\ r)$ or log tan $2\,q$, thence by tables of natural or log tangents to find $2\,q$. This, of course, gives q at once, and tan q, sin q, and other functions can then be found from tables. In some cases, notably in secondary work, it is, however, more convenient to have an expression for q, tan q, or other function, directly in terms of d and r. Such expressions generally take the form of a series. Several may be found in Czermak. The expression for tan q is the one in frequent use, and therefore this alone will be given here. It is derived from a development of tan q in series with ascending powers of tan $2\,q$, viz. :

$$\tan q = \frac{1}{2}\tan 2\,q\left[1 - \frac{1}{4}(\tan 2\,q)^2 + \frac{1}{8}(\tan 2\,q)^4 - \ldots\right].$$

Substituting $d\ /\ r$ for tan $2\,q$ gives

$$\tan q = \frac{1}{2}\cdot\frac{d}{r}\left[1 - \frac{1}{4}\left(\frac{d}{r}\right)^2 + \frac{1}{8}\left(\frac{d}{r}\right)^4 - \ldots\right].$$

This could be used directly, but may be modified into a more convenient form. Multiplying the terms of the parenthesis by d gives

$$\tan \varphi = \frac{1}{2\,r}\left[d - \frac{1}{4} \cdot \frac{d^3}{r^2} + \frac{1}{8} \cdot \frac{d^5}{r^4} - \cdots \right].$$

Then the quantity

$$-\frac{1}{4} \cdot \frac{d^3}{r^3} + \frac{1}{8} \cdot \frac{d^5}{r^4} - \cdots$$

being very small may be conveniently treated as a correction δ to be
applied to d. So we may write

$$\tan \varphi = \frac{1}{2\,r}\,[d + \delta].$$

The procedure would be to compute beforehand a table of values
of this correction for the given value of r and for successive values of
d ($= 200, 250, 300, \ldots 500$) at sufficiently short intervals over the
desired range. Then for a subsequent observed value of d', the corresponding value of δ' would be taken by interpolation in a table (or
upon a plot), and applied to d'. This corrected value of d' would
then be multiplied by $1/2\,r$ to obtain $\tan \varphi$. As the correction is
small, only an approximate value of r is required in computing it, and
slight changes in r during the progress of work would not vitiate the
table, although they must be allowed for in the term $1/2\,r$.

In secondary work any of the above methods of finding φ, $\tan \varphi$, etc.,
may be used, but for use in connection with tangent instruments the
application of the series expression is especially advantageous. This
will be illustrated by the tangent galvanometer. For this instrument
the current C is related to the steady, angular deflection φ which it
produces, by the expression

$$C = k \tan \varphi,$$

where k is a numerical factor which is constant so long as the magnetic
conditions and dimensions of the instrument are constant. Substituting the above expression for $\tan \varphi$ gives

$$C = \frac{k}{2\,r}\,[d + \delta]$$

or, as r is a constant throughout the work, we may write

$$C = K\,[d + \delta]$$

where K is a numerical constant $= k / 2 r$ and

$$\delta = -\frac{1}{4} \cdot \frac{d^3}{r^2} + \frac{1}{8} \cdot \frac{d^5}{r^4} - \cdots$$

But in the employment of such a secondary instrument it is calibrated, that is, its constant is found, by sending through it some known current C' and observing the steady deflection d'. Then denoting by δ' the known value of the correction δ computed for this deflection d', we have

$$K = \frac{C'}{d' + \delta'}.$$

This gives us the numerical value of K so that C can be at once computed for any subsequently observed values of d by the expression

$$C = K [d + \delta],$$

the proper value of δ being taken from a table or plot as above described. The correction δ will be a quantity to be subtracted from d, and therefore expressed in the same unit, *e. g.*, millimeters. The table must be carried out to the next place of significant figures beyond the last place obtained in reading the scale, *e. g.*, to hundredths of mm. if readings are taken to tenths of mm. It must also be computed for sufficiently short intervals to enable interpolation to be made with corresponding closeness, *i. e.*, to 0.01 mm. or 0.02 mm. in the above case. And if a plot is used it must be on a sufficient scale and with sufficiently frequent points to enable this same closeness to be obtained. Unless many readings of d are to be reduced, it is as well to compute δ for each observation as to make a table. Extensive tables of δ for graded values of r and d are given in Czermak. The number of terms to be retained in the series in computing δ must be determined by computing for the largest value of d the value of the successive terms until one is reached which is negligible, *e. g.*, less than about 0.01 mm. in the above case.

If the instrument is used always with deflections near the value d', observed in calibrating, it is evident that the values of δ both in obtaining K and subsequent measurements will be nearly the same. Hence if δ were omitted in computing K, and also in all subsequent computations of C, the errors thus introduced would nearly offset one another. To see how wide a range of deflections could thus be used without introducing into C a resultant error exceeding the limit which we have been using in the preceding discussion, we have merely to find the two deflections d_1 and d_2 at any desired part of the scale for which $(\delta_2 - \delta_1) / d_1 = 0.00\ 030$. For a rather extreme case of $d_1 = 400$ mm., $r = 1000$ mm.,

$$\delta_2 - \delta_1 = 0.12 \text{ mm.}$$

The value of d_2 for which δ_2 would be $\delta_{400} + 0.12$ could be found from tables, but in default of these we may compute as follows, neglecting the second term in δ,

$$\frac{1}{4} \cdot \frac{d_2^3}{r^2} - \frac{1}{4} \cdot \frac{d_1^3}{r^2} = 0.12$$

$$\therefore\ d_2^3 - d_1^3 = 0.48 \cdot 10^6$$

$$\therefore\ d_2^3 = 400^3 + 0.48 \cdot 10^6$$

$$= 64 \cdot 48 \cdot 10^6$$

$$\therefore\ d_2 = 401.0 \text{ mm.}$$

Thus the correction could be neglected in this case only over a range of 1 mm. Even with large values of r and smaller ones of d the correction becomes negligible over only a centimeter more or less, and thus is practically never negligible in 0.1 per cent. work.

SELECTION OF APPARATUS.

Rigidity of construction is a prime requisite. It is often impaired by unnecessary and weak adjustments or poor clamping devices. The scale may be, and generally is, carried on the same support as the telescope. But an entirely independent support is to be preferred, as facilitating adjustment and promoting stability. This will be the more obvious if it be remembered that the telescope need not be exactly over the center of the scale or in any determinate position relatively to it, although an approximately central position is usually more convenient and is presupposed in the preceding discussion of some of the corrections. A convenient method of supporting the scale would be by a bracket at each end clamped to the table by screws passing through slots in the bracket. This would permit the needed forward and back motion of the ends separately. Long and wide slots in the vertical arms of the brackets would afford the necessary vertical and endwise range of adjustment. With this arrangement the scale can be placed in final adjustment and clamped in position without employing the telescope, and the latter may be separately put in place with the minimum of trouble, and with no chance of disturbance of the scale.

The telescope must have rotation about both vertical and horizontal axes. It should be provided with a vertical adjustment through a range of six inches or more, preferably by a round rod, which gives at the same time the needed vertical axis. The base must be provided with screw holes for immovable attachment to the table. The draw tube of the telescope must be thoroughly firm, so that a moderate lateral pressure on the eye-piece shall produce no permanent shifting of the cross-wires over the scale. The tube must be long enough to focus on objects at a distance of somewhat less than one meter. The definition of the telescope should be tested as carefully as desired, but will usually be sufficiently good if a clear image of the scale is given under fair illumination at the usual distance.

The definition of the mirror on any instrument to be used with this method must be tested, for example, by the quality of the image of the scale when viewed through the telescope at the usual distance, but with no interposed cover-glass. The effect of the cover-glass on the definition must be carefully tested by alternately inserting and withdrawing it, noting the change in sharpness of the image of the scale. The best of thin plate glass is barely good enough, and selection from samples is often necessary, especially to secure sufficiently parallel surfaces. There is little advantage in having the diameter of the objective of the telescope more than double that of the mirror to be used; an inch to an inch and a half is usually abundant, and three-quarters of an inch is often enough. A magnifying power of twelve to fifteen diameters is desirable but is not common. The relation between size of scale-division, magnifying power, and distance will be briefly considered.

Let $OM = r$, $u =$ magnifying power, and $s_0 =$ the best size of division upon which to estimate tenths of a division by the unaided eye; this is about 1 mm. Then s_0 / v is, nearly enough, the angle in radians subtended by one scale division when seen directly at the distance v of most distinct vision. If at any other distance, as at $TM + MO$, reckoning TM from the eye-piece, the angle becomes $s_0 (TM + MO)$. This is the distance in the telescopic method. In order then that the telescope shall compensate for this removal, that is, in order that it shall make the arrangement as sensitive as a direct reading at the distance of most distinct vision (which is the condition attainable by the use of a spot of light proceeding from T and reflected from M to the scale at A), the telescope must have a sufficient magnifying power. This must be, for objects at a distance $TM + MO$, such that $s_0 / (TM + MO) = s_0 / v$, or $u = (TM + MO) / v$. For the case where $TM = MO = r = 1000.$ mm., and $v = 250.$ mm., this yields $u = 8$. Telescopes are often furnished for this use with smaller values of u than that just deduced, hence for ordinary use with low power instruments, the millimeter is about the proper size of scale division, but this is by no means the case with higher powers. The best value of s for any fixed value of u would be such that $us / (TM + MO) = s_8 / v$, or $s / s_0 = (TM + MO) / uv$. For a very good glass $u = 15$, so that with the values as before, the scale should be in half millimeters if used at the distance of $r = 1$ meter; since $s / s_0 = 2000 / (15 \times 25) = 0.53$.

With the mirror and spot of light method, the percentage or fractional precision with which a given angle can be measured with the scale at a distance MO is proportional to that distance; that is, the precision for MO is to that for a distance unity as $MO : 1$. With a telescope of magnifying power u the gain is further increased by the telescope in the ratio S / s, where S is the length of division actually employed, and s is the best length computed as just indicated, with the limitation, of course, that S is greater than s. Thus by the employment of unduly long divisions, the advantage derivable from high magnifying power may be sacrificed and most of the advantage over a much cheaper instrument rendered idle. Of course the superiority of the telescopic method over the spot of light does not lie wholly in the magnification, but partly in better definition, and in the avoidance of the necessity for screening. On the other hand, the latter method has the merit of simplicity and cheapness, as well as of facility of reading where there is much jarring and high accuracy is not demanded.

As to materials for the scale, a white metal surface with fine black rulings would be best, and is almost indispensable in accurate work, but is expensive and not generally, if at all, offered by makers. White porcelain or glass with fine black rulings is the next choice. Paper on wood, or celluloid, is not to be relied upon in careful work, and must be thoroughly tested for uniformity.

The warping of wooden scales may introduce serious error (cf. III).

The numbering usually extends from zero at the middle towards each end, but for many purposes a continuous numbering from one end is more convenient (cf. XIII). By the use of a circular scale with its center of curvature at M, the readings become directly proportional to twice the angle of deflection, and the focus is equally good throughout the length.

MASSACHUSETTS INSTITUTE OF TECHNOLOGY,
Boston, Mass., May, 1898.

RESULTS OF TESTS MADE IN THE ENGINEERING LABORATORIES.

X.

Received July 9, 1898.

STEAM.

DESCRIPTION AND RESULTS OF A 45-HOUR TEST MADE ON THE ENGINES, BOILERS, GENERATORS, AND AIR PUMPS AT THE HARVARD SQUARE POWER STATION OF THE BOSTON ELEVATED RAILROAD.

THIS test was made by students of the senior class under the direction of the instructors of the Engineering Laboratories at the Massachusetts Institute of Technology.

The test was divided into four watches of $11\frac{1}{4}$ hours, twenty-four students working at each watch. During the test 2,000 indicator cards were taken, and about 10,000 observations recorded. In the accompanying tables such summaries of these observations are given as were needed in the calculations of the results.

The power station has at present six Babcock & Wilcox boilers set in three batteries. The boilers all have extension furnaces. The gases are discharged into a common flue at the back of the boilers and pass from this through a Green economizer to the stack. The boilers are fed by power plunger pumps located in the basement of the engine room. These pumps are driven from a countershaft run by an electric motor supplied with current from the station. There are three main engines, $28'' — 56'' \times 5'$, of the cross-compound jet-condensing type, made by E. P. Allis & Company, of Milwaukee.

There is an independent condenser and air pump engine for each main engine. The main engines have Corliss valves with two wrist plates and eccentrics for each cylinder, thus making it possible to cut off later than half stroke on each cylinder.

The air pump engines are controlled by throttling governors, the valves being cylindrical D valves oscillating in cylindrical chests at right angles to the bore of the cylinder.

The main engines are set sufficiently high on their foundations to allow all the steam and exhaust piping to be placed below the engine room floor. The arrangement of this piping is similar to that at the Sullivan Square station at Charlestown, Mass. (See *Technology Quarterly*, Vol. IX, No. 4.) A separator is placed just before the throttle ; the exhaust from the high-pressure cylinder passes into a vertical receiver from which the low-pressure cylinder takes its steam. The exhaust from the low-pressure cylinder passes to the condenser or outboard.

On each exhaust pipe just before the condenser there is a feed-water heater. The feed-water, coming through a 6" pipe from the city main, passes through these heaters on the exhaust pipes, then through the economizer to the suction side of the power pumps, and from the pumps directly to the boilers. By this arrangement the economizer is under only such pressure as there may be in the city main.

The drip from the separators on steam mains is trapped into the suction pipe of the pumps. The drip from the receivers, between the cylinders, contains considerable oil and is thrown away.

The air pumps draw the condensing water from the Charles River through a 20" pipe about 300 feet long. About 100 feet from the station the suction and discharge pipes of the pumps are connected by cross-overs and valves, so that the outer end of either pipe may be used as suction or discharge. At this place there was also a second discharge, which could be used in case of accident to one of the other pipes.

The generators are direct connected, of the M. P. 12–1200–80 type, built by the General Electric Company. The armature is located upon the engine shaft. The nominal output of each machine is 2,180 amperes at 550 volts. Each generator is provided with a separate panel upon the switchboard gallery, on which is located a shunt-ammeter, wattmeter, circuit-breaker, switches, a rheostat, etc.

Upon a separate panel is a station shunt-ammeter that registers the total output of the station. A station voltmeter is also provided upon a swinging bracket, and connections arranged so that it can be connected to either generator.

The customary complement of feeder switches, ammeters, circuit breakers, etc., are provided and mounted upon slate panels.

ARRANGEMENT AND CHANGES MADE FOR THE TEST.

Lazy-tongs were attached to each crosshead to give the proper reduction of piston motion for 2″ diameter drums on the indicators. Indicators were attached to each end of the cylinders of main engines No. 2 and No. 3, the ones used during the test. Air-pump engines No. 2 and No. 3 were each piped up with three-way cocks. The motion for the drums of these indicators was taken from a pin screwed into the end of the shaft. Although this does not give a correct motion, it was considered to be sufficiently accurate, as the horse power of these engines is not considered in the subsequent calculation.

The drips from the receivers were caught in a tank on scales and weighed. The drip pipes from the separators were each disconnected from the traps and each connected with a steel reservoir about 16″ in diameter and 48″ in height, having on the side a gauge glass and scale giving the capacity between different levels. From time to time the levels were blown down by opening valves in the bottom, the discharge going into the suction pipe of the power pumps.

The feed-water pipe was broken at the pumps on the suction side, and the suction pipe changed so as to draw from three large barrels placed alongside the pumps. A check valve opening towards the pumps was placed in the suction pipe close to the barrels. The discharge from the separators was connected between this check valve and the pumps.

The regular suction pipe was carried up to the floor above and supplied four large barrels used for weighing. These barrels discharged into the three below connected with the suction of the pumps. The four weighing barrels were supplied from the city main with water which had passed through the heaters on the exhaust pipe. Ordinarily the water went through the economizer as well, before entering the pumps. During the test the pumps forced the water through the economizer into the boiler. It will be noticed that the temperature of the feed water entering the economizer is less, by a few degrees, than when leaving the heater on the exhaust pipe. This loss is due to the exposure to the air while weighing.   Wooden scales were attached to the water glasses on the boilers.

Calorimeters were placed on the main steam pipe from each battery and near the throttle of No. 2 or No. 3 engine.

Holes for flue thermometers and draught gauges were made at each end of the economizer, and connections for sampling flue gases were made at the entrance end.

The coal scales were tested.   The different gauges at the gauge boards in the engine room were tested and the corrections noted. With one exception, all connections on the feed water piping, where there was a possibility of leakage, were broken and blanked.

The place referred to as not being blanked was where an auxiliary steam pump connected with the line.   There were two Chapman valves here, one either side of the connection.   A pet-cock was tapped into the pipe between these valves and left open, so that any leakage would be noticed.

A weir box 10 feet long, and having a weir 3 feet wide, without end contractions, was placed at the end of the auxiliary discharge pipe for condensing water.   The pass valves were changed so as to send the water over the weir.   The height of water on the weir was measured by a hook gauge reading to $\frac{1}{1000}$ of a foot.   The water entering the weir box was quieted by straining through about 3 feet of brush and a screen of clapboards.

The other observations taken during the test may be noted by referring to the list of stations.

STATIONS.

(1) Electrical readings.

(2) Electrical readings.

(3) Coal, temperature, boiler room, barometer.

(4) Gas analysis every two hours.   Draught and temperature of flue at each end of economizer every half-hour, also temperature of feed at top of boilers every half-hour.

(5) Gas analysis every two hours.   Temperature of steam in main of batteries No. 1 and No. 3 every fifteen minutes; also calorimeters on boilers No. 1 and No. 3.

(6) Water barrel No. 6.

(7) Water barrel No. 7.

(8) Water barrel No. 8.

(9) Water barrel No. 9.

(10) Indicator, planimeter, length and per cent. cut-off, No. 2 H. H.

(11) Indicator, planimeter, length and per cent. cut-off, No. 2 H. C.

(12) Indicator, planimeter and length, No. 2 L. H.

(13) Indicator, planimeter and length, No. 2 L. C.

(14) Indicator, planimeter, length and per cent. cut-off, No. 3 H. H.

(15) Indicator, planimeter, length and per cent. cut-off, No. 3 H. C.

(16) Indicator, planimeter and length, No. 3 L. H.

(17) Indicator, planimeter and length, No. 3 L. C.

(18) Indicator, planimeter and length. Air pump No. 2 engine.

(19) Indicator, planimeter and length. Air pump No. 3 engine.

(20) Time, gong (every fifteen minutes) thus: . . . twenty seconds . . ten seconds . Counter engine No. 2, one-half minute after gong. Counter engine No. 3, one minute after gong. Counter air pump No. 3, one and one-half minutes after gong. Counter air pump No. 2, two minutes after gong. Pressure at gauge boards No. 2 and No. 3 engines. Hook gauge and temperature at weir every half-hour.

(21) Separator drip cans, No. 2 and No. 3 engines.

. (22) Temperature feed at city main, leaving heater No. 2 engine, leaving heater No. 3 engine, entering the economizer, and leaving the economizer ; also calorimeter near throttle No. 3 engine.

(23) Weight of receiver drips, No. 2 and No. 3 engines, also temperature hot and cold condensing water from air pumps, No. 2 and No. 3 engines.

(24) General log.

The circuit to the motor running the power pumps was broken, and an ammeter inserted, upon which readings were taken and the power required for feeding the boilers obtained.

The shunt field circuit of first one and then the other of the generators was connected through an ammeter and the current required for the fields was thus obtained. A standard voltmeter was connected to the circuit, and check readings obtained upon the station voltmeter.

During the night, in order to obtain a load for one engine after the load on the outside lines became too light, a large water rheostat was provided and connected to No. 3 generator on the first night, and to No. 2 generator on the second night, and a steady load of about the nominal capacity of the machine carried, the other engine being shut down.

A standard shunt and a portable ammeter were connected in this circuit and used when the rheostat was in circuit. Electrical readings upon all instruments were taken every five minutes, every third reading being taken upon the stroke of a gong.

The test on the boilers was continuous for 45 hours. The condition of the fires was noted at the beginning of the run, and the fires were brought to the same condition at the end.

If all the possible errors of the test on the boilers are assumed to be cumulative, the maximum error possible is 1.5 per cent. The tests on the engines and generators were divided up into five parts.

From 1-15 P.M. May 10, to 11-30 P.M. Engines No. 2 and No. 3 on regular station load.

From 12-45 A.M. May 11, to 5-45. Engine No. 3 alone on water rheostat with constant load.

From 6-30 A.M. May 11, to 11-30 P.M. Engine No. 2 and No. 3 on station load.

From 12-45 A.M. May 12, to 5-45. Engine No. 2 alone with constant load by water rheostat.

From 6-30 A.M. May 12, to 10-15 A.M. Engine No. 2 and No. 3 on station load.

The air pumps exhaust into the receivers between the high and low cylinders. During the first run No. 2 air pump exhausted into the condenser till 2-45, when the exhaust was turned into the receiver.

During the last run the exhaust of No. 3 pump was turned into the condenser at 8 A.M.

At the end of each run the levels in the boilers were noted, and if different from those at the beginning of the run, corrections were made for the difference.

The B. T. U. per horse power per minute of main engines was calculated by multiplying the steam per H. P. per minute by the heat in a pound of steam, of condition as determined by a calorimeter at the throttle, above the heat of the feed water leaving the heater on the exhaust pipe.

The coal per indicated horse power per hour of main engines was calculated by dividing the B. T. U. per horse power per hour by the B. T. U. taken up from a pound of dry coal. The probable error of work depending on the indicator may be assumed to be about 2 per cent.

SUMMARY OF DATA AND RESULTS OF A 45-HOUR TEST AT HARVARD SQUARE POWER STATION OF THE BOSTON ELEVATED RAILROAD.

ENGINE SIZES.

Main engines No. 2 and No. 3.
Diameter high, 28″. Diameter rod, 5½″.
Diameter low, 56″. Diameter rod, 6¼″.
Stroke, 5 feet.

AIR PUMP. SIZES NO. 2 AND NO. 3.

Diameter steam cylinder . . . . 16″
Diameter piston rod . . . . 2¹¹⁄₁₆″
Stroke . . . . 12″
Diameter bucket . . . . 36″

BOILER SIZES.

(4) B. and W. Boilers.
Heating surface (outside) one boiler . . . . . . 5340 sq. ft.
Grate surface, one boiler . . . . 84. sq. ft.
Total heating surface . . . . 21,360 sq. ft.
Total grate surface . . . . 336. sq. ft.

WEIR.

Weir 3 feet long without end contractions.
Crest of weir above floor of pit was 1.6 feet.
Calculate by formula of Fteley & Stearns.
$Q = 3.31 \, L \, H^{\frac{3}{2}} + .007 \, L$ first approximation.
Correct for velocity of approach, using for H (the height by hook gauge $+ \frac{1.5 v^2}{2 g}$).
$v =$ velocity of approach.
$L =$ length of weir.
$Q =$ cubic feet per second.
$g = 32.2$ feet.

BOILER TEST.

Duration of test . . . . 45 hours.
Kind of coal . . . . New River.
Boilers B. & W. with extended furnaces (2) batteries of two.
Total heating surface of the four boilers . . . . . 21,360 sq. ft.
Total grate surface of the four boilers . . . . . . 336 sq. ft.
Average absolute boiler pressure at boilers . . . . . 168.4 lbs.
Average quality of steam (at boilers) from both batteries (dry steam = 1), .989
Total weight of coal as fired from barrows . . . . . 159,116 lbs.
Total weight of dry coal (1 per cent. of moisture) . . . 157,525 lbs.
Total weight of ashes and clinkers . . . . . . 9,342 lbs.
Total combustible burned . . . . 148,183 lbs.
Dry coal burned per square foot of grate per hour . . . 10.42 lbs.
Total water weighed in barrels . . . . 1,490,621 lbs.
Total returns from separator drips . . . . . . 7,568 lbs.
Total feed water supplied to boilers . . . . . . 1,498,189 lbs.
Average temperature of feed water entering boilers . . . 209.1° F.
Average temperature of feed water entering economizer . 52.7° C. 126.86° F.
Average temperature of feed water leaving economizer . 98.5° C. 209.3° F.
Equivalent evap. from and at 212° per pound dry coal (boilers and economizer) . . . . 10.68 lbs
Total B. T. U. taken up by boiler and economizer per pound of dry coal . . . . 10,311 B. T. U.
Heat taken up by water in economizer per pound of dry coal . 786.4 B. T. U.
Heat gained in economizer in per cent. total heat acquired . . 7.62
Ash and clinkers in per cent. of total dry coal 5.9

ENGINE TESTS.

	No. 2.	Pump No. 2.	No. 3.	Pump No. 3.
Time of run	MAY 10, 1-15 P.M. TO 11-30 P.M.			
Engine number	No. 2.	Pump No. 2.	No. 3.	Pump No. 3.
Revolutions per minute during run	71,974	54,254	72,252	51,531
Average per cent. of cut-off on high pressure cylinder . . .	23.6	......	18.4	......
Mean effective pressure on H. H.	58.9	32.1	47.0	37.7
Mean effective pressure on H. C.	54.9	10.4	43.8	4.38
Mean effective pressure on L. H.	8.73	......	10.9	......
Mean effective pressure on L. C.	12.0	......	9.82	......
Horse power by indicator on H. H.	395.51	10.61	316.82	11.83
Horse power by indicator on H. C.	354.42	3.34	283.85	1.34
Horse power by indicator on L. H.	234.48	......	293.93	......
Horse power by indicator on L. C.	317.97	......	261.22	......
I. H. P. of Each Engine	1302 4	13.95	1155.8	13.17
Total I. H. P. Both Engines	......	2458.2		......
Total water supplied to boilers during run	......	366,824		......
Total steam used by three calorimeters during run	......	12,761		......
Steam supplied to engines and pumps	......	354,063		......
Steam per Hour per I. H. P. of Main Engines	......	14 05		......
Temperature of feed water { Entering heater on exhaust °F. .	50.8	......	50.8	......
Temperature of feed water { Leaving heater on exhaust °F. .	138.1	......	140.2	......
B. T. U. per I. H. P. of Main Engines per Minute . .	......	253.		......
Coal per I. H. P. of Main Engines per Hour	......	1.47		......
Temperature of condensing water { Cold °C.	......	13.1	......	12.8
Temperature of condensing water { Hot °C.	......	20.2	......	31.4
Weight of drip trapped from working side of receivers . . .	......	20,600		......
Quality of steam at throttle (Dry steam = 1).	......	......	.989	......
Pressure at throttle by gauge at board	149.0	......	148.4	......
Pressure in receiver	5.0	......	9.4	......
Vacuum in condenser. (Lbs)	13.2	......	12.3	......
Barometer. (Inches of mercury)	......	30''		......

Draught Pressures and Flue Temperatures.

	No. 2.	Pump No. 2.	No. 3.	Pump No. 3.
Average temperature of gases { Entering economizer °C. .	......	236.2		......
Average temperature of gases { Leaving economizer °C	......	123.3		......
Temperature of gases leaving economizer Maximum °C. .	......	129.0		......
Temperature of gases leaving economizer Minimum °C.	......	111.0		......
Av draught pressure in ins. of water { Entrance to economizer,	......	.5514		......
Av draught pressure in ins. of water { Leaving economizer . .	......	.4026		......
Draught leaving economizer { Maximum	......	.844		......
Draught leaving economizer { Minimum	......	.266		......

From Weir Measurements

	No. 2.	Pump No. 2.	No. 3.	Pump No. 3.
Average height of water on weir (Feet)	......	.533		......
Average temperature of water at weir ·C	......	31.7		......
Pounds of water over weir per hour	......	888,459		......
Pounds of steam condensed per hour	......	34,542		......
Condensing Water per Pound of Steam	......	24.7		......
Condensing Water per Hour per I. H. P. of Main Engines	......	344.		......

From Electrical Measurements

	No. 2.	Pump No. 2.	No. 3.	Pump No. 3.
Electrical H. P Output Each Engine	1156	......	1051	......
Electrical H P Output Both Engines	......	2207		......
Efficiency per cent Ratio of electrical to steam H. P. . . .	......	89 8		......
Coal per Electrical H P Ouput (Per hour). . . .	......	1.64		......
Electrical H P. to run feed pumps	......	15.0		......

ENGINE TESTS.

	May 11, 12-45 A.M. to 5-45 P.M.		May 11, 6.30 A.M. to 11·30 P.M.			
Time of run						
Engine number	No. 3.	Pump No. 3.	No. 2.	Pump No. 2.	No. 3.	Pump No. 3.
Revolutions per minute during run	71,967	50,420	70,828	55,123	72,075	55,659
Average per cent. of cut-off on high pressure cylinder	33.1	.....	23.43	....	20.60	....
Mean effective pressure on { H. H.	67.8	42.4	61.7	31.0	49.4	36.0
H. C.	68.4	3.27	58.8	11.0	47.7	3.3
L. H.	13.2·	.....	9.36	....	11.8	....
L. C.	12.6		10.40	....	10.5	....
Horse power by indicator on { H. H.	455.21	13.02	407.71	10.41	332.18	12.25·
H. C.	441.53	.98	373.55	3.59	308.37	1.09
L. H.	354.51	.....	247.40	....	317.38	....
L. C.	333.84		271.19	....	278.62	....
I. H. P. of Each Engine	1585	14.0	1299.8	14.00	1236.6	13 34
Total I. H. P. Both Engines	....	....	2536.4			....
Total water supplied to boilers during run	120,575	....	631,354	....		....
Total steam used by three calorimeters during run	6,225	....	21,165	....		....
Steam supplied to engines and pumps	114,350	....	610,189	....		....
Steam per Hour per I. H. P. of Main Engines		14.43	....		14.15	....
Temperature of feed water { Entering heater on exhaust °F.	50.1	....	50.9	....	50.9	....
Leaving heater on exhaust' °F.	162.1	....	133.3	....	142.0	....
B. T. U per I. H P. of Main Engine per Minute	254		....		254	....
Coal per I. H. P. of Main Engines per Hour		1.47	....		1.48	....
Temperature of condensing water { Cold °C.	....	10.6	....	13.3	....	11.8
Hot °C.		30.0	....	30.8	....	32.9
Weight of drip trapped from working side of receivers		7580	....	42,360		....
Quality of steam at throttle. (Dry steam = 1)	.990	....	....	....	.990	....
Pressure at throttle by gauge at board	150.5	....	149.0	....	148.2	....
Pressure in receiver	10.4	....	2.8	....	9.0	....
Vacuum in condenser. (Lbs.)	10.2	....	13.0	....	12.2	....
Barometer. (Inches of mercury.)		30			30	....

Draught Pressures and Flue Temperatures.

Average temperature of gases { Entering economizer °C.	225.8	....	240.6	....		....
Leaving economizer °C.	127.6	....	126.8	....		....
Temperature of gases leaving economizer { Maximum °C.	133.0	....	134.0	....		....
Minimum °C.	117.0	....	117.0	....		....
Av draught pressure in ins. of water { Entrance to economizer,	.4175	....	.6723	....		....
Leaving economizer.	.2902	....	.3632	....		....
Draught leaving economizer { Maximum	.824	....	.912	....		....
Minimum.	.308	....	.288	....		....

From Weir Measurements.

Average height of water on weir. (Feet.)	.459		.535			....
Average temperature of water at weir °C.	32.0	....	31.5	....		....
Pounds of water over weir per hour	708,851	....	893,429	....		....
Pounds of steam condensed per hour	22,870	....	35,893	....		....
Condensing Water per Pound of Steam	30.0	....	23.9	....		....
Condensing Water per Hour per I. H. P. of Main Engines	432.		338.			....

From Electrical Measurements.

Electrical H P. Output Each Engine	1409	....	1194	....	1092	....
Electrical H P. Output Both Engines	1409	....	....	2286		....
Efficiency per cent. Ratio of electrical to steam H. P.	88.8	....	....	90.1		....
Coal per Electrical H. P. Output. (Per hour.)	1.66	....	....	1.64		....
Electrical H. P. to run feed pumps	13.4	....	....	15.3		....

ENGINE TESTS.

	No. 2.	Pump No. 2.				
Time of run	May 12, 12-45 A.M. to 5-45 A.M.					
Engine number	No. 2.	Pump No. 2.				
Revolutions per minute during run	71,703	56,957	71,635	58,982	72,195	58,
Average per cent. of cut-off on high pressure cylinder . . .	36.1	....	19.0	....	20.2	
Mean effective pressure on { H. H.	81.1	38.4	56.7	19.0	45.1	27
H. C.	74.9	13.7	54.6	14.0	47.5	7
L. H.	11.7	....	8.20	....	11.0	
L. C.	11.8	....	8.96	....	9.70	.
Horse power by indicator on { H. H.	542.52	13.33	378.93	6.82	303.77	
H. C.	481.71	4.62	350.82	4.89	307.59	
L. H.	313.07	....	219.21	....	296.36	
L. C.	311.50	....	236.3	....	257.82	
I. H. P. of Each Engine	1648.8	17.95	1185.3	11 72	1165.5	1
Total I. H. P. Both Engines.	....			....	2390 8	.
Total water supplied to boilers during run	130,950			....	133,741	.
Total steam used by three calorimeters during run	6,225			....	4,669	.
Steam supplied to engines and pumps	124,725			....	129,072	.
Steam per Hour per I. H. P. of Main Engines . . .	15.13			....	14 64	.
Temperature of feed water { Entering heater on exhaust °F. .	52.0	....				
Leaving heater on exhaust °F. .	147.4	....				
B. T. U. per I. H P. of Main Engines per Minute . .	269.					
Coal per I. H. P. of Main Engines per Hour	1.56					
Temperature of condensing water { Cold °C.	....	11.8				
Hot °C.	....	25.8				
Weight of drip trapped from working side of receivers . . .	7170					
Quality of steam at throttle. (Dry steam = 1)	.989	....				
Pressure at throttle by gauge at board	149.9	....				
Pressure in receiver	3.7	....				
Vacuum in condenser. (Lbs.).	12.3	....				
Barometer. (Inches of mercury)	30					

Draught Pressures and Flue Temperatures.

Average temperature of gases { Entering economizer °C. . .	217.7			....	236.3	.
Leaving economizer °C. . .	126.4			....	132.3	.
Temperature of gases leaving economizer { Maximum °C. .	129.0			....	139.0	.
Minimum °C. .	121.0			....	124 0	.
Av. draught pressure in ins. of water { Entrance to economizer,	.6891			....	.4090	.
Leaving economizer . .	.4316			....	.3095	.
Draught leaving economizer { Maximum	.846			....	.682	.
Minimum	.298			....	.246	.

From Weir Measurements.

Average height of water on weir. (Feet.)	.451			....	.454	.
Average temperature of water at weir °C.	30.4			....	36.3	.
Pounds of water over weir per hour	691,108			....	696,585	.
Pounds of steam condensed per hour	24,945			....	34,418	.
Condensing Water per Pound of Steam	26.7			....	19 3	.
Condensing Water per Hour per I. H. P. of Main Engines	404.			....	281.	

From Electrical Measurements.

Electrical H. P. Output Both Engine	1499	....				
Electrical H P. Output Both Engines	1499	....				
Efficiency per cent Ratio of electrical to steam H. P. . .	90.9	....				
Coal per Electrical H P Output (Per hour.)	1.72	....				
Electrical H. P. to run feed pumps	14.7	....				

TESTS ON THE TRIPLE EXPANSION ENGINE, 9″–16″–24″ × 30″.

During the last six years quite an extended series of tests have been made on this engine to determine the effect of steam jacketing. The results have been published in the *Technology Quarterly*, and also in the *Trans. Am. Society of Mechanical Engineers.*

The following tests form the beginning of a new series, having for objects, first, the gain to be derived by using reheating surface on the receivers only ; second, the amount of reheating surface needed for best results. The tests are printed at this time because of the considerable interest taken in this work by designers of steam engines.

Up to the summer of 1897 the engine had double shell receivers made by the E. P. Allis Company, with jacket steam supplied to the space between the two shells. The reheating surface of these receivers was not very effective. During the summer of 1897 these two receivers were taken out and Wainwright reheaters of special design substituted. These new reheaters are so constructed that either one-third, two-thirds, or the entire amount of reheating surface can be used. The entire jacket surface of the heaters was used in the tests of which the results are given below.

A working drawing of the second reheater is shown in Figure 1.

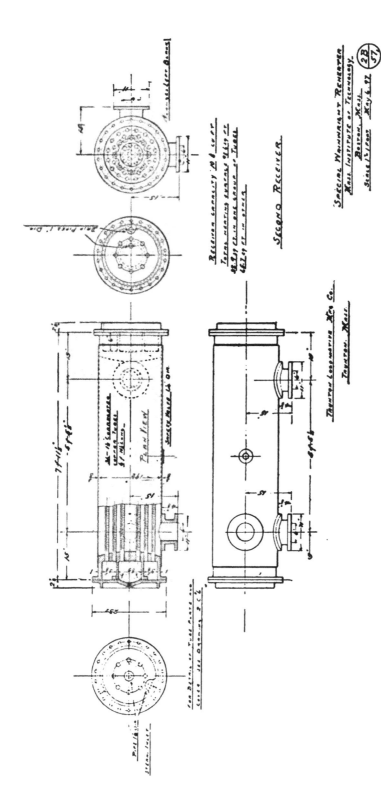

FIG. 1. — WAINWRIGHT REHEATING RECEIVER.

TESTS ON THE TRIPLE EXPANSION ENGINE AT THE MASSACHUSETTS INSTITUTE OF TECHNOLOGY.

Date.	Horse power.	Steam per horse power per hour.	Steam through cylinders per hour.	Water by Jackets Per Hour. First receiver.	Water by Jackets Per Hour. Second receiver.	Revolutions per minute.	Boiler pressure gauge.	Vacuum in condenser. (Inches of mercury.)	Barometer.
Mar. 8, 1898 . .	88.56	15.95	14 13	....	....	81.82	146.7	26.37	30.61
Mar. 7, 1898 . .	87.52	16.03	14.03	....	....	81.80	147.5	26.05	30.39
Mar. 4, 1898 . .	89.68	16.01	14.36	....	....	81.60	147 0	25.89	30.51
Mar. 29, 1898 . .	103.32	15.53	16.05	....	....	81.20	148.2	25.90	30.16
May 6, 1898 . .	66.45	15.71	9.21	....	123.3	85.54	147 3	25.48	29.98
May 9, 1898 . .	84.93	15.87	11.96	....	152.6	83.25	146.9	23.80	30.16
Mar. 22, 1898 . .	112.38	15.01	15.05	....	183.0	81.40	146.1	25.80	30.17
Apr. 8, 1898 . .	61.53	15.53	8.07	80.2	58.1	84.97	147.3	26.59	30.32
Mar. 14, 1898 . .	74.78	14.90	9.32	108.9	76 9	84.45	146.9	26.16	30.31
Mar. 15, 1898 . .	95.65	14.72	12.21	117.8	70.2	82.43	147.1	25.33	30.44
Apr. 1, 1898 . .	105.88	14.69	13.59	81.9	115.6	81.92	147.7	25.40	30.10
Mar. 28, 1898 . .	107.00	14.54	13.64	90.0	102.8	81.97	146.6	25.72	39.22

Date.	HIGH PRESSURE CYLINDER. Initial pressure.	Per cent. of cut-off.	Pressure at cut-off.	Pressure at release.	Pressure at compression.	Per cent. of steam in the cylinder at cut-off.	Per cent. of steam in the cylinder at release.	M. E. P. crank.	M. E. P. head.	Horse power.
Mar. 8, 1898 . .	142.1	27.0	138.4	37.4	40.6	76.3	84.3	48.7	47.8	36.86
Mar. 7, 1898 . .	142.5	27.3	132.4	37.3	39.3	73.1	82.2	50.1	49.2	37.92
Mar. 4, 1898 . .	142.2	29.2	128.5	36.7	41.4	75.9	82.4	51.9	49.3	38.50
Mar. 29, 1898 . .	144.3	35.5	135.6	48 8	50.8	81.6	88.9	52.1	51.5	39.25
May 6, 1898 . .	142.5	10.0	134.0	16.7	15.1	66.7	88.3	45.0	28.4	29.11
May 9, 1898 . .	139.5	18.9	128.8	27.0	30.0	69.8	83.6	49.9	36.3	33.30
Mar. 22, 1898 . .	143 8	31.1	134.3	42.0	44.3	78.3	86.2	51.0	49.9	38.31
Apr. 8, 1898 . .	145.4	7.8	125.4	12.2	15.9	69.9	83.7	34.0	27.1	24.10
Mar. 14, 1898 . .	146.1	10.0	136.2	17.1	19.8	63.4	84.6	39.2	30.7	27.43
Mar. 15, 1898 . .	146.2	20.5	135.0	29.4	32.3	72.4	85.1	46.3	42.8	34.20
Apr. 1, 1898 . .	144.8	26.6	132.2	38.7	45.6	75.3	89.0	44.6	43.4	33.62
Mar. 28, 1898 . .	145.0	27.8	130.6	37.8	43.4	76.6	87.0	46.9	46.9	35.91

TESTS ON THE TRIPLE EXPANSION ENGINE AT THE MASSACHUSETTS INSTITUTE OF TECHNOLOGY. — *Continued.*

| | INTERMEDIATE PRESSURE CYLINDER. | | | | | | | | | |
DATE.	Initial pressure.	Per cent. cut-off.	Pressure at cut-off.	Pressure at release.	Pressure at compression.	Per cent. of steam in the cylinder at cut-off.	Per cent. of steam in the cylinder at release.	M. E. P. crank.	M. E. P. head.	Horse power.
Mar. 8, 1898 . .	36.6	15.5	26.7	− 2.12	+ 0.37	55.1	76.3	11.7	11.7	28.85
Mar. 7, 1898 . .	34.1	17.5	24.1	− 1 93	+ 0.00	56 2	77.4	10.7	11.5	27.49
Mar. 4, 1898 . .	35.3	17.6	26.3	− 1.30	+ 0.75	57.3	78.5	10.8	11.9	27.84
Mar. 29, 1898 . .	43.2	15 9	35.1	+ 0.67	+ 1.51	58.2	79.3	15.6	14.2	36.52
May 6, 1898 . .	13.3	19.2	9 52	− 4.30	− 3.0	58.1	95.9	4.81	4.50	12.03
May 9, 1898 . .	25.7	17.5	19.5	− 0.80	+ 0.68	60.5	94.4	6.81	7.22	17.65
Mar. 22, 1898 . .	39.4	17.5	32.7	+ 1.39	+ 3.82	60.1	85.2	10.9	12.3	28.54
Apr. 8, 1898 . .	11.3	25.7	8.19	− 4.60	− 2.99	74.4	Super.	5.08	5.51	13.61
Mar. 14, 1898 . .	14.4	32.1	9.38	− 2.78	− 1.89	78.9	Super.	6.43	7.19	17.39
Mar. 15, 1898 . .	28.6	24.6	23.0	+ 0.95	+ 2.26	75.2	Super.	9.07	10.6	24.57
Apr. 1, 1898 . .	39.2	18 8	32.1	+ 1.05	+ 3.63	69.4	92.4	12.1	12.5	30.37
Mar. 28, 1898 . .	37.5	20.2	32.4	+ 1.07	+ 2.75	73.0	92.5	12.1	12.2	30.08

| | LOW PRESSURE CYLINDER. | | | | | | | | | | | |
DATE.	Initial pressure.	Per cent. of cut-off.	Pressure at cut-off.	Pressure at release.	Pressure at compression.	Per cent. of steam in the cylinder at cut-off.	Per cent. of steam in the cylinder at release.	M. E. P. crank.	M. E. P. head.	Horse power.	B. T. U. per H. P. per minute (actual).	B. T. U. per H. P. per minute reduced to 26" vacuum.
Mar. 8, 1898 . .	−2.6	22 6	−5.2	−11.1	−12 0	44.9	61.3	4.14	3.99	22 85	289.7	288.4
Mar. 7, 1898 . .	−2.3	19 5	−5.4	−11.1	−11.9	40.0	55 3	3.97	3.90	22.11	290.9	288.9
Mar. 4, 1898 . .	−1.5	18.7	−4 4	−11.2	−11.2	41 2	64.4	4.25	4.08	23.34	284.8	286 3
Mar. 29, 1898 . .	−0.7	22.5	−3 0	−10 2	−11.3	46 8	61.7	4.81	5.07	27 55	282.0	280.8
May 6, 1898 . .	−3.6	36.8	−5 7	−10.6	−11.3	72.1	98.8	4.06	4.54	25 31	277.1	272 7
May 9, 1898 . .	+0.7	35.5	−2.9	−10 0	−11.2	74 7	86.0	6.15	5.76	33 98	277.1	261.5
Mar. 22, 1898 . .	+2.3	39.4	−1.6	− 9.1	−11 4	82.9	81.9	7.97	8.32	45.53	265.8	264.4
Apr. 8, 1898 . .	−3.8	32.6	−6.0	−11.1	−12.0	92 1	Super.	3.92	4.24	23.82	268 6	274.4
Mar. 14, 1898 . .	−1.7	33.3	−5.3	−11.0	−12.1	86 7	90 2	5.19	5.14	29 96	260 7	259 9
Mar. 15, 1898 . .	+1.9	33 2	−2.5	−10 3	−11.6	83.7	80 8	6.46	6.57	36 88	257.8	252 0
Apr. 1, 1898 . .	+1.8	35 7	−1.4	− 9 4	−11.5	79.8	85.4	7.30	7.03	41.89	258.6	254.5
Mar. 28, 1898 . .	+1.7	36 0	−1.4	− 9.1	−11.6	86.1	89.6	7.14	7.43	41.01	250.4	254.5

Cranks set 120° apart, the high leading.

Total volume of working side of first reheater, 8.5 cubic feet.   Second reheater, 11 cubic feet.

Total reheating surface, first reheater, 80 2 square feet.   Second reheater, 102.4 square feet.

Volume in piping and ports between exhaust valves on high, and admission valves on intermediate, exclusive of reheater, 22.3 cubic feet.

Volume in piping and ports between exhaust valves on intermediate, and admission valves on low, exclusive of reheater, 20.4 cubic feet.

Steam used in the jackets was supplied at full boiler pressure.

Jackets on the cylinders were not in use.

FIG. 2. — APPARATUS FOR TESTING EXPLOSIVE MIXTURES OF
ILLUMINATING GAS AND AIR.

EXPLOSIVE MIXTURES OF ILLUMINATING GAS AND AIR.

During the last year a few experiments have been made upon explosive mixtures of illuminating gas and air. These experiments have formed a part of the regular work in the fourth year mechanical engineering laboratory.

The essential parts of the apparatus (Figures 2 and 3) are: (A) The cast-iron cylinder for holding the mixture. (B) Manometer tube used in obtaining the desired mixture. (C) Power pump used in clearing the cylinder of the products of explosion. (D) Exhaust pump for rarifying the air in the cylinder. (E) Indicator. (F) Tuning fork for obtaining a time line on indicator card. (G) Two storage batteries, one wired to terminals in cylinder to give the electric spark for the ignition of the mixture, the other being used in connection with an electro-magnet to keep the tuning fork in vibration while a card is taken; a metallic point attached to the end of one arm of fork tracing a wave line upon the card. (H) Induction coil.

The method of procedure in making an experiment is as follows: The cylinder is thoroughly cleared of the products of explosion by forcing air through it by means of the power pump. The connections between the cylinder and the atmosphere, the cylinder and the power pump, are then closed. To obtain a given mixture, for illustration, one part gas and five parts air, the barometer reading, say, 30 inches; the connections between cylinder exhaust pump and manometer tube are opened, and the exhaust pump is allowed to operate until there is a difference of level of 5 inches of the mercury in the manometer tube; then one-sixth of the original volume of air in the cylinder has been removed. The connection between the cylinder and the exhaust pump is then closed, and just as much gas is allowed to flow into cylinder as air previously removed, bringing the pressure in the cylinder back to atmospheric and also giving the desired mixture, one part gas and five parts air. The connection between the cylinder and the manometer tube is then closed and the gas is allowed to diffuse in the cylinder for a few minutes before ignition. The indicator card is placed on a flat circular disc which is driven from a shaft over the machine. · Metallic points are used for the indicator, and on the arm of the tuning fork. The atmospheric line, which in this case is a circle, is taken and a switch is thrown in, causing the tuning fork to vibrate. The indicator cock is opened and time and pressure lines are taken simultaneously upon the card.   The time line in this case is a wave line, from crest to crest corresponding to $\frac{1}{80}$ of a second.

A small projection from the frame carrying the tuning fork enters a spiral groove in a disc which is rotated while the card is taken; the object of this groove being to prevent the wave lines from being superimposed.

The indicator point is kept in contact with the paper for about two seconds.   The time of explosion, maximum pressure and pressure for each twelfth of a second, for one second, are measured and plotted upon coördinate paper to scale of ordinates $1'' = 40$ pounds, abscissæ $6'' = 1$ second. Pressures were also measured for each $\frac{1}{80}$ second from time of ignition to maximum pressure, so that the first part of the pressure line could be plotted more accurately. Areas under the curves were obtained by planimeters.

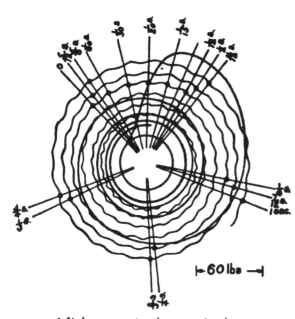

|- 60 lbs -|

Mixture:- 1 part gas, 9 parts air

FIG. 4. — A SAMPLE INDICATOR-CARD.

To determine which of the mixtures has the greatest capacity for doing work, suppose 1 cubic inch of illuminating gas to be used in each of the mixtures, 1–3, 1–4, 1–5, etc., then they would measure 4, 5, 6, etc., cubic inches.   Let them be placed in cylinders of 4, 5, 6, etc., square inches piston area, the pistons will be displaced 1 inch from the bottom of the cylinder.   If the pressure on the piston were the same, equal movements of the piston would give equal power; if, therefore, the mixtures gave equally good results, the mean pressure multiplied by the piston area will in all cases be the same.   The

FIG. 3.— THE SAME APPARATUS VIEWED FROM THE OPPOSITE SIDE.

mean pressures were obtained from the plot of the cards on coördinate paper, the same scale being used in all cases. The first fifth of a second only was considered, for in that time the ordinary gas engine has completed its working stroke. Column 6 of the results of tests shows the relative power of the mixtures. By a similar line of reasoning we obtain column 12 of the table, which shows the relative power of the mixtures to resist cooling. As the method here employed is applicable to other gases, we are enabled to compare the relative value of any similar explosive mixture.

RESULTS OF TESTS ON EXPLOSIVE MIXTURES OF ILLUMINATING GAS
AND AIR.

Mixture. (By parts.)	Maximum pressure. (Lbs. per sq. in.)	Time of explosion. (Seconds.)	FIRST ¼ SECOND.				¼ SECOND AFTER MAXIMUM PRESSURE.				
			Area. (Square inches.)	Mean pressure. (Lbs. per sq. in.)	Mean pressure. Divided by proportion of gas.	Final pressure.	Area. (Square inches.)	Mean pressure. (Lbs. per sq. in.)	Mean pressure. Divided by proportion of gas.	Final pressure.	Final pressure. Divided by proportion of gas.
1	2	3	4	5	6	7	8	9	10	11	12
Gas-Air.											
1 — 3	45	.49	0.32	11	44	26	1.30	43	172	40	160
1 — 4	86	.08	1.77	59	295	61	1.88	62	310	46	230
1 — 5	96	.05	1.86	62	372	52	1.93	64	384	44	264
1 — 6	88	.05	1.80	60	420	54	1.93	64	448	46	322
1 — 7	86	.06	1.97	66	528	58	1.93	64	512	48	384
1 — 8	87	.06	1.71	57	513	53	1.83	61	549	46	414
1 — 9	77	.08	1.60	53	530	57	1.86	62	620	46	460
1 — 10	71	.11	1.36	45	495	56	1.69	56	616	45	495
1 — 11	68	.14	1.21	40	480	60	1.66	55	660	43	516
1 — 12	39	.33	0.35	12	156	29	0.98	33	429	30	390
1 — 13	32	.42	0.18	6	84	16	0.79	26	364	24	336
1 — 14	9	.42	0.05	2	30	4	0.24	8	120	8	120

APPLIED MECHANICS.

TORSION TESTS ON COPPER WIRE. (Series 1.)

Date, 1897.	Diameter of cross section. (Ins.)	Length of specimen between jaws. (Ins.)	Maximum twisting moment. (In. lbs.)	Number of turns between jaws.	Apparent outside fiber stress. (Lbs. per sq. in.)	Average number of turns per inch.
Oct. 28 . . .	0.1658	12.0	36.6	52.1	40,900	4.34
Oct. 29 . . .	0.1644	12.0	37.3	53.8	42,800	4.48
Oct. 29 . . .	0.1643	12.0	38.3	42.1	43,900	3.51
Nov. 5 . . .	0.1664	12.0	37.6	46.4	41,600	3.87
Nov. 5 . . .	0.1678	12.0	37.4	51.5	40,300	4.29
Nov. 7 . . .	0.1663	12.0	37.0	43.4	40,900	3.62
Nov. 9 . . .	0.1669	12.0	38.0	57.8	41,600	4.82
Nov. 9 . . .	0.1666	12.0	38.0	16.8	41,900	1.40
Nov. 11 . . .	0.1656	12.0	36.5	49.2	41,000	4.10
Nov. 17 . . .	0.1663	12.0	35.7	46.3	39,500	3.87

TORSION TESTS ON COPPER WIRE. (Series 2.)

Date, 1897.	Diameter of cross section. (Ins.)	Length of specimen between jaws. (Ins.)	Maximum twisting moment. (In. lbs.)	Number of turns between jaws at fracture.	Apparent outside fiber stress. (Lbs. per sq. in.)	Average number of turns per inch.
Nov. 17 . . .	0.2586	10.0	130 7	37.7	38,500	3.77
Nov. 17 . . .	0.2584	10.0	126.8	37.8	37,400	3.78
Nov. 17 . . .	0.2582	12.0	118.6	43.2	35,100	3.60
Nov. 18 . . .	0.2576	10 0	116.3	33.1	34,600	3.31
Nov. 18 . . .	0.2572	10 0	127.6	31.5	38,100	3.15
Nov. 19 . . .	0.2575	10.0	122.3	34.2	36,400	3.42
Nov. 20 . . .	0.2578	10.0	125.0	43.2	37,200	4.32
Nov. 20 . . .	0.2581	10.0	131.5	46.5	38,900	4.65
Nov. 22 . .	0 2579	10.0	124.7	30.8	37,000	3.08

NOTE. — This material was a very soft ductile grade of copper.

TORSION TESTS ON COPPER WIRE. (Series 3.)

Date, 1897.	Diameter of cross section. (Ins.)	Length of specimen between jaws. (Ins.)	Maximum twisting moment. (In. lbs.)	Number of turns between jaws at fracture.	Apparent outside fiber stress. (Lbs. per sq. in.)	Average number of turns per inch.
Dec. 4 . . .	0.1037	10.0	9.1	28.1	41,400	2.81
Dec. 6 . . .	0.1038	8.0	9.5	29.6	43,300	3.70
Dec. 9 . . .	0.1025	8.0	11.6	24.4	54,700	3.05
Dec. 10 . . .	0.1034	8.0	12.9	5.2	59,400	.65
Dec. 11 . . .	0.1050	8.0	9.7	18.1	42,700	2.26
Dec. 18 . . .	0.1062	8.0	10.3	39.0	43,800	4.88
Dec. 18 . . .	0.1041	8.0	11.5	27.8	51,800	3.48
Dec. 28 . . .	0.1046	10.0	10.6	23.2	47,200	2.32

NOTE. — This material was a hard grade of copper.

TORSION TESTS ON BRASS TUBING.

Date, 1898.	Outside diameter of cross section. (Ins.)	Inside diameter of cross section. (Ins.)	Length of specimen between jaws. (Ins.)	Maximum twisting moment. (In. lbs.)	Number of turns between jaws.	Apparent outside fiber stress. (Lbs. per sq. in.)	Average number of turns per inch.
Apr. 7 . . .	0.246	0.186	10.0	81.5	1½	41,500	.15
Apr. 9 . . .	0.253	0.173	10.0	132.5	1¾	53,300	.18
Apr. 22 . . .	0.253	0.158	10.0	132.0	¼	49,200	.03
Apr. 22 . . .	0.253	0.158	10.0	131.0	¼	48,900	.03
Apr. 22 . . .	0.253	0.158	10.0	138.0	⅓	51,300	.03
Apr. 23 . . .	0.253	0.158	10.0	139.0	¼	51,800	.03
Apr. 23 . . .	0.253	.0158	10.0	135.0	¼	50,300	.03
Apr. 25 . . .	0.253	0.158	10.0	134.0	¼	50,000	.03
Apr. 25 . . .	0.253	0.158	10.0	115.0	⅜	42,900	.02

TORSION TESTS ON ROUND BRASS ROD.

Date, 1898.	Diameter of cross section.	Length of speci-men between jaws. (Ins.)	Maximum twisting moment. (In. lbs.)	Number of turns between jaws.	Apparent outside fiber stress. (Lbs. per sq. in.)	Average number of turns per inch.
Apr. 5 . . .	0.2518	8.00	17.5	12⅞	55,800	1.6
Apr. 5 . . .	0.2515	9.00	17.4	15	55,700	1.7
Apr. 5 . . .	0.2518	8.00	18.6	13¼	59,300	1.7
Apr. 5 . . .	0.2512	9.00	17.4	15	55,900	1.7
Apr. 5 . . .	0.2515	9.00	17.8	15½	57,000	1.7
Apr. 7 . . .	0.2510	8.00	17.0	11½	54,600	1.4
Apr. 9 . . .	0.2512	8.00	16.3	14	51,200	1.8
Apr. 12 . . .	0.2511	8.00	16.4	13	52,800	1.6
Apr. 12 . . .	0 2512	8.00	16.4	13	52,700	1.6
Apr. 14 . . .	0 2512	9.00	16.6	15	53,300	1.7
Apr. 14 . . .	0.2508	8.00	17.0	14	54,900	1.8
Apr. 15 . . .	0 2505	8.00	16.6	12	53,800	1.5
Apr. 16 . . .	0 2505	8.00	17.1	13⅞	55,400	1.7
Apr. 16 . . .	0 2505	8.00	16.9	14	54,800	1.8
Apr. 25 . . .	0 2512	10.00	16.5	16¼	53,000	1.6

TORSION TESTS ON GALVANIZED IRON WIRE.

Date, 1897.	Diameter of cross section. (Ins.)	Length of specimen between jaws. (Ins.)	Maximum twisting moment. (In. lbs.)	Number of turns between jaws at fracture.	Apparent outside fiber stress. (Lbs. per sq. in.)	Average number of turns per inch.
Oct. 18 . . .	0.1491	12.0	53.3	39.9	81,800	3.33
Oct. 18 . . .	0.1488	12.0	52.5	37.8	81,200	3.15
Oct. 20 . . .	0.1489	12.0	52.8	40.4	81,300	3.37
Oct. 20 . . .	0.1490	12.0	53.3	41.7	82,000	3.48
Oct. 20 . . .	0.1492	12.0	51.7	39.5	79,300	3.29
Oct. 27 . . .	0.1492	12.0	50.9	32.5	78,100	2.71
Oct. 27 . . .	0.1492	12.0	52.4	40.9	80,400	3.41
Oct. 27 . . .	0.1491	12.0	53.1	37.2	81,500	3.10
Oct. 28 . . .	0.1492	12.0	53 3	41.1	81,800	3.43
Nov. 24 . . .	0.191	12.0	107.0	35.4	77,900	3.0
Nov. 24 . . .	0.191	12.0	110.0	36.4	80,100	3.0
Dec. 2 . . .	0.191	10.0	108.9	29.7	80,100	3.0
Dec. 2 . . .	0.191	10.0	105.9	28.6	77,900	2.9
Dec. 4 . . .	0.191	10.0	107.8	27.6	78,600	2.8
Dec. 6 . . .	0.190	10.0	106.4	27.4	78,500	2.7
Dec. 9 . . .	0.190	10.0	106.5	31.7	79,600	3.2
Dec. 10 . . .	0.191	10.0	106.9	29.0	78,400	2.9
Dec. 11 . . .	0.191	10.0	105.0	28.0	76,500	2.8
Dec. 18 . . .	6.191	10.0	116.0	26.8	85,300	2.7
Dec. 20 . . .	0.191	10.0	109.7	30.0	80,700	3.0
Dec. 31 . . .	0.191	12.0	105.6	36.4	77,500	3.0
Dec. 31 . . .	0 191	12.0	103.0	33.9	75,500	2.8
Jan. 1 . . .	0.190	10.0	105.1	30.7	77,600	3.1
Jan. 1 . . .	0.190	10.0	103.5	27.2	76,600	2.7
Jan. 1 . . .	0.191	10.0	105.8	31.6	77,600	3.2
Jan. 1 . . .	0.191	10.0	104.2	28.5	76,400	2.9
Mar. 29 . . .	0.193	10.0	115.0	36.3	81,500	3.6
Mar. 31 . . .	0.193	10.0	120.0	37.5	87,000	3.8
Mar. 31 . . .	0.190	10.0	108.0	35.5	80,200	3.6
Apr. 4 . . .	0.194	10.0	116.0	....	80,900	...
Apr. 4 . . .	0.194	10.0	108.0	38.0	75,100	3.8
Apr. 21 . . .	0.193	10.0	117.0	37.5	82,900	3.7
Apr. 21 . . .	0.193	10.0	123.0	40.8	87,100	4.1

TRANSVERSE STRENGTH OF SPRUCE BEAMS.

Load applied at middle of span.

Number of test.	Width and depth. (Ins.)	Distance between supports. (Feet.)	Limits of loads for calculation of modulus of elasticity.	Deflection for difference of loads used in calculation of modulus of elasticity. (Ins.)	Modulus of elasticity. (Lbs. per sq. in.)	Breaking load. (Lbs.)	Weight of beam. (Lbs.)	Modulus of rupture. (Lbs. per sq. in.)	Maximum intensity of longitudinal shear. (Lbs. per sq. in.)	Manner of Breaking.	Remarks.
580	4 × 12	23.5	500–1500	.6155	1,310,000	5,010	180	4,430	95	Tension.	
581	5½ × 12	21	500–4500	1.3662	1,150,000	9,180	315	4,210	101	Tension.	
582	6 × 12	19	500–4500	.7361	1,600,000	14,685	350	6,050	138	Crushing and tension.	
583	6 × 11½	19	500–4500	.9713	1,200,000	12,880	310	5,300	140	Tension.	
584	5½ × 12	21	1000–4000	.8685	1,100,000	10,790	320	5,050	119	Tension.	
585	6 × 12	19	500–4500	.9586	1,100,000	8,155	335	3,490	88	Tension.	
586	6 × 12	19	500–4500	.9626	993,000	12,790	330	5,130	137	Tension.	
587	6 × 12	10	500–6500	.2519	1,210,000	22,230	155	4,650	233	Longitudinal shear.	
588	5½ × 12	9	500–6500	.1508		27,470	145	5,150	288	Longitudinal shear.	
589	5½ × 12	8.5	....		....	21,030	135	4,040	225	Shearing and tension.	
590	6 × 12	11	500–2500	.1688	1,120,000	23,930	135	4,340	208	Tension.	
591	3½ × 12	17	500–3500	.1017	1,500,000	11,110	105	4,170	371	Tension.	Part of beam No. 584.
592	4½ × 12	15	500–2500	.6280	801,000	21,380	100	8,310	110	Tension.	Part of beam No. 581.
593	4 × 12	15	500–3500	.4559	1,310,000	7,155	225	3,590	187	Tension.	Part of beam No. 583.
594	4 × 12½	15	500–4500	.6199	1,430,000	12,305	190	5,520	185	Tension.	
595	4 × 12	14	500–6500	.8346	1,500,000	12,310	185	5,460	139	Tension.	
603	3½ × 12½	15	500–6500	.688	1,540,000	8,930	185	4,060	237	Crushing and tension.	
604	4 × 12	17	500–6500	.820	1,480,000	14,625	155	5,940	199	Longitudinal shear.	
605	3½ × 11½	15	500–4500	.858	1,410,000	12,580	150	6,240	185	Tension.	
607	4 × 12	15	500–4500	.696	1,340,000	11,655	190	6,530	216	Crushing and tension.	
610	4 × 12	15	500–4500	.6518	1,310,000	13,130	190	6,610	221	Crushing and tension.	Well seasoned.
611	1½ × 12	14	500–4500	.5523	1,210,000	14,025	160	6,230	209	Longitudinal shear.	
612	4 × 12	14	500–2500	.5277	1,390,000	13,210	90	4,710	138	Crushing and tension.	
613	2 × 12		500–2500	.5973		4,980	95	5,560	200	Crushing and tension.	
614	1½ × 12	14		.6428	1,140,000	5,995	100	3,970	143	Tension.	
615	4 × 12					4,300					

TRANSVERSE TESTS OF YELLOW PINE BEAMS.

Load applied at middle of span.

Number of test.	Width and depth. (Ins.)	Distance between supports. (Feet.)	Limits of loads for calculation of modulus of elasticity.	Deflection for difference of loads used in calculation or modulus of elasticity. (Ins.)	Modulus of elasticity. (Lbs. per sq. in.)	Breaking load. (Lbs.)	Weight of beam. (Lbs.)	Modulus of rupture. (Lbs. per sq. in.)	Maximum intensity of longitudinal shear. (Lbs. per sq. in.)	Manner of Breaking.	Remarks.
596	4⅛ x 12	17	500–9500	1.4650	1,860,000	15,615	285	8,340	245	Crushing first, and then longitudinal shear.	
597	3⅞ x 12	17	500–6500	.9104	2,090,000	15,005	215	8,290	245	Tension.	
598	4½ x 12½	18	500–3375	.4085	2,410,000	7,475	285	4,070	116	Longitudinal shearing.	
599	4 x 12	17	500–6500	.8799	3,090,000	14,610	335	7,850	234	Longitudinal shearing.	
600	4½ x 12½	17	500–6500	.7939	2,120,000	17,475	300	8,710	264	Longitudinal shearing.	
601	4 x 12½	18	500–3500	.7123	1,360,000	11,480	305	6,030	177	Crushing and tension.	
603	4⅛ x 12½	14	500–8500	.7277	1,830,000	17,705	245	7,600	275	Tension.	
606	3⅞ x 12½	17	500–6500	.832	2,160,000	13,430	280	7,330	227	Tension.	
607	3⅞ x 12½	17	500–4500	.872	1,460,000	7,205	245	4,070	185	Tension.	
608	4 x 11½	17.5	500–4500	.895	1,550,000	8,230	265	4,670	234	Longitudinal shearing.	

THE PROTECTION OF STEAM-HEATED SURFACES.

By CHARLES L. NORTON.

Received July 18, 1898.

THE investigation, of which this is a partial report, has been under-taken at the request of Mr. Edward Atkinson, and has been pursued during a large part of the years 1896 and 1897, and is still incom-plete. The first object sought for was the relative efficiency of sev-eral kinds of steam-pipe covering now upon the market. The second object was to ascertain the fire risk attendant upon the use of certain methods and materials for insulation of steam pipes. Third, an attempt was made to show the gain in economy attendant upon the increase of thickness of coverings, and to show also the exact financial return which may be expected from a given outlay for covering steam pipes. Further information is given on many minor matters and con-ditions effecting the transfer of heat from a steam pipe to the sur-rounding air.

The method adopted is one which, so far as I know, is origi-nal. A piece of steam pipe is heated by electricity from the inside. The amount of electrical energy supplied is measured, and hence the amount of heat furnished is known. If the steam pipe is kept at a constant temperature by a given amount of heat, it is because that amount is just equal to the heat it is losing, for if the supply were not equal to the loss, the temperature would rise or fall. In other words, the heat put into the pipe is just equal to the heat lost from it by radiation, convection, and conduction. By measuring the electrical energy supplied I can determine the heat put in, and hence the heat given out or lost. It must be borne in mind that a given amount of electrical energy always produces the same definite amount of heat, the amount of heat furnished by one electrical unit of energy being known with greater accuracy than the amount of heat given out by a pound of steam in condensing.

The apparatus for making tests by this method comprises several pieces of steam pipe of different diameters and lengths, heated electrically from within by means of coils of wire in oil. The oil is stirred vigorously, and serves as a very efficient carrier of heat from the wires to the pipes. A brief description of the smallest tester may make the details of the apparatus more easily understood. A section is shown in Figure 1.

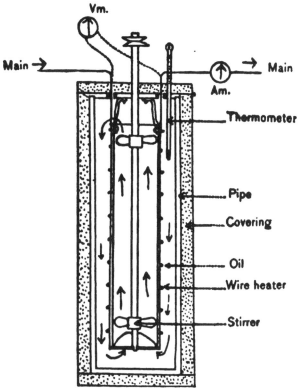

A piece of 4-inch steam pipe 18 inches long is closed at one end by a plate welded in, and at the other by a tightly fitting cover. This pipe is then filled with cylinder oil, and a coil of wire of sufficient carrying capacity, and a stirrer, are introduced into the oil. A thermometer is inserted in such a position as to record the temperature of the oil. An ammeter and voltmeter or a wattmeter may then be connected so as to record the amount of electrical energy supplied.

FIG. 1.—PLAN OF THE APPARATUS FOR TESTING STEAM-PIPE COVERINGS.

The stirring must be brisk, and if enough power is put into the stirrer to be comparable with the electrical energy supplied, such amount must of course be added, as it also is converted into heat. It is my custom to suspend the apparatus in the middle of the room on non-conducting cords, and read the thermometer with a telescope, so that no heat from the person of the observer may be added to the supply given to the cover from within, and also that care may be taken not to produce air currents by walking near the apparatus during a test.

In making a test the following operations are carried out, and observations are taken in the following order :

The current is turned on, and heat is generated in the wire coil until the wire, oil, and steam pipe have reached the desired temperature at which it is proposed to test. The current is then gradually diminished, until it is found to be of just the amount necessary to keep the pipe at this temperature without a rise or fall of one-tenth of a degree in 30 minutes. A reading of the voltage and current is now taken at intervals of 30 seconds, and the watts and B. T. U. are computed from their average. We then have the number of B. T. U. lost from the outside of this particular pipe at this particular temperature. If, now, we place a steam-pipe cover around the pipe, we shall find that a less amount of energy is sufficient to keep it at the required temperature, the difference being the amount of heat saved by the covering. The minimum length of time considered sufficient for the equalization of heat, or "soaking in" to the cover, is 6 hours. If, after a second heating of 6 hours, no change in the conducting power is noted, the cover is considered in a permanent condition, and is tested. Some covers, notably those composed wholly or in part of wool, can not be considered dry and constant until after an exposure upon a pipe at 200 pounds pressure for 6 or 8 days. Covers containing sulphate of lime are also slow in drying.

The three thermometers used were frequently standardized in naphthaline, and were examined to note any disagreement among themselves.

A discussion of the position of the tester and its exposure to air currents will be found in a later paragraph.

All tests were made at a temperature corresponding to 200 pounds steam pressure.

A comparative test was made in 1895 upon a number of steam-pipe covers on a 4-inch tester 16 inches long. The results obtained have been published in the circulars issued by the Boston Manufacturers Mutual Fire Insurance Company, and by the Steam Users' Association. The values given were stated to be purely *relative*, the specimen being too small to give reliable data on the *absolute* conduction, and the surrounding conditions not being controlled other than to maintain them constant during the several runs. The ends of the specimen were covered by massive heads, and the whole tester was situated within a few inches of a brick wall and a stone pier. It was

called to my attention that the heat loss was probably high, and I agree that the exposure was such as to make it so, being a rather harsh test, but one which was rigidly uniform in its requirements of the several covers. In short, the actual loss of heat per square foot of the pipe surface was correct for that particular piece under the conditions of the test, but was not sufficient for the estimation of the actual saving which might be expected from the general use of coverings. I deemed it wise, therefore, to construct new heaters 4 and 10 inches in diameter, and 36 inches long. These were suspended by non-conducting cords in the center of the laboratory, so as to hang freely and not be in contact with any conducting supports. Conduction up to the lead wires and stirring rod was found to be negligible.

It seems to me that I have approached more nearly the conditions of actual practice than can be obtained by any other method of testing, except the actual use of a long run of pipe, and the determination of the amount of heat put into such a pipe by the "condensation" method offers many difficulties and is open to much uncertainty. I feel, therefore, that in adopting this method I am using a reasonable exposure for the pipe, and have an exceptionally good opportunity to measure the heat supplied.

Results are given in English units, the British Thermal Unit or "pound — degree" being the measure ordinarily used by engineers. Its abbreviated form in the tables is B. T. U., it being the amount of heat necessary to raise 1 pound of water 1° F.

The general appearance of the testing apparatus is shown in Figure 2. Table I gives the relative conductivity of the various kinds of steam-pipe cover tested up to April, 1898.

It gives the results of the tests upon most of the samples used, some being omitted when found to be of such low efficiency as to be of doubtful value.

Specimen A. Nonpareil Cork Standard, consists of granulated cork, pressed in a mould at high temperature and then submitted to a fireproofing process.

Specimen B. Nonpariel Cork Octagonal, is similar in composition, but is made up of several strips of cork, instead of two semi-cylindrical sections.

Specimen C. Manville High Pressure Sectional Cover, is composed of an inner jacket of earthy material and an outer jacket of wool felt, the whole being 1¼ inches thick.

TABLE I.

Specimen.	Name.	Maker.	B. T. U. loss per sq. ft. pipe surface per minute.	Per cent. or ratio of loss to loss from bare pipe.	Thickness in inches.	Weight in ounces per ft. of length, 4" diameter.
A	Nonpareil Cork Standard . .	Nonpareil Cork Co. . .	2.20	15.9	1.00	27
B	Nonpareil Cork Octagonal . .	Nonpareil Cork Co. . .	2.38	17.2	.80	16
C	Manville High Pressure . . .	Manville Covering Co. .	2.38	17.2	1.25	54
D	Magnesia	Keasby & Mattison Co. .	2.45	17.7	1.12	35
E	Imperial Asbestos	H. F. Watson	2.49	18.0	1.12	45
F	" W. B."	H. F. Watson	2.62	18.9	1.12	59
G	Asbestos Air Cell	Asbestos Paper Co. . .	2.77	20.0	1.12	35
H	Manville Infusorial Earth . .	Manville Covering Co. .	2.80	20.2	1.50	..
I	Manville Low Pressure . . .	Manville Covering Co. .	2.87	20.7	1.25	..
J	Manville Magnesia Asbestos .	Manville Covering Co. .	2.88	20.8	1.50	65
K	Magnabestos	Keasby & Mattison Co. .	2.91	21.0	1.12	48
L	Moulded Sectional	H. F. Watson	3.00	21.7	1.12	41
O	Asbestos Fire Board	Asbestos Paper Co. . .	3.33	24.1	1.12	35
P	Calcite	Philip Carey Co. . . .	3.61	26.1	1.12	66
	Bare Pipe	. . . .	13.84	100.	....	..

Specimen D. The Keasby & Mattison Magnesia, is a moulded sectional cover, composed of about 90 per cent. carbonate of magnesia.

Specimen E. The Imperial Asbestos of the H. F. Watson Company, is essentially an air cell cover, being composed of sheets of asbestos paper which has been indented before being laid up, the indentations serving to keep the thin sheets of paper from coming in close contact with one another, thereby causing a considerable amount of air to be held throughout the body of the cover.

Specimen F. The "W. B." covering of the H. F. Watson Company, is composed of a wool felt with a lining of asbestos paper.

Specimen G. The Asbestos Air Cell Cover of the Asbestos Paper Company, is a cover made up of thin sheets of asbestos paper, fluted or corrugated, and stuck together with silicate of sodium.

Specimen H. A plastic covering made by the Manville Company of Infusorial Earth.

Specimen I. The Manville Low Pressure Covering is similar to Specimen F.

Specimen J. A plastic cover made by the Manville Company, and called by them Magnesia-Asbestos. It contains only a slight amount of carbonate of magnesium.

Specimen K. The Magnabestos of the Keasby & Mattison Company is a moulded cover, containing about 45 per cent. of carbonate of magnesia and a considerable percentage of carbonate of calcium.

Specimen L. The Moulded, Sectional Cover of the H. F. Watson Company is composed mainly of sulphate of calcium and some 20 per cent. of magnesium carbonate, and has upon its outer surface a thick sheet of felt board.

Specimen O. The Asbestos Air Cell Fire Board of the Asbestos Paper Company is similar to Specimen G, except that it has larger cells and contains much more silicate of sodium. It is very hard and strong.

Specimen P. The "Calcite," or Asbestos-Magnesia, of the Philip Carey Company, is a sectional, moulded cover, composed mainly of sulphate of calcium. It has an outer layer of felt board.

In regard to the compositions of Specimens C, J, L, and P, I desire to state that I have made no complete analysis, but have satisfied myself that the principal ingredient is sulphate of calcium and *not* carbonate of magnesium. Prospective purchasers of pipe covers should not be misled by names. Since the appearance of Professor Ordway's reports it has been recognized that carbonate of magnesium was of great value as a non-conductor of heat, hence the name "Magnesia" has been applied to a great many covers. It is to be observed that there is no virtue in a name. Asbestos is merely an incombustible material in which air may be entrapped, but when not porous is a good conductor of heat. Magnesia is a most effective non-conductor. This name has been applied to many compounds, of which the greater part consists of carbonate of calcium or of plaster of Paris, materials which are not good as heat retardents. The percentage of magnesium carbonate and plaster of Paris in several moulded, sectional covers is given in Table II.

I have made no investigation of the effect of the raw materials upon the metal of the pipe other than to satisfy myself that the Cork, Magnesia, Air Cell, and Imperial covers cause no corrosion.

TABLE II.

Specimen.	Name	Percentage Composition.	
		Mg.CO₃. Carbonate of Magnesium.	Ca.SO₄. Sulphate of Calcium.
D	K. & M. Magnesia	80 to 90	3
C	Manville H. P. Lining	Less than 5	65 to 75
L	Watson Moulded	20 to 25	50 to 60
P	Carey Calcite	Less than 5	75
J	Manville Magnesia Asbestos . .	10 to 15	None.

The conditions of testing were reasonably near the conditions of actual practice. The room temperature was kept at 72° F., and the openings into the room were carefully closed. It was found early in the series that variation in the amount of moisture present in the air altered the amount of heat lost from the covers, but no attempt was made to correct this. The error introduced is not greater than 1 per cent.

It was found that the heat loss per square inch of the flat surfaces at the ends of the pipes was less by several per cent. than the loss from the sharply curved sides, and as all pipe covers tested were used to cover both sides and ends, the figures given in the table show a loss less than would be shown were the pipe surface wholly cylindrical, and more than if it were all flat.

The pipes were suspended from the ceiling, as described in an early paragraph, and the circulation of the air about them was due only to their own convection currents. The variation in thickness in different places on the same specimen was considerable, but an average of twenty measurements was taken and results given in the table to the nearest ⅛ of an inch. Owing to these variations in thickness, the results of a measurement of the efficiency of any one cover cannot be used to predict the efficiency of a second cover of the same make with an accuracy greater than 2 per cent. Two specimens of each make were tested, and in some cases, four, the mean value being given in the table.

Table III gives the saving, in dollars, due to the use of the various covers.

TABLE III.

Specimen.	NAME.	Loss per sq. ft. B.T.U. at 300 lbs.	Saving B. T. U. per sq. ft.	Saving per year per 100 sq. ft.
A	Nonpareil Cork Standard	2.20	11.64	$37.80
B	Nonpareil Cork Octagonal	2.38	11.46	37.20
C	Manville Sectional High Pressure .	2.38	11.46	37.20
D	Magnesia	2.45	11.39	36.90
E	Imperial Asbestos	2.49	11.35	36.80
F	" W. B.".	2.62	11.22	36.40
G	Asbestos Air Cell	2.77	11.07	36.00
H	Manville Infusorial Earth	2.80	11.04	35.85
I	Manville Low Pressure	2.87	10.97	35.65
J	Manville Magnesia Asbestos . . .	2.88	10.96	35.60
K	Magnabestos	2.91	10.93	33.50
L	Moulded Sectional	3.00	10.84	35.20
O	Asbestos Fire Board	3.33	10.51	34.20
P	Calcite	3.61	10.23	33.24
	Bare Pipe	13.84	....	....

Table IV shows that at the end of ten years the best of the covers tested will have saved $46.00 more than the poorest. The difference between the several covers of the better grade is exceedingly small.

The money saving is computed on the following assumptions: Coal at $4.00 a ton evaporates 10 pounds of water per pound of coal; the pipes are kept hot 10 hours a day, 310 days a year. If computations are made, as is sometimes done, on an assumption that the pipes are hot 24 hours a day, 365 days in a year, the saving is nearly three times that shown in Table III.

Generally speaking, a cover saves heat enough to pay for itself in a little less than a year at 310 ten-hour days, and in about four months at 365 twenty-four-hour days.

It is evident that the decision as to the choice of cover must come from other considerations, as well as from the conductivity.

The question of the ability of a pipe cover to withstand the action of heat for a prolonged period without being destroyed or rendered

TABLE IV.

Net Saving Per 100 Square Feet.

Speci-men.	Name	1 year.	2 years.	5 years.	10 years.
A	Nonpareil Cork Standard	$12.80	$50.60	$164.00	$353.00
B	Nonpareil Cork Octagonal	12.20	49.40	161.00	347.00
C	Manville Sectional High Pressure .	12.20	49.40	161.00	347.00
D	Magnesia	11.90	48.80	159.50	344.00
E	Imperial Asbestos	11.80	48.60	159.00	343.00
F	" W. B."	11.40	47.80	157.00	339.00
G	Asbestos Air Cell	11.00	47.00	155.00	335.00
H	Manville Infusorial Earth	10.85	46.70	154.25	333.00
I	Manville Low Pressure	10.65	46.30	153.75	332.00
J	Manville Magnesia Asbestos . . .	10.60	46.20	153.00	331.00
K	Magnabestos	10.50	46.00	152.50	330.00
L	Watson's Moulded Sectional . . .	10.20	45.40	151.00	327.00
O	Asbestos Fire Board	9.20	43.40	146.00	317.00
P	Calcite	8.24	41.48	141.20	307.00
Q	Bare Pipe	....	....	....	......

less efficient is of vital importance. The increasing use of cork as an insulator has led to many questions as to its ability to remain "fireproof." I have exposed it to a temperature corresponding to 350 pounds of steam for three months, and to a temperature corresponding to 100 pounds for two years, and can detect no change ; and I am satisfied, as well as one can be without the actual experience, that any suspicion of its ability to withstand continued heating is groundless.

The magnesia covering is, of course, unquestionable on this ground, being almost indestructible by heating.

The Imperial Asbestos is also perfectly safe from any fire risks, as is the Air Cell and Fire Board.

The Manville Infusorial Earth, and also the Manville Magnesia-Asbestos are liable to no accident from fire, nor is the Carey Calcite.

It is to those covers, the " W. B." of the Watson Company, and the Manville Sectional and others which possess a composite structure, that I desire to call attention. I do not consider it safe to put upon a steam pipe, wool, hair felt, or woolen felt in any form. The causes of risk are two: First, the wool may become charred by heat from the pipe and finally ignited. However, this can hardly happen, even on high-pressure pipes, when the thickness of fireproof material (ashes. tos, magnesia, or whatever it may be) is as great as 1 inch. The second and most serious risk is from the presence in shops or mills of the long tubes of wool, dry as tinder, often connecting one room with another, and ready to flash at the slightest rise in the already too great temperature, I would even insist that the canvas jackets on the covers be fireproof. An accident in my own laboratory has proved the actual danger of these wool felts, and I should not be willing to allow their use again. Their efficiency is high as non-conductors, but not higher than any other perfectly safe covers. If the wool is separated by about 1 inch of fireproof material from the pipe, it is not kept so hot and dry, and the risks from outside ignition is less; but I do not endorse the practice of many engineers in wrapping hair felt outside of a sectional cover. The saving due to this practice is indicated in Table V.

The following assumptions have been made in computing Tables IV, V, and VI: First, that all the covers cost $25.00 per 100 square feet, applied. I realize that this is a high figure, perhaps too high, yet it is not far from the list price of several makers, and any attempt to get a definite price from them revealed a maze of discounts and double discounts, and flexible price lists too intricate for an unitiated mind to travel. In case the saving due to a cover which costs $20.00 instead of $25.00 is desired, the simple addition to the final saving of the $5.00 difference makes the necessary correction.

Secondly, by the advice of the makers, I have made an assumption that the cost is not nearly proportional to the thickness. As the thicker coverings are not now made in great quantities, the actual cost of their manufacture is uncertain.

Inspection of Table V shows the saving due to the use of hair felt outside a standard magnesia cover. In five years 100 square feet of hair felt saves $7.00 more than its cost, and in ten years it saves $20.00 above its cost. The further saving due to a second inch outside the first is $8.00 in ten years. Of course the well-known tendency of hair felt to deteriorate should be considered.

In the case of the Nonpareil cork, increasing the thickness from 1 to 2 inches raises the cost from about $25.00 to $35.00 per 100 square feet, and increases the net saving in five years by $10.00, and by $30.00 in ten years. In other words, the second inch of material in use about pays for itself in two years, while the first pays for itself in about one year. The third inch does not increase the saving even in ten years. The second inch, therefore, more than pays for interest and depreciation, while the third fails to do this.

TABLE V.

VARIATIONS IN THICKNESS, ETC.

SPECIMEN.	Saving in B. T. U. per sq. ft. per minute.	Saving in dollars per 100 sq. ft. per year.	NET SAVING.				Approximate cost.
			1 year.	2 years.	5 years.	10 years.	
Magnesia 1¼ in. thick	11.62	$37.75	$7.75	$45.50	$159	$347	$30
Magnesia 1¼ in. thick, and 1 in. of Hair Felt	12.38	40 22	5.22	45.44	166	367	35
Magnesia 1¼ in. thick, and 2 in. of Hair Felt	12.77	41.50	1.50	43.00	167	375	40
Nonpareil Cork :							
1 inch	11.64	37 80	12.80	50.00	164	353	25
2 inch	12.84	41 75	.....	48.50	174	383	35
3 inch	12 94	42 05	.....	34.10	160	370	50
Fire Board :							
1 inch	10 54	34 20	9.20	43.40	146	317	25
2 inch	11.48	37.25	2.25	39.50	151	337	35
3 inch	11 70	38 00	12.00	26.00	140	330	50
4 inch	11 83	38.40	26.60	11.80	127	319	65

In the case of the Asbestos Fire Board, a second inch in thickness causes a saving of $20.00 in ten years, the third and fourth inches showing a loss.

In general it may be said, therefore, that if five years is the length of life of a cover, 1 inch is the most economical thickness, while a cover which has a life of ten years may to advantage be made 2 inches thick.

In view of the custom which prevails to some extent of wrapping asbestos paper around a pipe and surrounding the whole with hair felt, I made tests as to the temperature of the bounding line of the asbestos paper and hair felt, using a LeChatelier Thermo-electric Pyrometer for this purpose. The different samples of asbestos paper give widely varying results, but a general idea of the protection afforded by the paper may be had from Table VI.

TABLE VI.

PROTECTION AFFORDED BY ASBESTOS PAPER. PIPE AT 200 POUNDS PRESS.

Thickness of Asbestos Paper.	Temperature of Pipe.	Temperature of Inside of Hair Felt.	Pressure Corresponding to the Temperature of the Inside of the Hair Felt.
$\frac{1}{32}$ inch.	384.7° F.	356° F.	146 pounds.
$\frac{1}{16}$ inch.	385.0° F.	329° F.	102 pounds.
$\frac{1}{8}$ inch.	384 6° F.	302° F.	70 pounds.
$\frac{1}{4}$ inch.	384.7° F.	266° F.	39 pounds.

I have had my attention called to the varying loss from bare pipes when their surfaces were in varying conditions as regard rust, dirt, paint, etc. I therefore made a few brief tests to satisfy my mind as to the chance of their being any large variation which might influence my figure for the loss from bare pipe, viz., 13.84 B. T. U. per square feet per minute. The results are shown in Table VII.

TABLE VII.

LOSS OF HEAT AT 200 POUNDS FROM BARE PIPE.

CONDITION OF SPECIMEN.	B. T. U. lost per sq. ft. per minute.
New pipe	11.96
Fair condition	13.84
Rusty and black	14.20
Cleaned with caustic potash inside and out . . .	13.85
Painted dull white	14.30
Painted glossy white	12.02
Cleaned with potash again	13.84
Coated with cylinder oil	13.90
Painted dull black	14.40
Painted glossy black	12.10

The rate of heat loss from a bare pipe is also affected by the air circulation and the temperature of the surrounding bodies. A few tests were made to indicate the magnitude of the errors likely to be caused by variation in these conditions, and a brief examination of some of the results may be interesting. They are given in Table VIII.

TABLE VIII.
EFFECT OF SURROUNDINGS.

CONDITION AND POSITION OF PIPE.	B. T. U. lost per sq. ft. per minute at 200 pounds.
1. Standard condition : hung in center of room .	13.84
2. Near brick wall, between windows	14.26
3. Hung horizontally in center of room . . .	12.06
4. Vertical 10-inch pipe { 36 inches long . . .	13.48
18 inches long . . .	14.42
5. Vertical 18-inch long { 10-inch diameter . .	14.42
4-inch diameter . .	15.20
6. 4-inch diameter in draft from electric fan . .	20.10

Table IX shows the varying loss from a bare pipe with the change in pressure.

TABLE IX.
VARIATION OF HEAT LOSS WITH PRESSURE.

PRESSURE.	Bare Pipe loss B. T. U. per sq. ft. per minute.
340	15.97
200	13.84
100	8.92
80	8.04
60	7.00
40	5.74

A very thorough test was made of the common method of judging a pipe cover by the sensation of warmth given the hand on touching it, and nothing too harsh can be said of this practice. The sensation is dependent to such an extent upon the *nature of the surface* that it fails utterly to give any idea of the actual temperature. I have been unable to devise any method of so attaching a mercury thermometer to the outside of a steam-pipe cover as to make use of it as a testing device in measuring heat loss.

I am desirous of calling attention to the advantages arising from the use of plastic, rather than sectional covers. The case of removal for repairs or alterations makes the sectional cover better for some work, but there is much pipe surface which might be covered securely with plastic, where a sectional cover is soon ruined by vibration. Of course, the plastic covers offer no possibility of leaky joints and long cracks. It should be borne in mind that in most cases about 20 per cent. of the entire surface to be covered is irregular, and must be covered by plastic or fittings. It will be well for prospective purchasers of pipe cover to see to it that their contracts call for fittings and plastic of as high an efficiency as the sectional cover shows.

I am now testing a considerable number of samples of non-conducting material, not perhaps, classed as pipe covers, but used for heat insulation. Table X gives some figures concerning them which may be of interest.

TABLE X.

MISCELLANEOUS SUBSTANCES.

SPECIMEN.	B. T. U. per sq. ft. per minute at 200 pounds.	Saving in 1 year per 100 sq. ft. pipe.
Box A :		
1. With sand	3.18	$34.60
2. With cork, powdered	1.75	39.40
3. With cork and infusorial earth . .	1.90	38.90
4. With sawdust	2.15	37.90
5. With charcoal	2.00	38.50
6. With ashes	2.46	36.90
Brick wall 4 inches thick	5.17	28.80
Pine wood 1 inch thick	3.56	33.80
Hair felt 1 inch thick	2.51	36.80
Cabot's seaweed quilt 1 inch thick . . .	2.78	35.90
Spruce 1 inch thick	3.40	33.90
Spruce 2 inches thick	2.31	37.50
Spruce 3 inches thick	2.02	38.50
Oak 1 inch thick	3.65	33.10
Hard pine 1 inch thick	3.72	32.90
Eider-down 1 inch thick, loose	*1.90 to 2.70	
Eider-down 1 inch thick, tightly packed .	*1.70 to 1.80	. . . . .

* Variable.

The box A, referred to in the table, is a $\frac{1}{8}$-inch pine box, large enough to surround the pipe, leaving a 1-inch minimum space at its four sides.   In it were tested several materials which I find are used in just this way for steam and cold storage insulation.

ROGERS LABORATORY OF PHYSICS,
MASSACHUSETTS INSTITUTE OF TECHNOLOGY,
April, 1898.

TECHNOLOGY QUARTER

AND

PROCEEDINGS OF THE SOCIETY OF ARTS.

VOL. XI.	DECEMBER, 1898.	No. 4.

THE PRODUCTION OF ILLUMINATING GAS AND COKE IN BY–PRODUCT COKE OVENS.[1]

BY H. O. HOFMAN.

IN June last the writer paid a visit to the gas works of Halifax, Nova Scotia. The illuminating gas is produced in by-product coke ovens, and the excess of heating gas not required for the ovens is collected in separate holders and piped to industrial establishments; the coke produced is sold for domestic use and boiler fuel.

The first by-product oven was that of Carvès, erected in 1867. Since then a great variety of coke ovens have been constructed for saving the by-products, which have thus become a valuable article of commerce, furnishing numerous important compounds used in the arts. Halifax is, so far as the writer knows, the first city in America to be lighted by gas made in the by-product coke ovens. In Europe, however, the gas from an Otto-Hoffmann plant has been sold to a gas works for several years. It has long been known from studying the analyses of coke-oven gases that those obtained at certain stages of the process were suited for illuminating purposes, but the first practical application of the knowledge here seems to have been made at Halifax, where it has proved an undoubted success.

Enormous works are now in process of erection at Everett for the purpose of supplying Boston in the same way with illuminating gas, heating gas, and coke. The plant will consist of 400 Otto-Hoffmann

---

[1] Reprinted from *The Engineering and Mining Journal* of October 8 and 15, 1898.

ovens, with a daily capacity of about 2,000 gross tons of coal. It will be equipped with modern mechanical appliances for the cheap handling of coal and coke, a complete system of condensation and purification and ammonia stills. The gas will be stored in a holder having a capacity of 5,000,000 cubic feet. The coke to be manufactured will, for the present at least, be principally of the kind suitable for domestic use and boiler fuel, there being little or no demand for blast-furnace coke. Probably, however, conditions will be changed extensively by the successful introduction of the new method of working and a market ultimately created for blast-furnace coke, especially in case the duty on iron ore should be removed, which would give Boston and the New England States the proper incentive to produce at least some of the iron and steel they consume. At present most blast-furnace coke is made in beehive ovens, which allow the by-products to go to waste. Only few kinds of coal make satisfactory coke in the beehive, thus restricting it to very small sections, while the by-product oven permits the use of a great variety of coal, and can be erected wherever coking coals occur. If it can produce illuminating gas, heating gas, and coke, varying with the character of the coal, there would seem to be no limit to its probable future usefulness in all parts of the country.

The great importance of the subject has led to the following outline, made with the knowledge of all parties concerned, of the work done at Halifax by the People's Light and Heat Company, and of the preliminary large-scale experiments made for the New England Gas and Coke Company at Glassport, Pennsylvania, with the coal that is to be used in the new Boston plant.

The People's Light and Heat Company at Halifax undertook the manufacture of illuminating gas for the city in March, 1897. They erected a Slocum[1] by-product coke oven. This furnace embodied a number of new ideas aiming to combine the advantages of the ovens of Hüssener[2] and of Semet-Solvay,[3] but it soon proved a failure and was remodeled according to the Semet-Solvay system. It has since been doing satisfactory work. A very interesting experience occurred in connection with the thickness of the partition walls supporting the

---

[1] See J. Fulton, *Coke.* Scranton, Pa., 1895. p. 215.

[2] *Stahl und Eisen*, 1883, p. 397.

[3] *Engineering and Mining Journal*, August 9, 1890.

roof. In the original Semet-Solvay they were 16 inches thick, in the modified oven 30 inches, which proved a permanent improvement in equalizing the temperature. The material first used was red brick, which had to be replaced by fire brick.

The plant consists of one block of 10 ovens, each being 30 feet long, 5 feet 6 inches high, and from 16 to 17 inches wide, and having 3 charging ports and 2 gas-collecting openings, and the necessary apparatus for condensing and extracting the tar and ammonia and for purifying the gas. An oven is charged with 5 net tons of coal, which is coked in 20 hours. It takes $2\frac{1}{2}$ hours to discharge the 10 ovens, and the same time to refill them, the latter work not being begun until they are all discharged. The time between the charging and refilling of the oven is given to heating up, so as to begin coking with a strong initial heat. The gas pressure in the ovens is kept at $\frac{1}{4}$ inch of water above the atmosphere, so as to prevent any air from entering, which would consume fuel and dilute the gases with nitrogen and carbon dioxide. The temperature in the ovens ranges from 1,800 to 2,000 degrees F.

The coal used is a mixture of washed slack from the International and Phelan seams of the mines of the Dominion Coal Company, Cape Breton. The coal charged contains approximately 60 per cent. fixed carbon.

The gas set free in the ovens is stored in two different sets of holders. All the gas above 16 candle power (measured by a Jones jet-photometer) goes to the illuminating-gas holders; as soon as it sinks below this figure it goes to the heating-gas holders. The illuminating gas, averaging 18 candle power, is supplied at $1.40, the heating gas, averaging $8\frac{1}{2}$ candle power, at $0.40 per 1,000 cubic feet. In 24 hours 37 short tons of coal is coked, furnishing 310,000 cubic feet of gas, of which 100,000 cubic feet (32.26 per cent.) is illuminating gas, and 210,000 cubic feet (67.74 per cent.) heating gas; of this 170,000 cubic feet (54.84 per cent.) is consumed in coking, leaving 40,000 cubic feet to be used as heating gas. A long ton furnishes on the average 5 pounds of ammonia gas and 12 U. S. gallons, 120 pounds, of tar. The ammonia liquor is distilled with milk of lime, and furnishes a shipping ammonia liquor with 17 per cent. ammonia. The tar is utilized in the summer in the manufacture of tar paper; in winter it is distilled, furnishing creosote, pitch, etc. Lastly, the commercial coke, forming 75 per cent. of the coal charged, is broken and sold at the rate of $4 a ton for domestic purposes and as boiler fuel, the price for anthracite being $4.25 a ton.[1]

This brief outline gives the leading technical results obtained so far. No doubt they will be improved upon by the time the normal method of working is reached, as, of course the first consideration must be to furnish regularly sufficient gas of a required candle power. Experiments in the direction of improvement can only be entered upon very gradually, so as not to interfere with the main product. The best ultimate method of working will depend to a considerable extent on the market for the two leading minor products, heating gas and coke. It may come to this, that with a larger demand for heating gas, the ovens will be heated with producer gas, and even some of the small coke burned in the producers, or stress may have to be laid on making a firmer coke in case a market for blast-furnace coke should open.

The working tests of the New England Gas and Coke Company for the new plant near Boston were made early in the present year at the works of the United Coke and Gas Company, Glassport, Pennsylvania, where the Otto-Hoffmann by-product oven is in operation, which is the one decided upon for the Boston plant. They were carried out by Dr. F. Schniewind with washed slack coal from the mines of the Dominion Coal Company, Cape Breton.

The works at Glassport were erected in 1896. They have four blocks of furnaces, each with 30 ovens. Each oven is 33 feet long, 5 feet 10 inches high and 20¼ inches wide. The coal usually treated is the run-of-mine of the Washington Coal and Coke Company's mine, on the upper Youghiogheny River, and has the following composition:

Moisture,	Fixed carbon,	Volatile matter,	Ash,	Total,
0.60	59.18	33.01	7.21	100.00.

Phosphorus, 0.0071; Total sulphur, 1.27; (fixed, 0.80; volatile, 0.47.)

A charge of 6¼ gross tons dry coal is coked in 34 hours 54 minutes, and yields 74.26 per cent. salable coke, the coke having the composition:

Volatile matter,	Fixed carbon,	Ash,	Moisture,	Total,
1.00	86.47	11.57	3.17	100.00

Sulphur, 0.96; Phosphorus, 0.0107.

is used in the iron blast furnace. A net ton of coal furnishes 10,000

---

[1] The record for the month of September shows an improvement on that of June given above. With nine ovens working, 1,416 net tons of coal were coked, which gave 12,889,900 cubic feet of gas, or 9,060 cubic feet per net ton of coal; of this 45 per cent. was surplus gas.

cubic feet of gas, of which 70 per cent. is required to heat the ovens, the rest piped 1½ miles to steel works for heating an open-hearth furnace. The other products are 5.27 per cent. tar and 1.230 per cent. sulphate of ammonia.

The tests were carried out in one of the ovens, the rest of the block being in use at the same time for their regular purpose, thus having actual working conditions. Of course, the oven was heated by the surplus gas from the ovens of the block. By measuring the volume of the gas used and determining its calorific power, the heat consumed by the oven is easily arrived at. The volatile compounds of the test-oven were collected in a separate main and examined as to quality and quantity, and the coke produced was kept apart from that of the other ovens. In all, 143 tons of coal from three different seams were coked. Of this, only the work on 40 tons of washed slack from the International seam will be discussed.

THE COAL. — As charged into the oven the coal contained 9.9 per cent. of water, this high figure being due to the coal having been exposed for several months to rain and snow. The hygroscopicity of the coal was determined at 4.01 per cent., so that under normal conditions the coal will not contain over 5 per cent. of water. An average of several ultimate analyses gave the composition :

C,	H,	N,	O + S,	Ash,	Total,
75.10	3.75	1.51	13.80	5.84	100.00.

and one of several proximate analyses :

V. H-C.	F. C.	Ash,	Total,
34.60	59.56	5.84	100.00.

It will be noted that sulphur and oxygen are given as a total of 13.80 per cent., which is unusual. In the destructive distillation of coal, sulphur has a harmful effect on the gas as well as on the coke, especially if the latter be used in the production of pig iron. From the gas it can be readily eliminated by an increased purification plant; from the coke the matter is more difficult. A great number of sulphur determinations have been made from the coal that is to be used. Instead of giving a mass of figures it will be better simply to tell what is being done at present with Cape Breton coal. It is supplied to the amount of 200 tons per day to the iron blast-furnace plant of the Nova Scotia Steel Company, at Ferrona, Nova Scotia, which uses it exclusively for the manufacture of its coke. The percentage of sulphur in the coal varies considerably; while in some parts of the bed it may occasionally reach 3 per cent., in others it is only a few tenths of 1 per cent., and on the average the percentage of sulphur is low. The coke made at Ferrona from unwashed Dominion coal contains as impurities 1.08 per cent. of sulphur and 8.20 per cent. of ash. Coals running higher in sulphur are first washed in the plant erected by Mr. W. M. Stein,[1] of Philadelphia, before they are coked. The iron ore used by the company comes in part from its own mines in the neighborhood, in part from Newfoundland.

Both basic pig and foundry iron are produced as seen in the following analyses:

Kind.	Si.	Ma.	P.	S.	G. C.	C. C.
Basic pig	0.50	0.87	1.23	0.017	....	....
Foundry iron	2.32	0.65	1.20	0.020	3.64	0.23

The percentage of phosphorus in the average coke analysis is exceptionally low, viz., 0.0028 per cent. The calorific power of the dry fuel with 5.84 per cent. of ash calculated according to the Dulong-Mahler formula, is 12,437 B. T. U. The coal is a good coking coal and has the advantage that the finished coke occupies a smaller space than the coal.

THE OPERATION. — In coking, the average weight of the four charges was 14,591 pounds of coal, which, deducting the moisture (1,348 pounds), corresponded to 13,602 pounds of dry coal. The average time required for complete destructive distillation was 33 hours 56 minutes, or 5 hours 35 minutes per long ton of dry coal. The regular working time for slack of the Youghiogheny mine with 1 per cent. of water is 5 hours 26 minutes. This proves that International coal compares favorably with it. With dryer coal the showing will be still better. By using an oven 18 inches wide instead of $20\frac{3}{4}$ inches, the coking time can be reduced to 26 hours, and if the coke is to be used only for domestic purposes or as boiler fuel instead of for the blast furnace, like the Glassport coke, the time required will be only 22 hours, and 24 hours will be a safe figure. The range of temperature in the four tests, measured by the Mesuré and Nouel optical pyrometer, was from 950 to 1,070° C. In a general way it is known that a high temperature increases the quantity, but diminishes the quality of the gas obtained, but the differences in temperature

---

[1] See Fulton, *Coke.* Scranton, Pa., 1895, p. 67.

were not sufficient to have any decided influence on the composition and candle power of the gas.

SUMMARY OF RESULTS. — A summary of the results obtained per long ton of coal, the volumes having been reduced to weights, is subjoined :

Products from 1 long ton of coal.	Pounds.	Per cent.
Coke, total { Large coke, $>$ 1″, 66.69 per cent. Small coke, $\frac{1}{4}$ — 1″, 1.64 per cent. Breeze, $<\frac{1}{4}$″, 2.80 per cent. }	1,593.4	71.13
Tar	75.7	3.38
Ammonia (= 1.373 per cent. sulphate)	7.6	0.34
Gas, total, 10,390 cubic feet, of 0.466 specific gravity	368.0	16.43
Sulphur compounds in gas : Hydrogen sulphide (H_2S), 0.98 pounds per 1,000 cubic feet . .	10.8	0.48
Carbon disulphide (CS_2), 0.13 pounds per 1,000 cubic feet . . .	1.6	0.07
Gas liquor and loss, by difference	182.9	8.17
Total	2,240.0	100.00

Of the 10,390 cubic feet of gas, 49.5 per cent. was surplus gas, that is, gas not required for the heating of the oven. This had the following composition :

Olefines, CmHn, 5.2,	Marsh gas, CH_4, 38.7,	Hydrogen, H_2, 38.4,
Carbon monoxide, CO, 6.1,	Carbon dioxide, CO_2, 3.6,	Oxygen, O_2, 0.3,
Nitrogen, N_2, 7.7,	Total, . 100.0.	

Its calorific power (H_2 burnt to liquid water) was 686 B. T. U., its candle power 14.7, and its specific gravity (air = 1) 0.510.

COKE. — The yield of dry coke, as shown above, was 71.13 per cent. After quenching and cooling it retained 3.67 per cent. of moisture, hence the figure 71.13 is increased to 73.91 per cent. As only the large and small coke are the marketable products, the salable coke will amount to 68.33 per cent. of the coal, or with the usual 2 per cent. of water of commercial retort coke, 69.70 per cent. of commercial coke. Most of the coke was what may be called a fair metallurgical fuel. It was hard, had a good cellular structure, a metallic ring and a silvery luster. Its composition is shown by the following approximate analysis :

Volatile matter,	Fixed carb. and S,	Ash,	Total,
1.27	69.82	8.91	100.00,
P.,	Moisture,		
0.0041	3.67.		

Regarding sulphur see above. The low percentage of volatile matter proves that the destructive distillation was complete. The subjoined analysis of the coke ash,

SiO_2,	Al_2O_3,	F_2O_3,	Mn_3O_4,	CaO.	MgO,
27.71	13.04	50.60	0.25	4.61	0.77

K_2O,	Na_2O,	SO_3,	P_2O_5,	Total,
0.85	0.18	2.62	0.10	100.73,

shows a high percentage of iron and small amounts only of alkali. The iron, due mainly to pyrite in the coal, can be greatly diminished, and, assuming it to be completely eliminated, the analysis will be changed to

SiO_2,	Al_2O_3,	Mn_3O_4,	CaO,	MgO,
55.29	26.01	0.50	9.19	1.53

K_2O,	Na_2O,	SO_3,	P_2O_5,	Total,
1.70	0.36	5.22	0.20	100.00,

giving a total of 2.06 per cent. of alkali. Such an ash is not liable to form clinkers on the grate.

TAR. — The yield of tar was 75.7 pounds per long ton of coal, or 3.38 per cent. ; its specific gravity was 1.170. The following table shows its behavior in fractional distillation as well as that from the Otto-Hoffmann oven at the Germania plant, and from German gas works :

Fractions.	Temperature, C°.	Otto-Hoffmann oven.		Average German gas-house tar.	
		International coal.	Germania plant.	I.	II.
Light oil	80—170	3.7	6.55	3.0	2.5
Middle oil	170—230	9.8	10.54	7.5	2.5
Heavy oil	230—270	12.0	7.62	33.5	25.0
Anthracene oil . . .	over 270	4.3 } 71.3	44.35 } 74.9	10.5 } 56.0	10 0 } 70.0
Pitch	. . . .	67.0 }	30.55 }	45.5 }	60.0 }
Water	. . . .	2.3	Trace.	. . . .	. . . .
Loss	. . . .	0.9	0.39	. . . .	. . . .
Total	. . . .	100 00	100 00	100.00	100.00
Specific gravity . .	. . . .	1.170	1 1198	1.155	1.155

The softening point of the pitch was 87 degrees C., and its specific gravity 1,350. There is considerable difference between the ratio of anthracene oil and pitch in the two tars of the Otto-Hoffmann oven. The reason for this is that with the Germania tar the distillation was carried on until all the vapors had been expelled. The softening point of the pitch was therefore much higher, viz., 165 degrees C. The table shows that the tar from the tests compares favorably with that from other sources. It has been thought that tar from by-product coking ovens was inferior in quality to that from retorts of gas works. The following is a comparative table [1] of tar analyses from gas works and from an Otto-Hoffmann plant using the same kind of Westphalian coal.

Fractions.	Gas-house tar.	Otto-Hoffmann coke-oven tar.
Water	2.9	2.2
Light oils to 200° C.	4.0	3.4
Aniline benzole	0.92	1.1
Solvent naphtha	0.20	0.32
Creosote oil	8.6	14.5
Crude naphthalene	7.4	6.7
Anthracene oil	17.4	27.3
Pure anthracene	0.6	0.7
Pitch	58.4	44.4
Totals	100.42	100.62
Free carbon in pitch	15—25	5—8

It shows that the coke-oven tar is better for the purposes of the tar distiller than gas-house tar. The yield of tar (3⅝ per cent.) is, however, low in comparison with that from gas works, which obtain about 5 per cent. The distillation of tar may be of importance with regard to the recovering of benzole as enriching material.

AMMONIA. — The gas-liquor obtained was 182.9 pounds per long ton of dry coal or 8.17 per cent., containing 7.6 pounds or 0.34 per cent. of ammonia. Tests showed that 96.6 per cent. was present as free, or volatile, ammonia, *i. e.*, ammonia that can be recovered by simple distillation; and only 3.4 per cent. was fixed ammonia, that is,

[1] Lunge, *The Mineral Industry*, V., p. 188.

ammonia in chemical combination as sulphate, chloride, rhodanate, etc., which can be recovered only by the addition of lime in the distillation. As the ultimate analysis given above shows 1.51 per cent. of nitrogen in the coal, and the 0.34 per cent. of ammonia requires 0.28 of nitrogen, it will be seen that 18.5 per cent. of the total nitrogen of the coal is converted into ammonia, which is a good showing, as usually only 13½ to 15 per cent. is thus changed.

GAS. — The total volume of gas per long ton of dry coal collected in the holders was 10,390 cubic feet. At the time when the tests were made the average temperature of the air was 37 degrees F., and the barometer showed a pressure of 750 millimeters. As the normal temperature for calculations is 60 degrees F., the volume of gas would be 4.7 per cent. less than under normal conditions. This is partly offset by the low atmospheric pressure, the normal being 760 millimeters; the volume, however, remains 3.2 per cent. less than when reduced to 60 degrees F. and 760 millimeters pressure. The error is so slight as to make a correction unnecessary. The samples for analysis were taken between the exhauster and the scrubber entrance and passed through a small iron-oxide purifier. A complete analysis was made about every 2 hours; the specific gravity was determined hourly as well as the calorific power and the candle power. Sulphur tests of the unpurified gas were made continuously, but only during the first 12 hours after charging.

Diagram 1 is a graphical representation of the average analysis of the gases from the four separate charges.

The percentage of marsh-gas (CH_4) shows a rapid decline, and this is especially the case toward the end of the coking operation. It is accompanied by a corresponding increase of hydrogen (H), which reaches nearly 67 per cent., or nearly 80 per cent. if figured for gas free from oxygen and nitrogen. The curve of carbon monoxide (CO) shows the percentage of this gas to vary between the narrow limits of 5.4 and 6.8 per cent., a very low figure. The olefines ($CmHn$) remain at a constant figure, 6½ per cent. for the first 5 hours, and then diminish at a uniform ratio to the end of the operation. The percentage of carbon dioxide (CO_2) is constant for the first 20 hours, ranging at from 3 to 4 per cent., and then decreases gradually to 1 per cent. Oxygen (O) and nitrogen (N) are accidental impurities. The oxygen gets into the gas through leaky joints and amounts to about 0.3 per cent. The average percentage of nitrogen during the first 14 hours and 46 min-

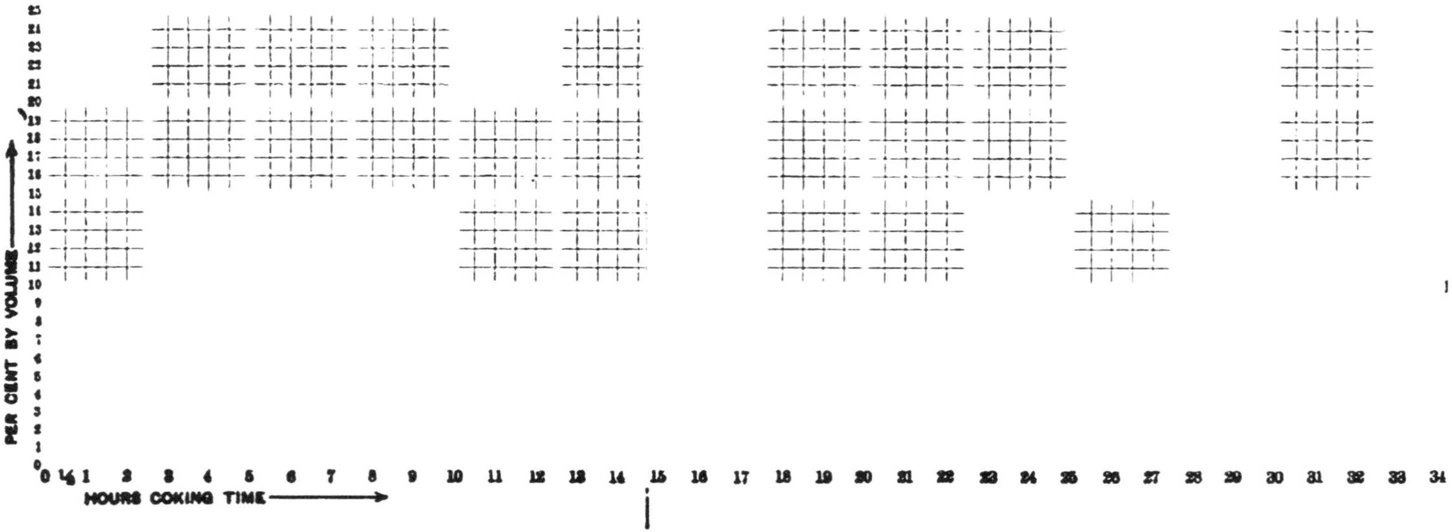

PER CENT BY VOLUME →

HOURS COKING TIME →

DIAGRAM No. 1.

Surplus gas produced during the first 14 hours, 46 minutes. Volume per long ton, 5,143 cu. ft. = 49.5 per cent. Average analysis:

C_2H_4	CH_4	H_2	CO	CO_2	O_2	N_2	Total.
5.2	38.7	38.4	6.1	3.6	0.3	7.7	100.0

Over-heating gas produced during the remaining 19 hours, 10 minutes. Volume per long ton, 5,247 cu. ft. = 50.5 per cent. Average analysis:

C_2H_4	CH_4	H_2	CO	CO_2	O_2	N_2	Total.
2.4	39.2	50.5	6.3	2.8	0.3	9.1	100.0

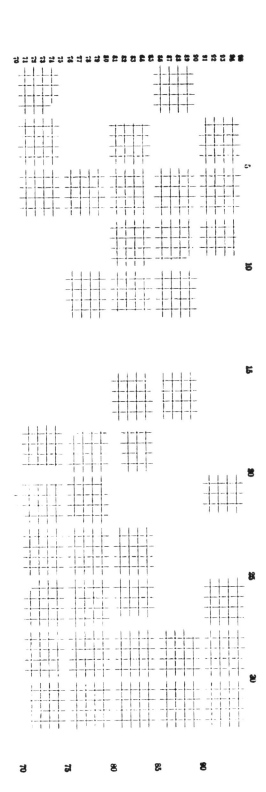

utes is seen to be 7.7 per cent. ; that of the last 19 hours and 10 minutes, 9.1 per cent. It remains pretty constant up to the 23d hour, and then increases rapidly. It has its origin in the destructive distillation, which accounts for 2–3 volumes, and in leakages which permit the entrance of air. The large increase toward the end finds its explanation in the diminished volume of gas generated in the oven and the consequent decrease of pressure. The percentage of nitrogen is high, and will under normal working conditions be reduced to 5 per cent.

Diagram 2 is a chart, the curves of which give the calorific power, the specific gravity, and the candle power of the gas. Disregarding a slight irregularity in the curves during the first two to three hours, due to the fact that the last portions of gas from a preceding experiment become mixed with the first gas of the following one, the calorific power at the beginning is about 685 B. T. U. It rises quickly to 775 B. T. U., then drops quickly until the 7th hour, and more·slowly until the 22d hour, after which it drops more quickly than at first until the end of the operation.

The curve of the specific gravity runs almost parallel to that of the calorific power, only that the drop is not quite so rapid. Toward the end the specific gravity is slightly higher, which is due to the increase in the percentage of nitrogen. The candle power is seen to drop rapidly during the first 7 hours, then slowly to the 24th hour, when the decline is again very rapid, so that toward the end the gas has no illuminating value whatever.

Diagram 3 gives a graphical representation of the total volume of the gas obtained from 1 long ton of dry coal, its calorific power, and the relation of this to the total amount of heat required for coking. The volume of gas obtained is seen to remain almost constant during the first 24 hours ; in the 22d and 23d hours there is an increase in volume, followed by a rapid decline. The increase is due to the last thin layer of coal in the centre of the oven being heated from both sides instead of only from one side, as was the case up to that time, the coal having received its heat chiefly from the side wall contiguous to it. The sudden drop is explained, of course, by the fact that the last of the coal having been heated to the coking temperature, the volume of gas liberated diminished very quickly.

The heat value of the gas is seen to keep pace pretty evenly with the volume. Diagrams 1, 2, and 3 give all the information required about the composition, the important physical properties, and the volume of the gas produced during the entire process. Combining the results of the three diagrams, three periods in the process can be distinguished.

The first period, lasting 9 hours, is one of declining value of gas. The percentage of marsh gas diminishes; that of hydrogen grows; the calorific power is reduced from 775 to 685 B. T. U.; the specific gravity drops from 0.550 to 0.490 and the candle power from 18 to 13$\frac{1}{4}$. The second period, reaching from the 9th to the 22d hour, produces gas of almost constant quality. The percentages of marsh gas and hydrogen are very little changed; the calorific power, specific gravity, and candle power are pretty constant.

The third period is again one of declining value of gas. The percentage of marsh gas diminishes rapidly, with a corresponding increase in hydrogen; the calorific power, specific gravity, and candle power show a rapid decline.

The decrease in volume is very gradual up to the 22d hour, when (for reasons given above) there is a small increase in the 23d hour which is followed by a rapid decline.

The gas of the first period, after purification, can be advantageously used for illuminating purposes; that of the second period for heating the ovens, and that of the third period, after having been purified and enriched with benzole or oil vapors, or with oil gas, can be mixed with that from the first period. This last gas, being so rich in hydrogen, is especially adapted to serve as a carrier of hydrocarbon vapors.

HEATING WITH OVEN GAS. — A question still to be settled is, How much of the gas is required to heat the ovens? It was stated at the beginning that the test oven had been heated with the gas from the other 29 ovens of the block. The amount of gas required per charge of 13,602 pounds dry coal was 36,169 cubic feet. As its calorific power was 499.2 B. T. U., the amount of heat required was 36,169 × 499.2 = 1,805,564 B. T. U. The heat consumption of 1 long ton for the entire coking time (33 hours, 55 minutes) was therefore 2,973,680 B. T. U., or of 1 long ton for 1 hour, 87,633 B. T. U., further the heat consumption per oven per hour, 532,090 B. T. U. This is graphically represented in Diagram 3 by the shaded area. This figure show sthat up to and including the 29th hour the heat value of the gas produced is greater than that required for coking, but after

AVERAGE CALORIFIC VALUE OF
SURPLUS GAS 685 B R T.U

CALORIFIC VALUE
OF HEATING GAS.
667 B.T.U.

AVERAGE SPECIFIC GRAVITY
OF SURPLUS GAS = .512

SPECIFIC GRAVITY
HEATING GAS .421

NITROGEN IN $\frac{1}{2}$ %

B. H. U. IN FULL NUMBERS.

AVERAGE ILLUMINATING VALUE
OF SURPLUS GAS = 14.7 C P.

TOTAL GAS IN 10 CU. FT.
CANDLE POWER IN $\frac{1}{10}$, SP. GR. IN $\frac{1}{1000}$

ILLUMINATING VALUE
HEATING GAS 9.0 C. P.

900 890 880 870 860 850 840 830 820 810 800 790 780 770 760 750 740 730 720 710 700 690 680 670 660 650 640 630 620 610 600 590 580 570 560 550 540 530 520 510 500 490 480 470 460 450 440 430 420 410 400 390 380 370 360 350 340 330 320 310 300 290 280 270 260 250 240 230 220 210 200 190 180 170 160 150 140 130 120 110 100 90 80 70 60 50 40 30 20 10 0

0 ½ 1
HOURS

Surplus gas produced during 6 minutes.
    Average calorific value . 667 B. T. U.
    Average illuminating value   9.0 C. P.
    Average specific gravity of   421
    Volume per long ton, 5,150.5 per cent

that time there is a lack of heat which would have to be supplied in case the ovens were exclusively heated by coke-oven gas and blast-furnace coke was to be produced. If the coke is to be used for domestic purposes or as a boiler fuel, a softer coke is more desirable and the coking can be advantageously interrupted at the 29th hour and the amount of gas represented by the area A B C saved. With ovens 18 inches wide instead of 20¾ inches, this can be done still sooner.

Above, three periods have been distinguished in the process of coking, furnishing three different kinds of gas. In practice it simplifies the arrangement of the parts of the plant to divide the gas produced into two parts, the first, richer in marsh gas, to be surplus or illuminating; the second, the heating gas, as is done at present at Halifax. The question is, Where must the line be drawn? The total gas produced per long ton of coal was found to be 10,390 cubic feet, having a heat value of 6,501,000 B. T. U. In order to coke 1 long ton of coal 2,973,680 B. T. U. were necessary; that is, the heat value of the gas produced in the last 19 hours and 10 minutes, leaving 14 hours and 46 minutes for surplus or available illuminating gas. This line of division is shown in the three diagrams. The data tabulated give the following heat balance:

Per long ton dry coal.	Volume.		Calorific Power.	
	Cubic feet.	Per cent.	B. T. U.	Per cent.
Gas used for heating oven	5,247	50.5	2,973,680	45.8
Surplus gas	5,143	49.5	3,527,320	54.2
Total gas consumed	10,390	100.0	6,501,000	100.0

The difference in calorific power, candle power, and specific gravity of the two gases is shown by the following table:

Per long ton dry coal.	Volume. (Cubic feet.)	Calorific power. (B. T. U.)	Candle power. (C. P.)	Specific gravity. (Air = 1.)
Surplus gas, 1st fraction	5,143	685.8	14.7	0.512
Oven heating gas, 2d fraction . .	5,247	366.7	9.0	0.412
Average of entire gas produced .	10,390	626.0	11.6	0.466

Finally, the difference in chemical composition is shown by the following analyses :

	Cm Hn	CH₄	H₂	CO	CO₂	O₂	N₂	Total.
Surplus gas, 1st fraction . . .	5.2	38.7	38.4	6.1	3.6	0.3	7.7	100.0
Oven heating gas, 2d fraction .	2.4	29.2	50.5	6.3	2 2	0.3	9.1	100.0
Average of entire gas produced .	3.8	33.9	44.5	6.2	2.9	0.3	8.4	100.0

The calculation of the volume of surplus gas has been based upon dry coal, while the coals tested contained 9.9 per cent. of water. This is equivalent to 222 pounds of water per long ton of coal. The heat required to bring 222 pounds of water from 60° F. to 212° F. is 33,744 B. T. U., and that required to convert 222 pounds of water having a temperature of 212° F. into steam of 212° F. is 214,452 B. T. U., which gives a total of 248,196 B. T. U. Leaving out the amount of heat required to raise the steam of 212° F. to the temperature of the gases of the gas-escape pipe from the ovens, the heat required for evaporation of the water is equivalent to 8.3 per cent. of the total heat required for coking 1 long ton of coal, or 2,973,680 B. T. U. This shows the importance of obtaining the coal in a drier state. Five per cent. may serve as an average figure. This decrease of water will give more surplus gas and shorten the coking time.

HEATING WITH PRODUCER GAS. — The amount of gas consumed by a city at different periods varies greatly. In making gas in small retorts a larger or smaller number can be put into operation, or the fires heating them can be urged in a greater or less degree. A by-product coking oven, however, has to run uniformly all the year round, which means producing the same amount of gas. If heated exclusively with its own gas it would not be well adapted for furnishing greatly varying amounts of illuminating gas. When the consumption is small the excess of the surplus gas would have to be disposed of at any price it could bring, or go to waste. In order to make the system elastic, the ovens will have to be heated by some other fuel than coke-oven gas during the period of greatest consumption of illuminating gas, and this fuel will be found in producer gas. In times when the consumption of illuminating gas is greatest the ovens will be heated exclusively with producer gas, and the whole coke-oven gas converted

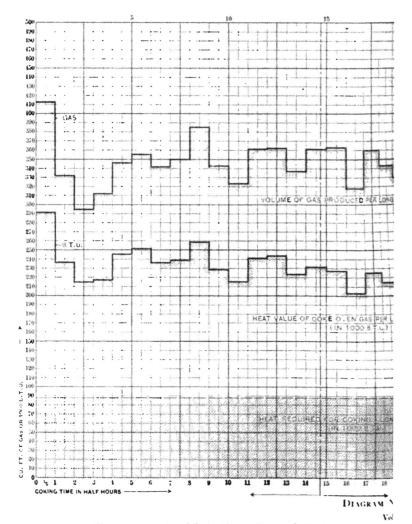

Surplus gas produced during first 14 hours, 46 minutes: 5.143
Surplus gas produced during remaining 19 hours, 10 minutes: 5.247

DIAGRAM V

Vol

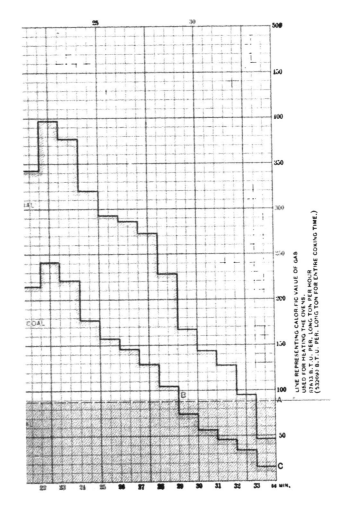

Calorific value.

9 5 per cent ; 3,527,320 B. T. U. = 54.2 per cent.
9.4 per cent; 2,973,680 B. T. U. = 45.8 per cent.

into illuminating gas ; in times of smallest consumption the ovens will be heated exclusively with coke-oven gas, illuminating gas being produced the first 14 hours of the coking period, heating gas the rest of the time. One net ton of bituminous coal furnishes 130,000 cubic feet of producer gas of a calorific power of 150 B. T. U. This figure comprises only the latent heat of the gas. The larger part of the sensible heat will be lost by freeing the gas from dust and ashes, as it is not advisable to introduce it into the checker-work of the Otto-Hoffmann oven while it still holds these particles in suspension. This loss, however, is not great, as the total heat of the gas being 100, the latent heat forms 82.8 per cent., the sensible heat 7.7 per cent., and the loss by radiation, etc., 9.5 per cent. The coke oven can be, so far as heat values are concerned, successfully heated with producer gas. As to the matter of cost, it is cheaper to heat with producer gas than with that from the coke oven. In regard to the flues and the checker-work, they will have to be larger than the size required for coke-oven gas, as producer gas contains about 60 per cent. of inert nitrogen.

The candle power of the oven gas will be reduced from 14.7 to 11.8 if all of it is collected in one holder. The additional cost of bringing it up again to the standard would more than be covered by the saving from the use of producer gas.

BY-PRODUCT COKE OVEN AND GAS HOUSE PRACTICE. — In this paper details as to cost have been intentionally omitted as being foreign to its purpose. However, by tracing the original calorific value of the coal through the products of destructive distillation a heat balance can be struck between the proposed production of illuminating gas in the by-product coke oven and the usual practice of gas works.

100 pounds dry coal yields —	B. T. U. (Per lb.)	Total calorific power. (B. T. U.)	Per cent. of calorific power of dry coal.
71.13 pounds coke	12,645	899,456	72 3
3.38 pounds tar	12,210	51,410	4.1
229.6 cubic feet surplus gas	686	157,504	12.7
234.2 cubic feet heating gas	567	132,835	10.7
Ammonia liquor, sulphur in purifier, and loss	. . . . . .	2,496	0.2
Total = 100 pounds dry coal	. . . . . .	1,243,700	100.0

The heat contained in 100 pounds dry coal is distributed as follows :	PER CENT. OF TOTAL HEAT VALUE OF COAL	
	German coal in gas retorts.[1]	International coal in Otto-Hoffmann ovens.
In coke, salable	46.4	72.3
In coke used for heating retorts	10.1	....
In tar	5.5	4.1
In gas, salable	21.0	12.7
In gas for heating ovens	....	10.7
In ammonia liquor, sulphur in purifiers, and loss . .	17.0	0.2
Total	100.0	100.0
Heat used and lost in distillation process	27.1	10.9
Heat contained in products	72.9	89.1
Total	100.0	100.0

The second table shows that in the coke oven practice 10.7 + 0.2
= 10.9 per cent. of the heat value of the coal is consumed in destructive distillation, while gas house practice requires 10.1 + 17.0 = 27.1
per cent.

Some minor details still remain to be discussed, such as the advantages alluded to above, of the premature pushing of the coke after 29
hours, instead of after 34 hours, when the coking is completely finished, and the disposal of coke breeze (use under boilers with mechanical stokers, manufacture of briquettes, recharging into coking ovens,
etc.) — questions having a direct financial bearing rather than one of
scientific interest. The purification of coke oven gas from carbon
dioxide and sulphur compounds and the enriching it to bring it up to
the standard of 20 candle power are matters of daily gas house practice, and need not be dilated upon here.

CONCLUSION. — The above summary will serve to show that from
the standpoint of chemical and thermal analysis the production of illuminating gas in by-product coke ovens is upon a very sound basis, and
much could be added to prove that this is equally true from a financial
point of view. The data have been obtained from the extensive report
of Dr. F. Schniewind, which is a model of its kind.

---

[1] Oechelhaeuser, W. von. Die Steinkohlen-Gasanstalten als Licht-, Wärme- und Kraftcentralen. Dessau, 1893

THE NORMAL CHLORINE OF THE WATER SUPPLIES OF JAMAICA.

By ELLEN H. RICHARDS, A. M., AND ARTHUR T. HOPKINS, S. B.

Read March 10, 1898.

JAMAICA, the leading English possession in the West Indies, lies 90 miles south of Cuba, and 100 miles west of San Domingo, or Hayti. Its somewhat irregular oval outline is 144 miles in length by 49 miles in width. It has an area of 4,190 square miles, or somewhat over two and a half million acres, rather less than one-sixth of this being arable, level land. This statement impresses one with the extent of the mountainous area of the island. One principal chain of mountains extends, under various names, the whole length of the island from east to west, and its numerous lateral branches, extending generally in a northerly and southerly direction, very effectually divide the land into hills and deep, narrow valleys. The island appears to be of an igneous formation, but is for the most part deeply overlaid with limestone. There are three extensive alluvial plains, all on the southerly side : that of the Black River district in the southwest ; the extensive Liguanean Plain, on which Kingston and Spanishtown are situated, in the south ; and the Plantain Garden River district in the southeast, containing the richest banana soil of the island. This Plantain Garden River valley will have a peculiar interest to us later. It is noticeable also, that while the outlines of the mountains and of the coast are bold and rugged on the northerly side, the mountains of the southerly side are more rounded and sloping, the shores gentle and shelving, and the rivers frequently end in swampy lagoons with the various formations that accompany undisturbed sedimentation.

Let us consider the factor which with the mountainous backbone determines the meteorology and climate of Jamaica, the trade winds. These winds are the " northeast trades," so called, and during the day from 9 to 5 o'clock their force and direction are very constant, being almost never interrupted, except by storms. The moisture-laden air of the Carribean Sea is driven forcibly against the hills and mountains of the eastern and northern shores ; the vapor is condensed and abundant showers result ; how abundant is indicated by the estimated rainfall for Port Antonio in 1891, nearly 200 inches, with an average for the entire island of about 70 inches. A curious variation from this rule is met with in the wind and rainfall of the Plantain Garden River district, which come from the *southeast.* A certain quantity of rain clouds come over the mountains and through Cuna Cuna Pass, but most of the rain comes from the undischarged "northeast trades" which are actually deflected by the John Crow Range into a southeasterly direction. The deflected moist air striking the cool mountain sides north and west of Bath gives up much of its rain to this district, and then passes on towards Kingston as a comparatively dry "southeast trade" wind.

As a result, then, of the topography of the country and of its position in the path of the trade winds, the eastern and northern sections have an abundant rainfall ; numerous mountain torrents and rivers, small and large, which change their depth and volume with almost every shower, and a most luxuriant and tropical vegetation. It is here, therefore, that the most varied and beautiful scenery is found ; here that the fruit plantations are most abundant and successful ; and here that sanitary problems are perhaps most interesting and fruitful. On the other hand, the Manchester and Santa Cruz Mountains in the southwest, with their elevation of three or four thousand feet, their New England-like scenery and cool, dry atmosphere, have a world-wide reputation as health resorts for the cure of pulmonary diseases ; while for those who desire a hot, dry climate there is the entire middle section, south of the central range, from Porus in the west to Yallahs River in the east. This portion is comparatively dry, except during the rainy seasons. Many of the rivers indicated on the map are nearly or quite evaporated during the dry season, obliging the inhabitants to rely largely for their water supply on vast cisterns, with the help of a few rivers which break through the mountain ; or, as in the case of Kingston, to tunnel through the mountain and tap the Agualta River. The variety of climate is convenient for the tourist, as it allows for a complete change within a few miles ; for the inhabitant it is less so, as his residence is usually fixed, by the exigencies of his occupation, within very narrow limits.

The peculiar topography of the country has another effect. On the north it limits the location of the important towns to the coast, not only on account of the heavy rains which seriously affect mountain roads, but also from the direction and the precipitous character of the mountain ranges, which force the lines of travel into the valleys, which here are usually at right angles to the coast; while in the south and west the towns are located away from the coast and on the inland plains and plateaus.

The soil is very rich, and usually of great depth in the valleys where it has accumulated from the constant scouring of the hillsides. It varies in color from a deep red to black, is rich in lime, iron, and phosphates, and is often somewhat stiff and clayey.

The air is exceptionally free from dust on the northern side, as might be expected from such a rain-washed atmosphere, and the sky is very blue; but on the dryer side there is naturally more dust, and in Kingston and on the Liguanean Plain there are frequent dust storms, only comparable to those of Boston in the height of the season. What effect the vast amount of decaying vegetation might have on the air, were there little or no atmospheric circulation, it is difficult to estimate; but the constant trades during eight or ten hours of the day, and the "doctor," as the reverse "land breeze" of evening is called, provide for constant and full oxidation of the organic wastes. As to the malarious exhalations of the swamps and low marshy lands, it is difficult to speak authoritatively, but the experience of the people leads them to avoid the evening dews and damps, and to aim at a residence as far above the level as circumstances or convenience will allow. Many parts of the island are comparatively free from malaria, notably the Manchester and Santa Cruz Mountains, before mentioned, and not every one who lives where the disease is prevalent is affected, but most people suffer to a certain extent who remain sufficiently long It does seem, however, that those whose vital forces are strongest, who are well nourished and clothed, and who avoid chill and dampness, are most immune to this, as to any other disease.

WATER.

In a chemical analysis of water there are three principal factors: first, nitrogen in the four forms of free ammonia, albuminoid ammonia (indicating the organic nitrogen), nitrites, and nitrates; second, lime and magnesia; and third, chlorine.

The nitrogen may come in the form of albuminoid ammonia from the living microscopic or visible organisms existing in the water, or from decaying vegetable matter, or in the form of free ammonia and nitrites from decaying animal matter or from animal or household wastes, and may vary from organic ammonia, suspicious ammonia, and nitrite to the inorganic and innocuous nitrate. These variations depend simply on the degree of oxidation of the organic matter, the ammonia being the first product of decay. Further, nitrogen is essential to plant life, and this cycle of change may be interrupted at any point and the nitrogen withdrawn from circulation. Nitrogen may or may not, therefore, be an exact indication of the extent of contamination.

The amount of lime and magnesia determines the degree of hardness, but does not necessarily affect the quality of the water except when excessive, that is, in a limestone region ; in a granitic country presence of lime and magnesia salts might indicate contamination from cesspools and the like.

In chlorine we have a factor of greatest importance in sanitary interpretation, when intelligently used. Common salt, of which chlorine is the active principle, is a necessity for man and beast. But little of it is retained in the system, however, and the chlorine, leaving the body with other wastes, is not absorbed by vegetation as are the nitrates, but being very soluble in all ordinary compounds, when once in the water is not lost, but remains there as a tell-tale of a most useful character. This demand for salt is so universal and the amount consumed is so constant, that the quantity of chlorine found in a sample of water becomes an excellent index of the degree of contamination, *provided* we know the amount of salt rightfully belonging to that locality, coming from that great repository of chlorine, the ocean. The importance of this knowledge of the normal amount of chlorine for a locality may be easily illustrated. The normal amount for low, sandy, ocean-washed Nantucket is 2.16 whole parts per 100,000 parts of water, while North Adams in the western part of the State, 140 miles from the coast, has a normal of .06 parts per 100,000. Now the last published analyses for the water of the Hoosac River at North Adams give about .40 parts per 100,000. The sewage-polluted Hoosac River has, then, but one-fifth as much chlorine as the well-filtered and perfectly safe water of Nantucket, and any one unacquainted with the facts would, on the basis of the chlorine, give an absolutely erroneous decision. The State Map of Normal Chlorine, however, shows us that the chlorine of Hoosac River is seven times the normal for that locality, and the question is at once decisively settled. The great importance of the normal chlorine chart to the sanitarian will be evident from the foregoing. The information upon which was based the first normal chlorine map of Massachusetts included thousands of analyses, made during two years and cost thousands of dollars. This large expense was necessitated by the widely scattered population in Massachusetts and consequent variation of the contamination of the waters. A very limited number of analyses for a single locality might have answered, had the region been uninhabited and uncontaminated.

In Jamaica the circumstances which, as before described, limited the towns to a narrow line of coast, leaving the interior but sparsely inhabited, made it possible to obtain what may be considered a fairly accurate chart, at least for the northern side of the island, with but seventy-seven samples of water, mostly taken within a few miles of the coast. The water taken from rivers at a distance of one mile from the coast was usually as pure as if taken many miles inland. Even had it been otherwise we should have attacked the problem, as it was probably the first attempt at such an investigation of a large island, and there was hope that with these unusually favorable surroundings we might be able to obtain some light as to the principles underlying certain features of the normal chlorine map of Massachusetts.

EQUIPMENT.

Through the kindness of the Boston Fruit Company we were furnished with two carriages and mules to transport apparatus and luggage. A dozen four-liter demijohns, and six dozen quart bottles (in crated cases) carefully sterilized, and packed to avoid breakage, were carried, as well as a bacteriological outfit.

Our plan was to circle the island on the main road, to bottle a sample of the purest river or brook water obtainable at intervals of five to ten miles, and to give a full description of the appearance and surroundings of each sample, with our impression as to its purity; and besides to obtain, where possible, the cistern rain waters, which would give the chlorine in the rain and atmosphere only, except as affected by the cement linings of ground cisterns. The case when filled with samples was sent to Boston. So successfully was this collection done that but one sample had to be completely rejected as being mixed with sea water. That particular sample was accompanied by the notation that it was "possibly brackish," and was bottled because it was the only one obtainable in several miles. Some few showed evidence of contamination, but in each case, fortunately, we had an undoubtedly pure brook or river water in the near vicinity to check with. The analyses were made entirely at the Institute of Technology, and the results check admirably with our knowledge of the surroundings. It is another evidence of the truth of the saying that "the sanitary engineer *must* collect his own samples." Had these samples been taken by an inexperienced person, or had they been submitted with no data as to localities and surroundings, the results would in some cases have been entirely uninterpretable.

ANALYSES.

The exhaustiveness of the analyses depends, primarily, on the amount of water available. The quart bottle samples were tested for turbidity, sediment, color, hardness, chlorine, nitrogen as nitrate, and in a few cases for the ammonias. The demijohn samples were given a further examination for nitrogen as ammonia and nitrite and for iron, while some five samples were further analyzed for silica, sulphuric acid, aluminum, calcium, and magnesium. Then on an outline tracing of the map of Jamaica the results of the analyses for chlorine were written in, on the spot from which the sample was obtained, and iso-chlors, or lines of equal chlorine, drawn with due allowance for evident contamination from adjacent villages and from salt springs, of which there are several. The chlorine of cistern waters cannot be used directly as a basis for normal chlorine, being always too low. Cistern water, generally speaking, contains only the chlorine from the rain and from its own particular watershed, generally a roof, and is but little concentrated by evaporation. Pure river water chlorine under the usual climatic conditions generally represents that which is normal to the locality after the usual evaporation has been effected.

While strict accuracy cannot be claimed (for this would require repeated examinations of a wider range of territory, extending over a considerable period of time and during the different seasons of the year), we are satisfied as to the general reliability of the results of our studies.

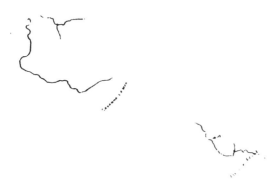

THE NORMAL CHLORINE CHART OF JAMAICA, JANUARY, 1898.

The places where the samples described in the tables were taken are indicated by the corresponding numbers on the chart. The decimals upon the iso-chlors show the parts of chlorine per 100,000 parts of water.

I. TABLE OF ANALYSES.

| Number | Appearance | | | Hardness | Ammonia | | Chlorine | Nitrogen as nitrates | Nitrogen as nitrites |
	Turbidity.	Sediment.	Color.		Free.	Albuminoid.			
1	None.	Slight sandy.	0.00	21.8	.0000	.0000	1.53	.0800	.0000
2	None.	Slight sandy.	0.00	14.4	.0000	.0800	.79	.0050	.0801
3	Slight green.	Considerable green.	0.05	92.0	.0012	.0044	7.48	.1600	.0006
4	Very slight.	Considerable rusty.	0.12	24.5	.0052	.0106	1.90	.0050	.0080
5	None.	Slight rusty.	0.05	6.4	.0000	.0000	.72	.0050	.0081
6	Slight.	Considerable earthy and rusty.	0.03	18.2	.0000	.0016	.70	.0280	.0001
7	Very slight.	Slight.	0.03	8.3	....	....	.51	.0050	....
8	Very slight.	Slight.	0.03	10.1	....	....	.77	.0050	....
9	Very slight.	Slight.	0.05	10.1	....	....	.80	.0050	....
10	Very slight.	Slight.	0.12	16.9	....	....	1.16	.0050	....
11	Very slight.	Slight.	0.05	13.3	....	....	.87	.0050	....
12	Very slight.	Slight rusty.	0.03	33.4	....	....	3.39	.0050	....
13	None.	Slight.	0.03	19.9	.0000	.0080	1.87	.0070	....
14	None.	Slight.	0.04	12.3	....	....	1.31	.0070	....
15	None.	Slight.	0.03	16.4	....	....	.89	.0600	....
16	None.	Slight sandy.	0.01	12.4	....	....	.84	.0450	....
17	None.	Slight sandy.	0.01	13.3	....	....	.79	.0620	....
18	None.	Slight earthy, cyclops.	1.30	2.2	....	....	.46	.0120	....
19	Very slight.	Considerable earthy and rusty.	0.00	18.3	.0000	.0000	.71	.0700	.0000
20	Slight clayey.	Considerable clayey and sandy.	0.00	13.6	....	....	2.87	.0400	....
21	Slight clayey.	Considerable clayey.	0.40	12.3	....	....	3.30	.0200	....
22	None.	Very slight.	0.00	18.0	.0008	.0000	.95	.0400	....
23	Very slight.	Slight organisms.	0.00	18.6	....	....	1.65	.0400	....
24	Very slight.	Slight rusty.	0.10	19.9	....	....	1.79	.0280	....
25	Very slight.	Slight sandy.	0.05	16.7	....	....	1.01	.0220	....
26	None.	Very slight.	0.00	23.0	....	....	.96	.0550	....
27	None.	Very slight.	0.08	27.2	....	....	1.12	.0320	....
28	Very slight.	Slight.	0.04	16.9	....	....	.84	.0150	....
29	None.	Slight.	0.01	12.1	....	....	.74	.0200	....
30	Very slight.	Slight.	0.02	14.2	....	....	3.05	.0070	....
31	Very slight.	Slight rusty.	0.03	74.0	....	....	135.0	.0050	....
32	None.	Very slight.	0.01	18.3	.0000	.0000	1.56	.0500	....
33	None.	Slight gelatinous.	0.00	19.6	....	....	3.70	.0850	....
34	None.	Slight rusty.	0.30	16.9	....	....	2.09	.0070	....

I. TABLE OF ANALYSES. — *Continued.*

Number.	Appearance.			Hardness.	Ammonia.		Chlorine.	Nitrogen as nitrates.	Nitrogen as nitrites.
	Turbidity.	Sediment.	Color.		Free.	Albuminoid.			
35	None.	Very slight.	0.00	16.7	....	....	.45	.0170	....
36	None.	Very slight.	0.00	22.7	....	....	.52	.0250	....
37	None.	Slight.	0.18	22.0	....	....	.67	.0050	....
38	None.	Slight.	0.05	19.6	....	....	.51	.0050	....
39	None.	Very slight.	0.00	21.2	.0016	.0012	.47	.0220	....
40	None.	Slight rusty.	0.07	22.9	....	....	.90	.0400	....
41	None.	Very slight.	0.01	18.2	....	....	.80	.0400	....
42	None.	Slight.	0.01	14.8	....	....	.50	.0400	....
43	None.	Slight.	0.01	14.8	....	....	.67	.0400	....
44	None.	Slight.	0.05	20.7	....	....	6.00	0.170	....
45	None.	Slight rusty.	-0.20	17.9	....	....	.37	.0500	....
46	None.	Very slight.	0.00	22.0	....	....	.75	.0650	....
47	None.	Very slight.	0.00	12.7	....	....	.37	.0400	....
48	None.	Slight.	0.03	16.3	....	....	.51	.0400	....
49	None.	Very slight rusty.	0.33	5.4	.0012	.0040	.14	.0200	....
50	None.	Slight rusty.	0.42	9.1	.0000	.0156	.32	.0200	....
51	Distinct milky.	Slight rusty.	0.60	5.0	.0000	.0042	.17	.0050	....
52	None.	Very slight gelatinous.	0.50	4.9	....	....	.19	.0350	....
53	None.	Considerable rusty.	0.05	8.0	.0000	.0128	.67	.1000	....
54	None.	Slight rusty.	0.07	18.5	....	....	.57	.0070	....
55	None.	Very slight.	0.05	16.1	....	....	.65	.0300	....
56	None.	Very slight rusty.	0.80	6.3	....	....	.40	.0250	....
57	None.	Slight.	0.00	11.9	.0000	.0000	.87	.0050	....
58	None.	Very slight.	0.00	19.9	....	....	1.80	.0050	....
59	Distinct milky.	Slight clayey.	0.20	20.2	....	....	.89	.0580	....
60	None.	Slight.	0.05	15.0	....	....	.89	.0320	....
60a	None.	Slight.	0.00	14.9	.0000	.0012	.88	.0250	.0001
61	None.	Slight.	0.02	10.00	.0000	.0000	.32	.0050	.0000
62	None.	None.	0.02	12.90	.0000	.0000	.37	.0030	.0000
63	None.	Very slight.	0.07	13.6	....	....	.72	.0030	....
64	None.	Very slight.	0.02	59.5	....	....	2.50	.0030	....
65	None.	Very slight.	0.00	11.9	....	....	.60	.0050	....
66	None.	Slight.	0.05	22.2	.0034	.0016	1.12	.0070	....
67	None.	Very slight.	0.02	30.0	....	....	2.90	.0420	....

I. TABLE OF ANALYSES. — *Concluded.*

Number.	Appearance.			Hardness.	Ammonia.		Chlorine.	Nitrogen as nitrates.	Nitrogen as nitrites.
	Turbidity.	Sediment.	Color.		Free.	Albuminoid.			
68	None.	Slight.	0.03	21.7	....	....	1.47	.0500	....
69	Very slight.	Slight earthy and rusty.	0.03	19.1	.0000	.0056	1.35	·0330	.0005
70	None.	Very slight.	0.08	19.6	....	....	1.16	.0520	....
71	None.	Slight.	0.30	12.4	....	....	.67	.0090	....
72	None.	Very slight.	0.03	26.6	.0000	.0004	1.22	.0090	.0000
73	None.	Slight.	0.03	15.0	....	....	.97	.0100	....
74	None.	Slight.	0.00	15.9	....	....	.61	.0070	....
75	None.	Very slight.	0.03	21.7	....	....	.86	.0230	....
76	Very slight.	Considerable earthy and rusty.	0.00	18.3	.0000	.0000	.71	.0700	.0000

II. TABLE OF REMARKS.

Number of sample.   Source.

1. From town supply; Morant Bay; reservoir well located.
2. Irrigating canal from Johnson River; one mile inland.
3. Well 15 feet diameter, 22 feet deep; surface water; slight scum on top; 100 feet from sea.
4. Phillips field; well located reservoir; heavy growth on shore; three miles inland.
5. Town supply; Port Antonio; reservoir small; peasant houses three hundred yards above.
6. Wharf supply; reservoir larger; no houses nearer than five hundred yards.
7. Swift river; deep and rapid; rises ten miles away in mountains; one and a half miles inland.
8. Buff Bay River; deep water; river rises fifteen miles back; one mile inland.
9. Buff Bay River; three miles above (8).
10. Dry River; low water; rises twenty miles distant; few villages on banks; one mile inland.
11. Agualta River; one mile inland; under bridge women washing clothing; rises fifteen miles inland.
12. Crawl Pond Spring; three miles inland; taken above fording.
13. Town supply; Port Maria; reservoir fairly well protected.
14. Sambo River; near sea; few settlements on river.
15. Bungalo River; near shore.
16. Roaring River; one-half mile inland.
17. Landovery River; near shore; a mountain stream.
18. Brownstown; cistern water from roof.
19. Dornack River Head; a sink hole; water comes underground several miles.
20. Roseall River (Little River); taken near coast; only one mile in length.
21. White Gut; small brook near coast.

Number of sample.	Source.
22.	Town supply, Montego Bay; from Barnet River.
23.	Small stream; one mile from shore on Arwin estate.
24.	Barnet River; taken under bridge; villages near; rises in the mountains.
25.	Jerry Duncan River; rises in interior; taken one mile inland.
26.	Redding River; a mountainous stream near coast.
27.	Great River; near shore; still water; perhaps some chlorine.
28.	Flint River; a large stream; houses near on banks; comes great distance.
29.	Brook on Tryall estate; used for water power; probably unpolluted.
30.	Mr. Davis's brook; a small stream one mile from coast.
31.	Kew Bridge; "backwater" near coast; possibly brackish.
32.	Town supply, Lucea; from Prospect; two miles inland.
33.	Great well; unfavorable surroundings.
34.	Morgan's Brook; small; probably no contamination.
35.	Picket's River; a deep stream; rises up country.
36.	Town supply; Savanna La Mar; from spring five miles distant.
37.	Smithfield Brook; small shallow stream one and a half miles inland.
38.	Lochiel Spring; small brook.
39.	Deans Valley River; fair size; but few inhabitants.
40.	Cave Brook; small; comes from back country; near coast.
41.	Clear running brook.
42.	Bluefield River; rapid brook from hills.
43.	Akindown River; comes from foot hills through estate; cattle near; washing place.
44.	Scott's Cove Brook; small flowing stream; clear; comes from hills; no houses near; cattle abundant.
45.	Black River; tap; reservoir six miles from coast; good conditions.
46.	Spring Head; a small spring brook.
47.	Shaw's River; deep, quietly flowing; villages on banks.
48.	Black River at Lacovia; deep river; many villages on banks.
49.	Santa Cruz; cistern water from roof of constabulary station; cedar shingle roof.
50.	Elginton; water from cemented limestone cistern nine feet deep, ten feet diameter.
51.	Spur Tree Pen; cistern forty by one hundred feet; twenty feet deep; water flows down hillside; cattle on hills.
52.	Mandeville; water from concreted cistern twenty feet diameter, fourteen feet deep; Mrs. Halliday's lodgings.
53.	Williamsfield; cistern water used on railway; contaminated by street flowage.
54.	Belvidere; small brook across road; comes from hills; probably uncontaminated.
55.	Clarendon Park; small running brook.
56.	Four Paths; cistern water at constabulary station; rain from roof.
57.	May Pen; town supply; no particular care of reservoir.
58.	Old Harbour; town supply from mountain springs.
59.	Dry Harbour; water from Irrigation Canal.
60.	Spanishtown Irrigation Canal; Rio Cobre River.
60a.	Town supply at Spanishtown; water from Rio Cobre River; fairly good conditions.
61.	Constant Springs Hotel; water from tunnel to Wag Water River; good conditions.
	Town supply of Kingston at Myrtle Bank, from Wag Water River.
62.	Cane River near mountains; no settlement visible.
63.	Chalk River; small milky white stream from mountains; no apparent contamination; limestone.

Number of sample.	Source.
65.	Albion River from mountains; very little contamination.
66.	Buff Bay; well in yard; dug in limestone; low ground; contamination of surface water.
67.	Mundicott River; contaminated by washing, etc.; fairly large river.
68.	White Horses; brook comes over cliffside; uncontaminated.
69.	Morant estate; small reservoir; cattle about.
70.	Small brook from hills; probably very good.
71.	Devil's River; from the mountains; fair.
72.	Swift's Spring, Golden Grove; six inches deep, four feet diameter; exposed to surface contamination.
73.	Small river from mountains; somewhat muddy.
74.	Muirton River; large; from mountains; very fair.
75.	Mulatto River; small running stream from mountains; good.
76.	Falmouth; town supply from river two miles inland.

INTERPRETATION.

The iso-chlors drawn on the map are those for two, one, nine-tenths, eight-tenths, five-tenths, and three-tenths parts per hundred thousand. The first thing noticeable is that the iso-chlors for two whole parts per hundred thousand begin and end in the seacoast and inclose rather small areas, that those for one and nine-tenths parts are similar but of greater extent, while the remaining iso-chlor areas cover the central portions of the island, more or less evenly spaced between the northern and southern coasts.

There are three distinct areas covered by the iso-chlors for two parts per hundred thousand, the districts about Rio Nuevo, Little River, and Mosquito Cove. Now, as a matter of fact, each of these three districts consists of a more or less narrow plain backed by a wall of mountains extending to the sea on either side, which effectually shuts in the rain and salt spray from the Carribean Sea blown against that part of the coast. Consequently we have, condensed in a small area, an amount of salt which but for the mountain barrier would be spread over a territory perhaps several miles in depth and with a lower normal chlorine.

It will be noticed that the Port Morant and Plantain Garden River district has an isolated iso-chlor for one part per hundred thousand; this results from the diversion of the northeast trade by the John Crow Mountains previously mentioned, and is a necessary accompaniment to the generous rains which have made this limited district so important in the sugar cultivation of the past and the banana plantations of the present.

We have already mentioned the curious and important fact of the generally equal spacing of the iso-chlors for eight-tenths parts and less, with respect to the northern and southern coasts. This was, we confess, entirely unexpected by us. We had looked to find a high chlorine along the northern coast, somewhat greater than on Cape Cod, because of the greater degree of concentration of the ocean in that latitude. Our disappointment in this regard was probably due to the general moderation of the winds in the Carribean Sea, for investigations in Massachusetts indicate that the amount of chlorine carried depends on the force of the winds bearing inland the salt spray from the coasts. We had also expected to find a very low chlorine on the southern side generally, reasoning that the high mountain wall which shut out so much of the rain would also prevent the entrance of chlorine. Why, then, this discrepancy between theory and fact? It is due to the neglect of a very important factor obtaining in tropical countries having excessive precipitation and evaporation. This factor we may term "the dilution and concentration of chlorine." The rainfall for the northern and eastern districts is very great, ranging from 100 to nearly 200 inches in a year (the showers coming at frequent intervals, but being especially concentrated in the May and October seasons), and this gives a large amount of water to a given amount of chlorine. It must be remembered that the chlorine should be a very constant factor, being directly dependent on the proverbially constant trade winds. This large if somewhat variable amount of water, then, dilutes the chlorine to an apparently small quantity, although even here the excessive evaporation serves to concentrate the chlorine in river or surface water to an amount considerably in excess of that in cistern (or rain) waters. On the dry side of the mountains we have the reverse conditions, the amount of rainfall is so moderate, about one-quarter to one-third that of the weather side, and it is further so reduced by evaporation that there is a considerable concentration of the chlorine, giving the results referred to.

This theory tallies favorably with the results obtained from certain cistern waters in both regions, north and south. Rain water contains only the chlorine in the atmosphere, and when it is collected in uncontaminated and non-evaporating cisterns, represents this atmospheric chlorine very exactly. The cistern water at Brownstown in the north contains .46 parts of chlorine, about two-thirds that of the near-by uncontaminated Dornack River. The cistern water at Mandeville just over the mountains to the south shows .18 parts of chlorine, while the nearest uncontaminated brooks show .57 and .65 parts respectively. In short, while the cistern water on the weather side showed nearly three times as much chlorine as that from behind the mountains, the surface waters were almost alike.

The case of Port Antonio is a curious one. It has, perhaps, the greatest rainfall of any place in the island, is surrounded by mountains, and is directly in the path of the trade winds, yet its chlorine is very moderate, coming in the third group. It is easily understood, however, when the facts are considered. Two supplies were examined : one, the general town supply; the other, that belonging to the fruit company and supplying the ice factory and the ships. Both waters come from streams in the mountains back of the town, at an elevation of some 1,500 feet. Both reservoirs have an excessive supply of water, fully as much water running to waste as is consumed. One reservoir is probably above the level where salt spray might reach it ; the other is possibly within its reach during heavy blows. Chlorine simply does not reach these waters to any extent except during unusual weather. A slight preponderance of chlorine in the water from the town supply over the other is due to the people living on its watershed. This reminds us of Mr. Stearns's statement, that an increase of 28 persons per square mile in any section would raise the chlorine for that section one-tenth part. It is an interesting fact that on these two adjoining watersheds, each containing about one square mile, there are probably two dozen more persons living on one than on the other, while the actual difference in chlorine is .07, nearly a tenth.

Our Normal Chlorine Map for Jamaica has an important point of similarity with that for Massachusetts, in the corresponding distances of the iso-chlors from the coast. Our lowest line, .30, and probably the one most free from disturbing local influences, varies between ten and twenty miles from the coast ; that for Massachusetts is about twenty miles inland, with the advantage of a smoother topography. The iso-chlor for .80 is at an average distance of about five miles inland, closely corresponding with the Massachusetts line. This is valuable corroboration of the belief that the amount of chlorine varies inversely with the distance from the coast.

In view of these facts, we think the Normal Chlorine Map of Jamaica may justly be regarded as an interesting contribution to the knowledge on this important subject. Not only do its correspondences with the map for Massachusetts confirm the theories of the distribution of chlorine, but in an even more marked degree are its differences a witness to the correctness of those conclusions.

The analyses for hardness, nitrates, and color follow in a general way those for chlorine, with, however, higher figures for the leeward side. It is reasonable that there should be more lime in rivers with a slight current and little water, and that such rivers should be more favorable to the growth of· algæ, which tends to increase the nitrogen and color. The greatest degree of hardness obtained was, naturally, at "Chalky River," a river milky white with lime. The temporary hardness equalled or exceeded the permanent for most of the waters examined. The highest color was found in cistern waters, most of the river waters having almost none.

A word as to the cisterns. Those in rainy districts were mostly of wood attached to the roof gutters by spouts, but those in the vicinity of Mandeville, where great capacity is required, were dug and cemented cisterns, 20 to 50 feet in diameter and 20 feet deep, sometimes connected with roofs, but more frequently so built that the hillside could be used as a watershed. One such cistern was rectangular, 40 by 150 feet and 20 feet deep, situated at the foot of a considerable slope, and having the ground covered with cement for over 100 feet around. We have examined a few wells also in the course of the investigation, but the less said of them the better.

This investigation has included, with two or three exceptions, all the public supplies of the island, all the rivers of any importance, and a large proportion of the smaller rivers and brooks. Had it been possible, we would have obtained several samples from each river, at increasing distances from the coast. This task we must leave for some future occasion. Seventy-seven samples in all were collected, and four hundred miles traveled over in carriages.

Of the public supplies examined, none could be absolutely condemned on the evidence of but one analysis and without a careful inspection of the reservoirs and watersheds. It should be stated, however, that we found several waters, mostly on the south or dry side, suspiciously high in chlorine and in nitrites. Some few of the brooks and rivers, as also the wells examined, showed evidence of contamination, but most of the waters were really very good. A serious attempt was made to check the chemical analyses by bacteriological analyses made on the spot. Experience quickly showed the impracticability of this on account of the very limited period of time available, and the lack of ice.

AN IMPROVED FILTER FOR MICROSCOPICAL WATER ANALYSIS.

` BY DANIEL D. JACKSON, S. B.

THE greatest percentage of error in the microscopical analysis of water by the Sedgwick-Rafter method is found in the examination of those waters which contain a large number of delicate Protozoa. There is hardly a surface water which does not contain some of these organisms, and the plankton of some waters is often very largely made up of them.

A careful study of the method of analysis will point out the reason for this great loss in Protozoa. The measured quantity of water passes entirely through the sand, and very little moisture is left. This is quite sufficient in itself to break up a large percentage of the more delicate forms in water, but the organisms are then washed down into the tube by water of a different specific gravity (often distilled water), and the result is that some of the microscopic growths which survive the partial drying process on the filter are collapsed by the change of media. Occasionally when a water is swarming with some delicate form of Protozoa, only a few, or perhaps none at all, are discovered after filtration.

This very serious difficulty is now obviated by an attachment to the filter funnel which holds back a definite quantity of the original water in the funnel. No change in the method of analysis is necessary, and the original filter funnel, as devised by the author and described in a previous article,[1] is used, but to the bottom of this funnel is fitted an attachment by means of a ground glass connection.

The attachment consists of a prolongation of the filter tube, and is closed at the bottom by means of a solid rubber stopper. A smaller tube is connected with the main tube by a T joint, and rises just to the level of the six-cubic-centimeter mark on the funnel. This tube is

---

[1] *Technology Quarterly*, Vol. IX, No. 4, Dec., 1896.

in consequence of such a height that, when the solid rubber stopper is in place, all the water may filter through the sand except five cubic centimeters.

The ground glass on the bottom of the funnel has a well-marked upper border, and the perforated rubber stopper (with a wet pellet of No. 20 silk bolting cloth over the upper opening) is inserted into the bottom of the funnel up to this mark. Then exactly two cubic centimeters of sand is measured out and poured into the top of the funnel. If the sand is 60-120 mesh Berkshire [1] (quartz), it will hold one cubic centimeter of water, and therefore, when covered with water will displace exactly one cubic centimeter. Hence, from the top of the rubber stopper to the mark corresponding to the top of the small upright tube, the capacity of the tube should be exactly six cubic centimeters.

The perforated rubber stopper must be of such a shape that the attachment may be put on and taken off quickly without interference. The bottom of the attachment is open so that the solid rubber cork which fits into it may be left out, if desired, until the latter part of the operation, thus gaining in the rapidity of filtration. Some of the filtrate may be used to wash down the sides of the funnel after the water has fallen sufficiently in height.

Without suction the process of direct filtration requires half an hour. Filtration without suction, but with the solid rubber stopper in place, requires over an hour, but the latter part of the process will take much less time if slight suction be applied by means of the mouth and an intercepting bottle.

The dimensions of the filter funnel as described in a previous article are as follows: "The inside diameter of the top of the funnel is 2 inches, the distance from the top to the beginning of the slope is 9 inches, and the length of the slope is 3 inches. The tube of small bore at the bottom is $2\frac{1}{2}$ inches long and $\frac{1}{2}$ an inch in inside diameter." [2]

The illustration accompanying this article shows a battery of filter funnels, and, in front, a funnel with the new attachment at the bottom. The main part of the attachment is $1\frac{1}{2}$ inches long, $\frac{1}{2}$ an inch

---

[1] Berkshire Glass Sand Co., Chester, Massachusetts.

[2] The most satisfactory funnels, with or without the attachment, have been obtained from Richards & Co., 30 East 18th Street, New York City.

FILTER FUNNELS FOR SEDGWICK-RAFTER METHOD.

in inside diameter, and has a ground glass connection with the filter funnel. The small upright tube is $\frac{1}{4}$ of an inch in bore and is joined to the main tube $\frac{3}{4}$ of an inch below the top.

When the funnel is ready and the attachment is in place, 250 cubic centimeters of the water to be examined is poured into the top of the funnel. Filtration is rapid at first, but toward the end it becomes very slow, and then slight suction is applied to the small upright tube and the water drawn down till the bottom of the meniscus just reaches the six-cubic-centimeter mark. Just before the analyst is ready to examine the concentrated water with the microscope, the attachment is quickly removed, and a wide test tube is placed beneath the perforated rubber stopper. This stopper is then carefully withdrawn from the bottom of the funnel, and all of the sand and water is allowed to run into the test tube. The test tube is then shaken by a rotary motion in order to distribute the organisms equally. Then the sand is allowed to drop to the bottom of the tube suddenly, and the water, without the sand, is quickly decanted into another smaller test tube from which, after careful mixing, one cubic centimeter is taken for examination, as described in previous articles upon the subject.

Very careful experiments have been performed upon the use of the centrifuge in the microscopical examination of water, and the results obtained in this manner first pointed out the large error in the estimation of Protozoa by the Sedgwick-Rafter method. The centrifuge may be used with a considerable degree of success in the volumetric determination of plankton in water for the fish industry, but its use in the accurate qualitative and quantitative microscopical examination of drinking water is open to serious objection. The chief errors by this method come from the matting together of the material to be examined, thus preventing even distribution on the slide, and from the fact that the Cyanophyceæ and also other microscopic organisms of low specific gravity are precipitated only in small percentages.

Exhaustive experiments have been carried on also to find a filtering material which would replace the quartz sand. Occasionally a water contains a considerable number of very minute forms, hardly larger than bacteria, and quite a large percentage of these will pass through the sand filter. In order that the microscopic organisms may be properly collected for examination a very small filtering area must be used. It has been found that pellets of such filtering material as is used in the Pasteur and Berkefeld filters will hold these forms back, but will filter so slowly, even with strong suction, that their use is entirely impractical. When these very minute organisms are found in the filtrate, they may be concentrated by means of the small Berkefeld filter using suction, and the figures thus obtained from the filtrate may be added to the analysis by the Sedgwick-Rafter method. If these minute forms are in sufficient numbers, they may with at least equal accuracy be examined directly from the original sample of water.

The following table gives a partial analysis of a representative surface drinking water, in which the results of filtration with and without the attachment are compared. Only those forms are given which might be affected by the use of the attachment. It will be seen that the figures corresponding to the more hardy forms of Protozoa are not increased by its use.

	Without attachment. No. per cc.	With attachment. No. per cc.
CHLOROPHYCEÆ.		
Zoospores	8	56
PROTOZOA.		
Mallomonas	4	4
Peridinium	2	2
Dinobryon	16	16
Bursaria	0	?
Halteria	0	?
Cryptomonas	8	40
ROTIFERA.		
Asplanchna	1	2

A comparison of results with and without the attachment to the filter funnel shows practically no change in the number of Diatomaceæ, Chlorophyceæ (except in zoospores), Cyanophyceæ, Fungi or Crustacea, but in the Rotifera, and especially in the case of the delicate forms of Protozoa, a decided increase in number is noted. In other words, the hardy microscopic organisms, or forms not easily destroyed, have given little error in the analysis without the attachment, but the more delicate forms, most of which are included under the Protozoa, require that the attachment should be used in order that an accurate analysis may be obtained.

It will be seen also that the error due to an unknown quantity of water left in the sand after filtration is entirely eliminated by the use of the attachment.

MT. PROSPECT LABORATORY,
    Brooklyn, N. Y.

THE ACETYLENE STANDARD OF LIGHT.

BY H. E. CLIFFORD AND J. S. SMYSER.

IN an investigation of some of the photometric properties of the acetylene gas flame recently carried on at the Institute, results were obtained which will be briefly stated, together with an explanation of the diagram of the apparatus used. The diagram shows merely the arrangement of the apparatus, and is not a scale drawing.

The generator, *d*, Figure 1, is made of heavy galvanized iron. Calcium carbide is introduced in the form of large lumps, and the top is then closed by a cast-iron plate *a*, faced smooth, and having a rubber gasket between it and the top of the generator. The plate is held in place by means of a wrought-iron yoke *b*, through the top of which is a screw bolt *c*, which forcés the gasket tight against the plate and the top of the generator, thus preventing any leak of gas at this point.

The generator stands in a galvanized iron cylindrical reservoir *e*, and is surrounded by water up to within three or four inches of the top, means being provided for keeping a circulation of .cool water through this outer reservoir, so that the heat which is developed within the generator can be carried off easily. A small glass reservoir *f* is attached to the top of a pipe *g*, leading to the generator, provided with a stop-cock *h* and a tube *i*, which connects the space above the surface of the water within it with that below, thus insuring a flow whenever it is needed. A piece of rubber tubing *j* forms part of this by-pass, so that it can be closed by a pinch-cock *k* when it is necessary to refill the reservoir.

Another pipe *l*, leading from the generator, branches, one branch *m* leading to a gas holder *o* of about 3 cubic feet capacity, the other *n* to the air outside the building, so that in case the gas is generated too rapidly a part of it can be sent out of doors. A weight *p* is arranged to balance the movable cylinder *q* of the gas holder, and serves to keep it concentric with the other cylindrical parts of the holder, in which it is free to move in a vertical direction.

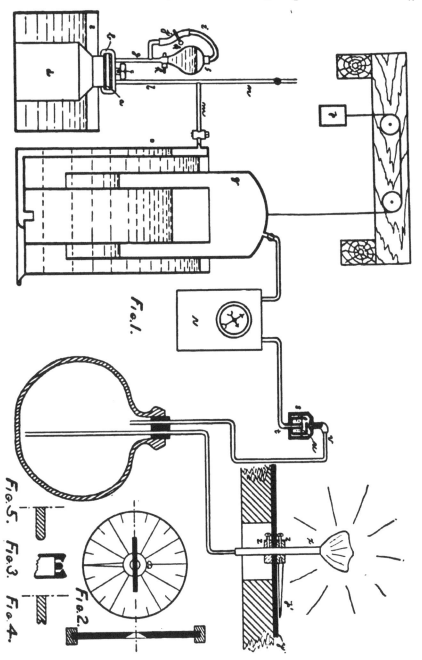

Fig. 1.

Fig. 5. Fig. 3. Fig. 4. Fig. 2.

From the gas holder connection is made by rubber tubing and lead pipe to a one-light acetylene gas meter *r*; and from here to a small rheometer or pressure regulator *s*, shown in section. This is merely a small brass reservoir into which the pipe *t* projects, and which is partly filled with oil. Within this space and resting in the oil is the regulator proper *u*, which resembles an inverted gun cap. Near the top there is a small hole through which the gas passes into the outer space. The top is conical, and is directly below the opening leading from the outer space. When the gas is supplied at a certain rate it passes through this cap-shaped piece, and then out through the pipe *v*. If the pressure increases, the cap rises and partly closes the exit pipe.

Any such regulator as this, however, would tend to produce fluctuations in pressure, even though the average pressure were maintained constant. To reduce as much as possible any such fluctuations before the gas is burned, it is passed into the top of a large empty glass carboy *w*, and then conducted through a pipe leading from the bottom of the carboy to a Bray 1-foot acetylene burner *x*. The latter is fastened to the end of a short piece of brass tubing, which is free to turn in the hole bored in the center of the brass plate *y*. It is capable also of vertical adjustment, the two brass collars *z, z*, serving to hold it at any given height. The pipe which attaches to its lower end is made of rubber tubing, so that the burner may be turned freely. A pointer *p'* is rigidly attached to the upper collar, and serves to indicate the angular position of the burner by reference to a graduated circle immediately below it, and concentric with the axis about which it revolves. The stop is located very near the burner, the diameter of the aperture used in these tests being 8.566 mm.

A Methven 2-candle standard was used, and a Lummer-Brodhun photometer, great care being taken to screen the disk from all stray light. Twelve tests were made, each consisting of eighteen double settings (direct and reversed), the mean of each double setting being taken as the setting of the photometer. The flame occupied a different angular position in each test, so that the horizontal intensity was studied at intervals of 30° around the entire circle.

The tests were made by two observers, one making the settings, taking the temperature of the air in the room, and reading the gas meter; and the other keeping the time, recording the settings, and changing the angular position of the flame at the end of each test.

The candle power corresponding to each observation was computed, also the mean candle power for each test, and the average deviation of a single observation — the latter being a measure of the precision with which the settings were made. A series of curves was also drawn, showing the relation between candle power and time for each of the eighteen minutes during which the test lasted. Then another curve was drawn showing the relation between the mean candle power, in each position of the flame, and the angle which the plane of the flame made with the plane of the stop.

In studying the curves from 1 to 12 inclusive, it must be remembered that they are plotted on a very large vertical scale, so that the variations can be more easily seen. A table is given in which the more important facts brought out in the tests are collected. It will be observed that the variation in intensity in each of the twelve tests is, on the average, about 0.5 of a candle power.

Particular attention is called to the columns of mean candle powers, angular positions of the flame, and rates of gas consumption. It will be seen at once that the figures in the latter are very uniform,

Number of test.	Maximum candle power.	Minimum candle power.	Difference.	Angle between the flame and plane of the stop.	Mean candle power.	Gas consumed in cubic feet.	Rate of gas consumption in cubic feet per hour.	Temperature of the air in the room. (Centigrade.)	Average deviation of a single observation. a. d.	Percentage change above minimum candle power.
1	7.38	6.9	.48	0°	7.12	0.355	1.18	20°.0	0.076	6.9
2	10.80	10.40	.40	270°	10.63	0.318	1.06	20°.3	.098	3.8
3	8.06	7.32	.74	330°	7.71	0.308	1.03	20°.4	.170	10.0
4	11.20	10.50	.70	300°	10.90	0.283	0.94	20°.7	.200	6.7
5	11.80	11 20	.60	240°	11.40	0.283	0.94	20°.8	.100	5.0
6	7.96	7.58	.38	210°	7.71	0.308	1.03	25°.7	.080	5.0
7	7.10	6.62	.48	180°	6.83	0 307	1.02	25°.7	.110	7.2
8	8.18	7.64	.54	150°	7.74	0.283	0.94	22°.0	.073	7.0
9	11.18	10.76	.42	120°	10.96	0.313	1.04	22°.2	.086	3.9
10	10.74	10.34	.40	90°	10.46	0.342	1.14	24°.5	.0075	3.8
11	12.48	11.84	.64	60°	12.14	0.318	1.06	24°.5	.148	5.4
12	8.48	7.90	.58	30°	8.24	0.308	1.03	25°.7	.146	7.3

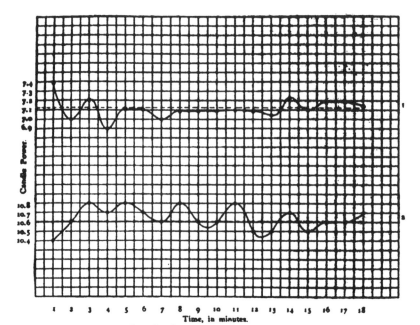

Time, in minutes.

FIG. 6. CURVES FOR TESTS 1 AND 2.

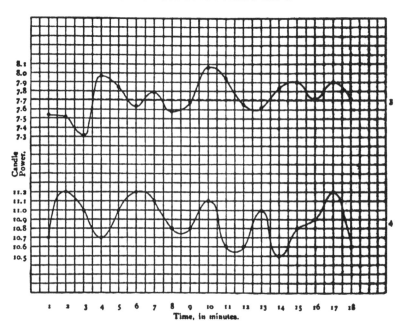

Time, in minutes.

Dotted line in each curve shows mean candle power.

FIG. 7. CURVES FOR TESTS 3 AND 4.

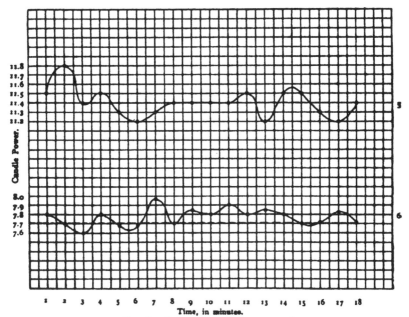

FIG. 8. CURVES FOR TESTS 5 AND 6.

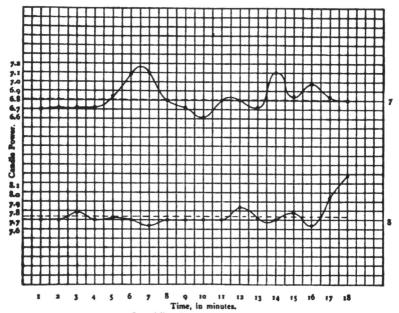

Dotted line shows mean candle power.

FIG. 9. CURVES FOR TESTS 7 AND 8.

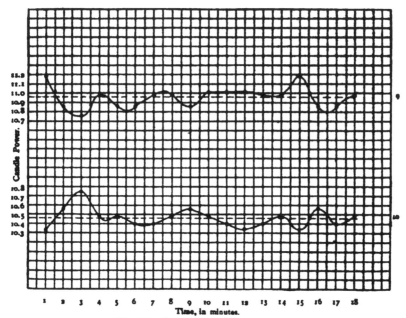

FIG. 10. CURVES FOR TESTS 9 AND 10.

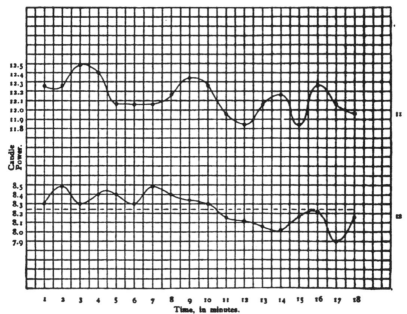

Dotted line shows mean candle power.

FIG. 11. CURVES FOR TESTS 11 AND 12.

while the mean candle powers show a wide variation, proving beyond question that these variations are not due to unequal rates of gas consumption ; for if we look at the table and compare tests Nos. 5 and 8, we see at once that, while their rates of gas consumption are exactly the same, their mean candle powers differ greatly; and it is instructive to note that the angular positions of the flame in these two cases differ by 90°. By referring to Curve 13 this relation is shown graphically.

The intensity of light emitted in different directions in the same horizontal plane seems to follow a certain law. It would appear as though the horizontal intensities were greatest in the directions which coincide with the plane of the flame. The curve, however, shows that the directions of greatest intensity lie a little either side of this plane —a phenomenon which would lead us at once to the belief that the cross section of the flame, instead of having, as we would naturally suppose, the form

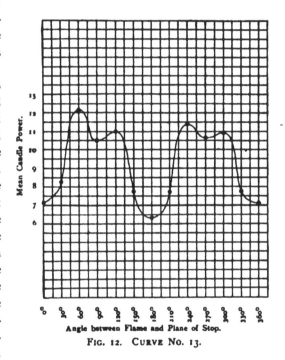

FIG. 12. CURVE No. 13.

shown in Figure 5 in half section, has more the form similarly shown in Figure 4.

This seems quite reasonable when we remember that the gas issues from the burner in two small jets which impinge against each other at an angle of about 90°, and that these jets flatten out into a fan-like flame. By looking at the edge of the flame we could not, of course, hope to see this constriction at the edges, supposed to exist, for the reason that the external surface of the flame being, of course, incandescent, could give us no indication as to the shape of the surface emitting the light.

It rarely happens that a flame is symmetrical with respect to the axis of the burner, and it is quite likely that the lack of symmetry of the two crests in Curve 13 is due to the fact that the flame was non-symmetrical with respect to this axis. · The same effect might, however, have been produced if the axis of the stop had failed to coincide with the horizontal axis of the flame.   We are not inclined to the use of the acetylene flame as a standard of light.

ROGERS LABORATORY OF PHYSICS.
November, 1898.

A MODIFICATION OF BISCHOF'S METHOD FOR DETER-MINING THE FUSIBILITY OF CLAYS, AS APPLIED TO NON-REFRACTORY CLAYS, AND THE RESIST-ANCE OF FIRE CLAYS TO FLUXES.[1]

By H. O. HOFMAN.

IN determining experimentally the fusibility of clays, two kinds of methods may be distinguished — the direct and the indirect. Of the direct methods, that of Seger has found much favor. It consists in placing in a crucible the clay to be tested with Seger cones[2] (graded mixtures, the melting-points of which are known), and heating in a suitable furnace until a cone is found which shows the same behavior in the fire as the clay. Different apparatus is used for refractory and non-refractory clays, the standard for a refractory or fire clay being that it shall not melt before Seger cone No. 26, or at an approximate temperature of 1650° C. The details of Seger's method of compar-ing fire clays with his standard refractory cones Nos. 26–36 have been given in a previous paper.[3] For determining the fusibility of non-refractory clays, a special form of gas furnace[4] is required, in which, with a good pressure of gas, Seger cone No. 25, the highest of the non-refractory mixtures, can be melted down in from 4½ to 5 hours. It may be said, however, that the melting down of cones Nos. 20–25 is often difficult.

Of the indirect methods, that of Bischof[5] has been extensively used. It consists in toning up weighed samples of the clay to be tested with increasing quantities of an intimate mixture of equal parts of chemically pure silica and alumina, and forming them into

---

[1] Reprinted from Transactions of the American Institute of Mining Engineers, Buffalo Meeting, October, 1898.

[2] Thonindustrie-Zeitung, 1893, p. 1252; Berg- und Hüttenm. Zeitung, 1894, p. 119; The Clay Worker, August, 1897.

[3] *Technology Quarterly*, 8, 63.

[4] Thonindustrie-Zeitung, 1896, No. 63; Berg- und Hüttenm. Zeitung, 1897, p. 21.

[5] Dingler's Polyt. Jour., cxcvi., pp. 438, 525; cxcviii., p. 396.

small prisms, to be heated with a prism of Saarau fire clay (equal to Seger cone No. 36) to above the melting-point of wrought iron. The sample which shows the same behavior in the fire as the Saarau prism is the critical mixture, and the amount of toning-up substance needed forms the criterion of the fusibility of the clay.

Beyond question, the direct method is the simpler, since the clay need only be ground, moulded, dried and heated with standards which are always uniform and can be bought in the market at a low price. The first test is made with Seger cone No. 26, to see if the clay is refractory or not, and the work is then continued with the Deville or the Seger gas furnace. There is no drying and igniting, no weighing out of clay and of fluxes, and no intimate mixing — all of which are tedious operations. If many tests have to be made, the direct method will always be followed. The indirect method, however, has the advantage that only one standard and one furnace are required. Considering the difficulties often encountered in melting down in the gas-furnace clays that are near the refractory line, and the advantage in this respect of using only one furnace (the one for refractory clays, which is cheap, and which anyone can build for himself), and simply determining how much refractory material is necessary to bring a non-refractory clay up to the required standard, we must confess that there are conditions under which the indirect method can hold its own.

The work to be described in this paper formed part of a thesis of Messrs. J. L. Newell and G. A. Rockwell, of the Class of '95, who with much care carried out the large number of tests required to verify the method.

In these experiments the Bischof standard (Saarau clay, corresponding to Seger cone No. 36) was changed to Seger cone No. 26, which, as previously observed, forms the line of separation between refractory and non-refractory clays, the non-refractory clays being toned up until they showed the same behavior in the fire as Seger cone No. 26. This was done because it was of more interest to find out how far the non-refractory clay stood below the point of being a fire clay than how much refractory material would have to be added to bring it up to the Saarau or Seger cone No. 36 standard. Moreover, it was thus possible to work at a lower temperature, with a saving in time and gas carbon, and a prolongation of the life of the furnace lining.

The method may of course be varied. For example, the writer has toned up low-grade fire clays with bauxite until they showed the same behavior as certain high-grade fire clays, or their equivalents in terms of Seger cones.

The silica used in the 'experiments was quartz, ground to pass a 100-mesh sieve and purified by boiling with nitrohydrochloric acid. Upon analysis, it showed 99.88 per cent. of SiO_2, and was assumed to be pure. The alumina was obtained from the Solway Process Company, Syracuse, New York. An analysis furnished by the makers showed Al_2O_3, 98.46; SiO_2, 0.25; Na_2O, 0.50; Fe_2O_3, 0.04; loss by ignition, 0.75 per cent. As the substance readily absorbs moisture, a sample was ignited, which gave a loss of 6.42 per cent., and an allowance for this loss was made in all the work. The clays tested were kindly furnished by Professor Edward Orton, Jr. Their composition and the results of the tests are shown in the accompanying table:

ANALYSES OF CLAYS AND RESULTS OF TESTS.

Sample No.	26[1]	25[1]	3[1]	22[1]	24[1]	33[1]	1,982[2]
	Per cent.	Per cent.	Per cent.	Per cent.	Per cent.	Per cent.	Per cent.
SiO_2	64.10	55.60	57.10	57.45	57.15	49.30	43.94
Al_2O_3	21.79	24.34	21.29	21.06	20.26	24.00	11.17
H_2O comb.	6.05	6.75	6.00	5.90	5.50	9.40	3.90
Total	91.94	86.69	84.39	84.41	82.91	82.70	59.01
Fe_2O_3	2.51	6.11	7.31	7.54	7.54	8.40	3.81
CaO	0.10	0.43	0.29	0.29	0.90	0.56	11.64
MgO	0.58	0.77	1.53	1.22	1.62	1.60	4.17
K_2O	2.62	3.00	3.44	3.27	3.05	3.91	2.90
Na_2O	0.03	0.09	0.61	0.39	0.58	0.17	0.71
Total	5.84	10.40	13.18	12.71	13.69	14.63	23.23
Moisture	1.10	2.65	1.30	1.90	2.70	1.20	15.66[3]
Grand total	98.88	99.74	98.87	99.02	99.30	98.54	98.00[4]
Stiffening ingredient, p. c.	20	40	60	80	80	100	180

[1] Analyzed by N. W. Lord.
[3] Includes CO_2.
[2] Analyzed by E. Orton, Jr.
[4] Includes P_2O_5, 0.10 per cent.

The method of operation was to weigh out samples of 1 gram of the clay to be tested (the moisture being allowed for); mix them severally with 0.1, 0.2, 0.3, etc., gram of the silica-alumina flux in small porcelain dishes; turn out each mixture upon a glass plate, moisten it with a 10 per cent. dextrine solution; work it with a spatula until it has acquired the right consistency, and mould it into the form of the small-size Seger cone.[1] When dried, three of them were placed in a crucible with a Seger cone, No. 26, and so heated in the Deville furnace as to melt down the Seger cone. In addition to the usual 30 grams of paper and 200 grams of charcoal, from 920 to 925 grams of gas carbon were required, with a pressure of blast of about 1 inch of water. A fusion required about 35 minutes.

In the table above, the clays are arranged according to their degree of refractoriness. Sample No. 26 requires 20 per cent. of flux to raise its melting-point to that of Seger cone No. 26; sample No. 25 requires 40 per cent., and so on.

In order to check the method, tests were made in two ways with large-size Seger cones. In the first, cones Nos. 1 to 25 were pulverized and toned up with the silica-alumina flux. It was found that 0.036 gram of the flux added to 1 gram of Seger cone substance, would raise its melting-point to that of the next higher number. In the second, two clay samples, Nos. 22 and 23, which according to indirect preliminary tests, ought to melt down at the same time as Seger cone No. 4, were placed in a graphite crucible, the bottom of which had been tamped with refractory clay, and were heated in a coke furnace with under-grate blast. In two hours the test was finished. The two clays showed in the fire approximately the same behavior as the Seger cone, thus again proving the accuracy of the method.

This modification of Bischof's indirect method may also be used for determining the resistance of fire clays or fire bricks to the corroding influence of sodium chloride, sodium sulphate, sodium carbonate, potassium carbonate, etc., to which they are exposed in glass pots, or to calcium carbonate, lead oxide, iron oxides, etc. To illustrate the operation, samples of 1.5 grams of clay are mixed severally with 5, 10, 15, etc., milligrams of flux, formed into small size Seger cones, and heated in the Deville furnace with Seger cone No. 26 in such a way

---

[1] *Technology Quarterly*, 8, 68.

that the Seger cone will melt. Here, again, the sample which shows the same behavior as the Seger cone will be the critical mixture, and the percentage of flux it contains will form the criterion of the clay's resistance to corrosion. It is true that the results obtained do not altogether determine the suitability of a clay for the manufacture of glass pots, as the requirements [1] of such a clay are not only that it shall resist heat and fluxes, but that it shall be highly plastic and burn dense at a comparatively low temperature. Nevertheless, the results obtained by the method form a valuable guide in the making up of mixtures. This was proved to the writer by the results with certain clays selected on account of their good behavior under the tests, which turned out a true prediction of what occurred on a large scale afterwards.

The method can, of course, be applied to a number of cases where an acid furnace material is to be exposed to the corrosion of a basic charge, and *vice versa*.

---

[1] Seger-Cramer Thonindustrie-Zeitung, 1897, p. 47.

DOES THE SIZE OF PARTICLES HAVE ANY INFLUENCE IN DETERMINING THE RESISTANCE OF FIRE CLAYS TO HEAT AND TO FLUXES?[1]

By H. O. HOFMAN and B. STOUGHTON.

BEFORE examining a fire clay in the laboratory for its resistance to heat or to fluxes, the sample is always ground to an impalpable powder. But when the clay is actually used for the manufacture of bricks, blocks, pots, etc., it is not ground to a uniform size, the particles varying from coarse grains to the finest slimes. The natural inference is that the tests with finely ground substances will give lower results than if the materials are tested just as they are going to be used. The following experiments were made to find out how far this inference is justified. The method employed for fusion was Seger's direct method,[2] and, for fluxing, the modified Bischof method described in the paper on that subject presented at the present meeting. The first question to be decided was, how large the test cones ought to be made to include representative proportions of the different sized particles composing the mixture. This was subjected to a screen analysis with the following results :

TABLE I.
RESULTS OF SCREENING.

	Grams.	Per cent.
Before screening	74.05	....
On 8-mesh	16.1	22.1
On 12-mesh	16.6	22.8
On 16-mesh	6.9	9.5
On 20-mesh	6.6	9.0
On 30-mesh	7.4	10.1
On 40-mesh	2.5	3.4
On 60-mesh	4.0	5.5
Through 60-mesh	12.8	17.6
Lost in screening	1.15	....

[1] Reprinted from Transactions of the American Institute of Mining Engineers, Buffalo Meeting, October, 1898.
[2] *Technology Quarterly*, 6, 301; 8, 63.

TABLE II.

WEIGHTS AND PROPORTIONS OF DIFFERENT SIZES IN A LARGE SEGER CONE.

	Sizes.	No. 1. Weight (Grms.)	No. 1. Per cent.	No. 2. Weight (Grms.)	No. 2. Per cent.	No. 3. Weight (Grms.)	No. 3. Per cent.	No. 4. Weight (Grms.)	No. 4. Per cent.	No. 5. Weight (Grms.)	No. 5. Per cent.	No. 6. Weight (Grms.)	No. 6. Per cent.	No. 7. Weight (Grms.)	No. 7. Per cent.	No. 8. Weight (Grms.)	No. 8. Per cent.
Before disintegrating.		7.950	....	7.670	....	6.833	....	8.715	....	7.434	....	7.283	....	7.396	..	8.462	....
After disintegrating there / Remained on sieve.	8-mesh.	1.605	20.4	2.062	27.2	1.840	27.1	1.872	21.6	2.162	29.1	1.657	22.8	1.305	17.9	2.277	27.3
	12-mesh.	1.857	23.6	1.502	19.9	1.372	20.2	1.939	22.4	1.535	20.7	2.127	29.3	1.942	26.7	2.207	26.5
	16-mesh.	0.950	12.0	0.685	9.0	0.514	7.8	0.855	9.8	0.725	9.8	0.718	9.9	0.875	12.0	0.750	9.0
	20-mesh.	0.820	10.4	0.645	8.5	0.587	8.8	0.742	8.8	0.640	8.8	0.637	8.8	0.665	9.2	0.707	8.5
	30-mesh.	0.825	10.5	0.782	10.3	0.669	9.8	0.860	9.9	0.745	10.0	0.680	9.3	0.774	10.7	0.762	9.0
	40-mesh.	0.277	3.5	0.292	3.9	0.254	3.8	0.312	3.6	0.245	3.3	0.252	3.5	0.264	3.6	0.249	3.0
	60-mesh.	0.442	5.7	0.430	5.8	0.422	6.2	0.572	6.6	0.380	5.1	0.352	4.9	0.417	5.9	0.407	4.9
Passed through sieve.	60-mesh.	1.095	13.9	1.168	15.4	1.117	16.4	1.498	17.3	0.982	13.2	0.832	11.5	1.023	14.0	0.980	11.8

After a number of preliminary trials it was decided to choose as a standard the large-size Seger cone,[1] $\frac{3}{4}$ inch at base and $2\frac{3}{4}$ inches in height. The results given in Table II show that the choice was a correct one.

This table represents eight cones made up from a sample of the mixture which had been moistened with water containing a small amount of dextrine. Each cone when dried was first weighed, then disintegrated by rubbing in a Wedgwood mortar and sifted through the same screens with which the first sizing-test had been made; the resulting different sized particles were then all separately weighed and their percentages calculated. The table shows the figures for 8-mesh and coarser sizes to be irregular. This was to be expected, as a few grains being a little larger or smaller than the average would cause a considerable difference in the percentage in view of the small weight of the cone. What went through an 8-mesh and remained on a 16-mesh screen showed more regularity; and the sizes from 16-mesh to 60-mesh give very uniform data. That the material finer than 60-mesh should again show some differences was to be expected, as the fine clay substance, adhering more or less to the coarser grains, could not be evenly separated by mere rubbing, which is all that can be used.

As the Seger cones for refractory clays, Nos. 26 to 36, are only $\frac{3}{8}$ inch at base and $1\frac{3}{16}$ inch high, large-size refractory cones had to be specially made as standards for the tests. The Deville furnace, and the crucibles, with their lids and supports, were of the same general character as those described in a previous paper,[2] only the dimensions were larger. The furnace, of $\frac{1}{8}$-inch sheet iron, was 25 inches high and 12 inches in diameter; the cast-iron plate, $3\frac{3}{4}$ inches from the bottom, was 1 inch thick, had a central opening 2 inches in diameter surrounded by four rows of $\frac{1}{4}$-inch holes. The lining, of sintered magnesite from the Fayette Manufacturing Company, Pittsburg, Pennsylvania, was $3\frac{1}{4}$ inches thick at the bottom, and $2\frac{3}{4}$ inches at the top, making the inner dimensions of the furnace: diameter at bottom $5\frac{1}{2}$ inches, at top $6\frac{1}{2}$ inches, height 20 inches. The inner diameter of the air-inlet pipe was 1 inch. The outside dimensions of the crucibles were: diameter $2\frac{1}{2}$ inches, height 3 inches, thickness of wall $\frac{1}{4}$ inch;

---

[1] *Technology Quarterly*, **6**, 313.
[2] *Technology Quarterly*, **8**, 63.

the thickness of the lids ¼ inch; the size of the supports: diameter 2½ inches, height 2⅝ inches. The method of firing differed slightly from that pursued with the small furnace. With the latter the blast is slowly started, 30 grams of paper are ignited and pressed down into the furnace, and then 200 grams of charcoal are charged, to be followed by the required amount of gas carbon. In the large furnace the space around the crucible support was filled with small-size gas carbon (about 40 grams) before the paper (50 grams) and charcoal (250 grams) were introduced to kindle the increased amount of gas carbon. Table III gives the leading details in regard to the manner of working with the furnace.

TABLE III.

OPERATION OF THE TESTING FURNACE.

SHORT CONE.		FUEL CHARGE.			Blast pressure. (Inches water.)	Furnace. (Initial condition.)	Time required for experiment. (Minutes.)
No.	Form after fusion.	Paper. (Grams.)	Charcoal. (Grams.)	Gas carbon. (Grams.)			
26 . . .	Lenticular.	50	250	2900	2	Cold.	60
31 . . .	Globular.	50	250	3750	3	Hot.	50
32 . . .	Globular.	50	250	4000	3	Cold.	55
33 . . .	Globular.	50	250	2300	3½	Hot.	35
34 . . .	Globular.	50	250	2400	3½	Hot.	40

The charcoal was passed through a small Blake crusher, set to 1½ inches, and then sifted through a 3-mesh screen, and the fines discarded. The gas carbon was broken in the same way, and screened through a 1-mesh sieve. The under-size was passed through a 3-mesh screen to remove the fines. What remained on the 1-mesh sieve was passed through the Gates laboratory-crusher, set to ¾ inch, and was then used for filling the space around the crucible support, or reserved for the small Deville furnace.

Table IV gives the results of fusing- and fluxing-tests of mixtures just as they are to be used in making bricks, blocks, pots, etc., and of the same mixtures ground fine, as is the usual custom in laboratory tests. It shows only a slight difference in the results, and proves that, in testing fire-clay mixtures, the samples can be safely ground fine and compared with the small-size Seger cones in the ordinary Deville furnace. Exceptional conditions only would make it necessary to test the mixtures just as they are going to be used.

TABLE IV.

COMPARATIVE TESTS OF THE SAME MATERIAL IN NATURAL CONDITION AND FINELY GROUND.

MIXTURE.		MATERIAL AS RECEIVED.		SAME MATERIAL FINELY GROUND.	
Mark.	Tested for resistance to.	Fusing test. Equal to Seger cone No.	Fluxing test. Grams of $CaCO_3$ to 1.5 of mixture.	Fusing test. Equal to Seger cone No.	Fluxing test. Grams of $CaCO_3$ to 1.5 of mixture.
A . . .	Heat.	34	....	34	....
B . . .	Heat and fluxes.	33	0.155	33	0.160
D . . .	Heat.	34	....	34	....
E . . .	Heat and fluxes.	34	0.180	34—33	0.180
H . . .	Heat and fluxes.	33—34	0.160	33—34	0.160
No. 1. .	Heat and fluxes.	31	0.100	30	0.120

REVIEW

American Chemical Research.

VOL. IV. 1898.

Contributed by Members of the Instructing Staff of
the Massachusetts Institute of Technology.

ARTHUR A. NOYES, Editor.

HENRY P. TALBOT, Associate Editor.

REVIEWERS.

ANALYTICAL CHEMISTRY·········· H. P. Talbot and W. H. Walker
BIOLOGICAL CHEMISTRY·········· ···············A. G. Woodman
CARBOHYDRATES··································· G. W. Rolfe
GENERAL CHEMISTRY ·····························A. A. Noyes
GEOLOGICAL AND MINERALOGICAL CHEMISTRY·· ······W. O. Crosby
INORGANIC CHEMISTRY ·······························H. Fay
METALLURGICAL CHEMISTRY AND ASSAYING·········H. O. Hofman
ORGANIC CHEMISTRY································J. F. Norris
PHYSICAL CHEMISTRY··········· ······················ H. M. Goodwin
SANITARY CHEMISTRY·············· ··········· E. H. Richards
TECHNICAL CHEMISTRY ··············A. H. Gill and F. H. Thorp

EASTON, PA.:
THE CHEMICAL PUBLISHING CO.
1898.

[Contribution from the Massachusetts Institute of Technology.]

REVIEW OF AMERICAN CHEMICAL RESEARCH.

Vol. IV. No. 1.

Arthur A. Noyes, Editor; Henry P. Talbot, Associate Editor.
Reviewers: Analytical Chemistry, H. P. Talbot and W. H. Walker; Biological Chemistry, W. R. Whitney; Carbohydrates, G. W. Rolfe; General Chemistry, A. A. Noyes; Geological and Mineralogical Chemistry, W. O. Crosby; Inorganic Chemistry, Henry Fay; Metallurgical Chemistry and Assaying, H. O. Hofman; Organic hemistry, J. F. Norris; Physical Chemistry, H. M. Goodwin; Sanitary Chemistry, E. H. Richards; Technical Chemistry, A. H. Gill and F. H. Thorp.

INORGANIC CHEMISTRY.

Henry Fay, Reviewer.

A New Explosive Compound Formed by the Action of Liquid Ammonia upon Iodine. By Hamilton P. Cady. *Kan. Univ. Quart.*, 6, 71–75.—When 25–30 cc. of liquid ammonia are added to 4–5 grams of iodine in a vacuum-jacketed tube protected from moisture, the iodine is dissolved, forming a dark opaque liquid which changes to olive-green and deposits a dark-green, crystalline precipitate. This substance, dried over sulphuric acid, dissolves in ether, alcohol, and chloroform; but it is insoluble in dilute acids, and is decomposed by them, generally with explosive violence. It is instantly decomposed by strong acids. Potassium iodide solution dissolves the crystals, and they are decomposed and dissolved by hydrogen sulphide, sulphurous acid and potassium hydroxide. The composition of the substance, as shown by analysis, agrees well with the formula HN_2I; and the formation of the compound is supposed to take place according to the equation : $11NH_3 + 9I = HN_2I + 8NH_4I$.

Note on Extemporaneous Chlorine Water. By F. B. Power. *Pharm. Rev.*, 15, 108.—The author criticises the statement made by Griggi that chlorine can be prepared according to the equation :

$$2H_2C_2O_4 + PbO_2 + CaCl_2 = CaC_2O_4 + PbC_2O_4 + 2H_2O + Cl_2.$$

It was shown that what was taken for chlorine was probably ozone; and that while the oxalic acid was mostly oxidized to carbon dioxide, practically all of the calcium chloride was unacted upon.

The Action of Carbon Dioxide upon Sodium Aluminate and the Formation of Basic Aluminium Carbonate. By William

C. DAY. *Am. Chem. J.*, 19, 707–728.—The author has analyzed a series of preparations made by precipitating sodium
aluminate by carbon dioxide. The sodium aluminate was prepared from Connetable phosphate rock, by treating it with
sodium carbonate and quicklime, which leaves all iron compounds and calcium phosphate undissolved. Before working
with the samples prepared, a commercial product, "Alumina
soluble in acids," was analyzed, and found to contain 26.38 per
cent. CO_2, and 21.30 per cent. Na_2O. Some of the original substance was then washed with water until no alkali could be detected by litmus or phenol-phthalein. The dried residue from the
washing effervesced with acid, and gave on analysis 17.35 per
cent. CO_2, and 12.22 per cent. Na_2O, amounts corresponding
closely to that required for sodium bicarbonate. Several preparations were then made and washed with hot water until
apparently free from alkali; they were then dried on the water-
bath and analyzed; but it was found that all portions contained
sodium oxide and carbon dioxide. This dried sample showed
alkali on washing, and continued to show it after apparently
washed free from it a second time. Preparations washed with
cold water by decantation until apparently free from alkali
always showed it on analysis, but gave off more after drying
and washing again. In two samples it was possible to wash the
sodium out until only 0.23 and 0.34 per cent. remained; and in
the same preparations there were left respectively 9.06 and 3.99
per cent. of carbon dioxide, which is more than that needed for
the formation of sodium bicarbonate, indicating that a basic
aluminum carbonate of varying composition might exist.

On the Chloronitrides of Phosphorus. (II). BY H. N.
STORES. *Am. Chem. J.*, 19, 782–796.—In a previous paper
(this *Rev.*, 3, 4), the author described the phosphonitrilic chlorides having the composition $P_3N_3Cl_6$ and $P_4N_4Cl_8$, and their corresponding acids. This paper is a continuation of the work de-
scribing the polymers of the same series. The following table
shows the additional members which have been isolated, together
with their properties :

	Melting-point.	Boiling-point. 13 mm.	760 mm.
Triphosphonitrilic chloride, $(PNCl_2)_3$	114°	127°	256.5°
Tetraphosphonitrilic chloride, $(PNCl_2)_4$	123.5°	188°	328.5°
Pentaphosphonitrilic chloride, $(PNCl_2)_5$	40.5–41°	223–224.3°	Polymerizes
Hexaphosphonitrilic chloride, $(PNCl_2)_6$	91°	261–263°	Polymerizes
Heptaphosphonitrilic chloride, $(PNCl_2)_7$	liquid at --18°	289–294°	Polymerizes
Polyphosphonitrilic chloride, $(PNCl_2)_x$	Below red heat.	Depolymerizes on distillation.	

There was also obtained a liquid residue of the same empirical composition of a *mean* molecular weight corresponding nearly to $(PNCl_2)_n$; but it has not yet been prepared in a pure condition. The method that gives the best results in the preparation of these compounds is to heat a mixture of four parts of dry phosphorus pentachloride with one part of ammonium chloride. As the mixture liberates hydrochloric acid, it is necessary to open the tube occasionally to relieve the pressure; and this is best accomplished by heating to 150° C., allowing to cool to 100°., and opening the tubes while in the furnace. This process is repeated several times, each time heating 1°–2° higher. After most of the hydrochloric acid has been removed in this way, the tube can be safely heated to 200° C. The contents of the tube are then distilled off, and the distillate is found to consist of a crystalline mass impregnated with a yellow oil, amounting to 95 per cent. of the theoretical quantity of phosphonitrilic chlorides, of a small amount of phosphorus pentachloride, of some chloronitride of the composition $P_4N_2Cl_3$, and of traces of an unknown compound. The melted distillate is poured into cold water to remove phosphorus pentachloride; and it is then heated for about two hours. The product is then fractionated up to 200° C. at 13–15 mm., using an Anschütz flask. The distillate at this temperature consists of about 70 per cent. $P_3N_3Cl_6$ and $P_4N_4Cl_8$; the residue of $P_5N_5Cl_{10}$ and the higher members. If the distillate is not desired, it is best to take advantage of the peculiar properties of the substance and polymerize it to the compound $(PNCl_2)_x$ which can in turn be depolymerized, giving the different lower members of the series. The residue of $P_5N_5Cl_{10}$ and higher members is fractionated above 200° C. Polymerization often takes place during the heating, as is indicated by the frothing; but this may be overcome by heating with water and removing the small amount of oil formed, and again continuing the fractionation. It has not been found advisable to heat above 370°, as polymerization takes place rapidly at that temperature. Pentaphosphonitrilic chloride at its melting-point is soluble in all proportions of benzene, ether, carbon bisulphide, and gasoline; but it cannot be crystallized from any of them. It shows a decided tendency to superfusion. The next two members of the series are soluble in benzene, ether, carbon bisulphide and gasoline; hexaphosphonitrilic chloride possessing a strong crystallizing power. When any of the lower members are heated polyphosphonitrilic chloride is formed slowly at 250° C., and rapidly at 350° C. As the change is reversible, complete transformation cannot be effected; but it reaches 90 per cent. The pure chloride is colorless, transparent, and elastic; it is insoluble in all neutral solvents; but it readily absorbs benzene, swelling to a rubber-like compound of many times its own volume. Hot water slowly dissolves it with decomposition; it swells, gelatinizes, and finally dissolves in ammonia. Depolymerization begins at 350° C., and this change is rapid below a red heat.

On the Solubility of Ammonia in Water at Temperatures below o°. By J. W. MALLET. *Am. Chem. J.,* 19, 804–809.— The author has extended the determination of the solubility of ammonia made by Roscoe and Ditmar from o° to —40° C. Strong aqueous ammonia was placed in a burette-like tube surrounded by a freezing mixture, and gaseous ammonia, previously cooled by being passed through tubes placed in freezing mixtures, was passed into the already strong solution. The temperature was recorded by means of a standardized alcohol thermometer. Measured portions were drawn off, beginning at —40° C. at intervals of 10° into water cooled to o° and titrated with sulphuric acid. The solution of enough ammonia to form ammonium hydroxide was not attended with any apparent change of behavior; and continuous liquefaction went on smoothly to the lowest temperature reached, without separation of any solid product. The table expresses the amount of ammonia dissolved in one gram of water at a pressure of 743–744.4 mm. :

°C.	Grams.	°C.	Grams.
—3.9	0.947	—25	2.554
—10	1.115	—30	2.781
—20	1.768	—40	2.946

The amount dissolved at —3.9° C. corresponds almost exactly to that calculated for ammonium hydroxide.

The Action of Nitric Acid on Aluminum and the Formation of Aluminum Nitrate. By THOMAS B. STILLMAN. *J. Am. Chem. Soc.,* 19, 711–717.—The author shows that aluminum in the form of coarse turnings is readily acted upon by nitric acid, hot or cold, of specific gravity 1.15 or 1.45, the weaker acid acting the more rapidly. The reaction is retarded if the aluminum is in the form of plates, the retardation being very slight with cold acid, but considerable if the acid is warm. The solutions deposited crystals of the composition $Al(NO_3)_3.9H_2O$.

Notes on Selenium and Tellurium. By EDWARD KELLAR. *J. Am. Chem. Soc.,* 19, 771–778.—It is shown that ferric hydroxide carries down both selenium and tellurium, the amount needed for the precipitation of the former element being much greater than for the latter. To obtain both elements quantitatively in precipitating from hydrochloric acid solutions, it is necessary to have the acidity of the solution between definite limits, 30–50 per cent. Figures are given for the partial precipitation of both elements, showing that the amount of precipitate varies with the time, temperature, and the acidity of the solutions. There is complete precipitation of tellurium when the solution contains 0.5 per cent. acid, but then none of the selenium is precipitated. When the acidity has reached 80 per cent., all of the selenium, but no tellurium, is precipitated. The reducing action of ferrous sulphate is almost identical with that of sulphurous acid. In precipitating copper solutions containing selenium with hydrogen sulphide, some of the latter is always carried down, probably in combination with copper as selenide.

Further Study on the Influence of Heat Treatment and Carbon upon the Solubility of Phosphorus in Steel. By E. D. CAMPBELL AND S. C. BABCOCK. *J. Am. Chem. Soc.*, 19, 786–790.—This is a continuation of a previous paper (this *Rev.*, 3, 2) greater care having been taken here to measure temperatures by means of a Le Chatelier pyrometer. The general results reached are : that phosphorus may exist in at least two forms in steel ; that carbon and the different heat treatment, annealing and quenching, considerably affect the solubility of the phosphorus in acid mercuric chloride solution, the amount going into solution apparently increasing with increase of the temperature at which the metal is quenched, and also with the percentage of carbon.

Contribution to the Chemistry of Didymium. By L. M. DENNIS AND E. M. CHAMOT. *J. Am. Chem. Soc.*, 19, 799–809. —The mixture of earths used in this work was extracted from monazite sand from Brazil. The sand was decomposed by sulphuric acid, and the rare earths separated from the other elements in the usual way. Cerium was separated from the didymium group by the chlorine process in apparatus specially devised for the purpose. The oxalates were then decomposed by heat and converted into nitrates, their solution being treated with an excess of potassium sulphate for the purpose of removing the yttrium group. The double sulphates were then dissolved in dilute ammonium acetate, converted into the hydroxides and then into nitrates. Fractionation of the double ammonium nitrates by the methods of Welsbach and Bettendorf did not yield satisfactory results ; and it was found that the best results were obtained by slow spontaneous evaporation. In this way there was more rapid separation of neodidymium from praseodidymium. A large portion of lanthanum salts was separated in the first fractions, disappearing for a time, and again appearing in later fractions. It was found that the presence of the lanthanum salts aided much in the splitting of the two constituents of didymium. Detailed directions are given for the method giving the most satisfactory results.

A Study of the Mixed Halides and Halo-Thiocyanates of Lead. By CHARLES H. HERTY AND J. R. BOGGS. *J. Am. Chem. Soc.*, 19, 820–824.—Lead chloride and lead bromide form isomorphous mixtures in all proportions, while lead iodide with lead chloride or bromide forms mixed crystals with the latter in excess. Lead thiocyanate forms true compounds with both lead chloride and lead bromide, but with lead iodide forms neither mixed crystals nor a chemical compound.

On the Existence of Orthosilicic Acid. By T. H. NORTON AND D. M. ROTH. *J. Am. Chem. Soc.*, 19, 832–834.—Silicic acid obtained by passing silicon tetrafluoride into water, was thrown on a cloth filter, washed quickly with ether or benzene, dried by pressure between folds of filter paper, and analyzed. The results obtained agree well with the formula H_4SiO_4. The substance thus obtained readily loses water.

The Solubility of Stannous Iodide in Water and in Solutions of Hydriodic Acid. By S. W. YOUNG. *J. Am. Chem. Soc.*, 19, 845–851.—Portions of saturated solutions of stannous iodide in water and in hydriodic acid protected from the air by a layer of oil, were drawn off at different temperatures, weighed and titrated with iodine solution. The results are normal with water, but the solubility in hydriodic acid decreases until the solution contains 6–7 per cent. of acid, and thereafter increases. When the concentration has reached 25 per cent. of hydriodic acid there is a marked increase in the solubility with decreasing temperature (see next abstract). Tables and curves are given expressing the solubility values.

On Iodostannous Acid. By S. W. YOUNG. *J. Am. Chem. Soc.*, 19, 851–859.—When the solution of stannous iodide in hydriodic acid, containing 25 per cent. acid, is cooled in ice water, light yellow needles crystallize out. Unsuccessful attempts were made to dry these crystals in hydrogen and in hydriodic acid at different temperatures, but decomposition always resulted. Their composition was established indirectly by analyzing for total stannous iodide, iodine, and free hydriodic acid, before and after cooling to separate the crystals. The difference in the amounts found represented what had left the solution to form the crystals, which were found to have the composition represented by the formula $SnI_2.HI$.

On the Reactions between Mercury and Concentrated Sulphuric Acid. By CHARLES BASKERVILLE AND F. W. MILLER. *J. Am. Chem. Soc.*, 19, 873-877.—When mercury is in excess almost pure mercurous sulphate and sulphur dioxide are formed, with no evidence of hydrogen as hydrogen sulphide ; but when the sulphuric acid is in excess, there is formed at 150° a compound having the composition $Hg_2SO_4.HgSO_4$. With increasing temperature there was corresponding decrease in the amount of mercurous sulphate formed. This method for preparing sulphur dioxide is not convenient on account of the volatility of mercury.

ORGANIC CHEMISTRY.

J. F. NORRIS, REVIEWER.

1. On the Oximes of Mucophenoxychloric and Mucophenoxybromic Acids. By HENRY B. HILL AND JOHN A. WIDTSOE. 2. On the Action of Aluminic Chloride and Benzol upon Mucochloryl Chloride, Mucobromyl Bromide, and the Corresponding Acids. By HENRY B. HILL AND FREDERICK L. DUNLAP. *Am. Chem. J.*, 19, 627–650.— Hill and Cornelison (*Am. Chem. J.*, 16, 188, 277) found that mucochloric and mucobromic acids yielded with hydroxylamine oximes of normal character, while the bromanhydrides of these acids could easily be converted by reduction into the corresponding substituted crotonolactones. The former reaction indicated the presence of an aldehyde group, while the latter was evidence of a different structure. The two views are expressed by the following formulæ :

$$
\begin{array}{ll}
\begin{array}{l} ClC.CHO \\ \parallel \qquad (1) \\ ClC.COOH \end{array} &
\begin{array}{l} ClC.CH.OH \\ \parallel \quad >O \quad (2) \\ ClCCO \end{array}
\end{array}
$$

The action of hydroxylamine on the methyl esters of mucochloric and mucobromic acids was studied in order to see if they, too, exhibited tautomerism. They did not form an oxime and accordingly were derivatives of formula (2). The present paper contains the results of a similar investigation of mucophenoxybromic and mucophenoxychloric acids and their esters. The methyl esters readily gave oximes identical with those prepared by the action of methyl iodide on the silver salts of the oximes of the acids. The marked difference in behavior which was shown by mucochloric and mucobromic acids and their esters disappears as soon as one atom of halogen is replaced by the phenoxy group. No explanation of this is offered. In the course of the investigation the following compounds were prepared and are described fully : the silver salt and methyl and ethyl esters of mucophenoxybromoxime, methylmucophenoxy-bromate, mucophenoxybromoxime anhydride, phenoxybrom-maleïnimide and its silver salt, and the corresponding deriva-tives of mucophenoxychloric acid.

2. The conversion of mucochloric and mucobromic acids, through their bromanhydrides, into substituted crotonolactones, finds its simplest explanation in the assumption that these acids them-selves are oxylactones (formula 2).   This conclusion has been reached by the study of a reaction of an entirely different char-acter.   The two possible structures for the chloranhydrides of mucochloric acid are :

$$\begin{matrix} \text{ClC.CHO} \\ \| \\ \text{ClC.COCl} \end{matrix} \quad \text{and} \quad \begin{matrix} \text{ClC.CHCl} \\ \| \quad {>}\text{O}. \\ \text{ClC.CO} \end{matrix}$$

The first should give an aldehyde ketone while the second should yield dichlordiphenylcrotonic acid when treated with benzene and aluminum chloride.   Both the acid and its chloranhydride when so treated gave dichlordiphenylcrotonic acid.   With muco-bromic acid the analogous acid containing bromine was formed. The structure of the latter was determined by conversion through oxidation into diphenylacetic acid.   The following substances are described : mucochloryl chloride ; dichlordiphenylcrotonic acid and its barium and calcium salts ; diphenylbutyric acid, formed by the reduction of diphenyldichlorcrotonic acid, and its silver salt; dibromdiphenylcrotonic acid, its barium, calcium, and silver salts, and methyl ester ; and diphenylbromallylene dibro-mide, formed by the action of bromine on the product resulting from the decomposition of dibromdiphenylcrotonic acid in alka-line solution.

On the Absorption of Oxygen by Tetrabromfurfuran. BY HENRY A. TORREY.   *Am. Chem. J.,* 19, 668–672.—Hill and Hartshorn (*Ber. d. chem. Ges.,* 18, 448) found that α-dibrom-furfuran was readily oxidized to maleic acid upon exposure to the air, but that tetrabromfurfuran, which also contains two bromine atoms in the α-position, under ordinary conditions, was not affected.   It has lately been discovered that exposure to direct sunlight causes a similar decomposition of tetrabromfur-furan.   This change is most readily effected when the substance is exposed to strong sunlight in an atmosphere of dry oxygen. In 7 or 8 hours the increase in weight is equivalent to about 85 per cent. of one molecular weight of oxygen.   The main prod-uct was shown to be dibrommaleyl bromide.

On Certain Derivatives of Brommaleic and Chlormaleic Acid-Aldehydes. BY HENRY B. HILL AND EUGENE T. ALLEN. *Am. Chem. J.*, 19, 650–667.—Limpricht (*Ann. Chem.*,

165, 287) obtained with great difficulty, by the action of aqueous bromine upon pyromucic acid, a compound which he considered to be the half aldehyde of fumaric acid, as it yielded the latter when oxidized with silver oxide. Since the authors were unable to isolate the compound from the viscous reaction-product, it was treated with hydroxylamine hydrochlorate and an aldoxime was obtained in small quantity, which contained bromine. By using pyromucic acid and bromine in the molecular proportions of one to three the yield of the oxime was greatly increased. From the brommaleïc acid-aldoxime so prepared the barium, lead, and silver salts and methyl ester were formed. The anhydride was made by the action of concentrated hydrochloric acid and by dissolving the oxime in concentrated sulphuric acid and precipitating with water. In the same way the anhydrides of mucochlor- and mucobromoxime were prepared. Brommaleïn-imide could not be formed by heating directly the brommaleïc acid-aldoxime, but was obtained in small yield by sublimation from the carbonaceous residue left when the anhydride was cautiously heated. Hydrobromic acid reacts with brommaleïc acid-aldoxime forming brommaleïc acid, or, on long standing, dibromsuccinic acid. By dissolving the oxime in a mixture of acetic acid and acetic anhydride and saturating with hydrochloric acid gas, bromchlorsuccinic acid was obtained. By the action of hydrochloric acid, hydrobromic acid, and bromine on the methyl ester of brommaleïc acid-aldoxime direct addition-products were formed. Tribromsuccinic acid-aldoxime was changed into dibromacroleïnoxime by warming with water at 40°. It was impossible to isolate dibromacroleïn from its oxime, but by boiling the latter with dilute hydrochloric acid a solution was obtained, which showed the reactions of aldehydes and from which an $\alpha.\beta$-dibromacrylic acid was obtained in small yield by oxidizing with silver oxide. As the substituted acrylic acid did not melt at the melting-point of dibromacrylic acid obtained from tribromsuccinic acid, it was probably an isomer of it. Chlormaleïc acid-aldoxime was prepared by the action of chlorine on pyromucic acid and was studied in the same way that the corresponding bromine compound was.

Derivatives of Eugenol. BY F. J. POND AND F. T. BEERS. *J. Am. Chem. Soc.*, 19, 825–832.—Benzyl eugenol, prepared by the action of benzyl chloride on the sodium salt of eugenol, was transformed into its isomer isobenzyl eugenol,

$$CH_3O.C_6H_3\Big\langle{{OCH_2=C_6H_5}\atop{CH=CH.CH_3,}}$$

by boiling with alcoholic potassium hydroxide. This yielded a dibrom addition-product, which was converted by sodium methylate into the ketone $CH_3.C_4H_3\left\langle\begin{array}{l}OCH_3.C_4H_3\\CO.CH_3.CH_3\end{array}\right.$ The oxime of the latter was also prepared.

Derivatives of Benzenesulphonic Acid. By T. H. NORTON. *J. Am. Chem. Soc.*, 19, 835–838.—A detailed description is given of the preparation and properties of benzenesulphonbromide and the sodium, potassium, ammonium, and lithium salts of benzenesulphonic acid.

On a Soluble Compound of Hydrastine with Monocalcium Phosphate. By T. H. NORTON AND H. E. NEWMAN.. *J. Am. Chem. Soc.*, 19 ,838–840.—A large excess of hydrastine was agitated with a saturated solution of monocalcium phosphate. At the end of 80 hours the solution was filtered and evaporated to dryness, when an amorphous residue was left, which was soluble in about ten parts of water. The analyses corresponded to the formula $2Ca(H_2PO_4)_2.3C_{21}H_{21}NO_6$.

Aliphatic Sulphonic Acids. By ELMER P. KOHLER. *Am. Chem. J.*, 19, 728–752.—The work described is the result of an effort to find general methods for the preparation of unsaturated aliphatic sulphonic acids, and particularly of isomeric sulphonic acids analogous to fumaric and maleic acids. A large number of sulphonic acids were studied with reference to their behavior with phosphorus pentachloride and the behavior of the resulting chlorides toward water. As it was found that the main product of the action of water on the chloride of a (1, 2)-disulphonic acid was an unsaturated acid, (1, 2)-ethanedisulphonic acid was carefully studied. The work now published is the result of this part of the investigation. Sodium ethanedisulphonate was prepared by boiling ethylene bromide with a saturated solution of sodium sulphite. The author's method gives 95 per cent. of the theoretical yield. From this a nearly quantitative yield of the chloride was obtained by the action of phosphorus pentachloride. As it melted sharply, the possibility of the formation of isomers was excluded. From the close analogy of (1,2)-ethanedisulphonic acid, succinic acid, and orthosulphobenzoic acid, it seemed probable that the former, like the latter, would yield both a symmetrical and an unsymmetrical chloride. But one, however, was obtained, although the conditions under which it was prepared were varied. As the same compound was also formed by the action of carbonyl chloride on ethanedisulphonic acid, it has probably the symmetrical structure $(CH_2SO_2Cl)_2$. The chloride reacts with water very slowly at 0°. At 100° it decomposes rapidly, forming sulphur dioxide, hydrochloric acid, and ethylenesulphonic acid, according to the following equation: $(CH_2SO_2Cl)_2 + H_2O = CH_2 = CH.SO_3H + SO_2 + 2HCl$. The yield of unsaturated acid is about 90 per cent. of the theory. A small amount of ethanedisulphonic acid is formed at the same time. Alcohols produce a similar decomposition, the yield of unsaturated acid decreasing as the number of carbon atoms in the alcohol increase; *e. g.*, methyl, propyl, and amyl alcohols, give 73, 56, and 51 per cent. respectively. Ammonia does not give an amide, but ammonium chloride, ammonium sulphite, ammonium ethanedisulphonate, and anhydrotaurine. With aniline

anhydrophenyltaurine, $\begin{matrix} CH_2 \\ | \\ CH_2SO_2 \end{matrix} \Big\rangle N.C_6H_5$, and the hydrochlorate

of anilidoethanesulphone anilide, $\begin{matrix} CH_2NHC_6H_5.HCl \\ | \\ CH_2SO_2NHC_6H_5 \end{matrix}$, were obtained. From the latter the free base was isolated, and from this the monoacetyl derivative was made, which had the structure $\begin{matrix} CH_3CO \\ \\ C_6H_5 \end{matrix} \Big\rangle N.CH_2CH_2SO_2NHC_6H_5$, as it was soluble in alkali. Since the amide of ethanedisulphonic acid could not be prepared by the usual reactions, an effort was made to get it by the action of acetamide on the chloride. The reaction did not take place in the manner expected. The hydrochlorate of acetamide, ammonium ethanedisulphonate, and ammonium chlorethanesulphonate were obtained. The chloride, dissolved in glacial acetic acid, was decomposed by sodium acetate, the products being, as before, ethylene sulphonic acid and ethane disulphonic acid. With zinc dust a 53 per cent. yield of (1,2)-ethanedisulphinic acid was obtained. As a result of the work the conclusion is drawn that the best way to prepare unsaturated aliphatic sulphonic acids is to decompose with water the chlorides of polysulphonic acids, which have the sulphonic acid residues in the (1,2)-positions with respect to each other.

Some New Derivatives of Diacetyl. By HARRY F. KELLER AND PHILIP MAAS. *J. Franklin Inst.*, 144, 379–385.—As hydrogen peroxide converts diacetyl quantitatively into acetic acid, its action was studied on dibrom and tetrabromdiacetyl in order to determine their structure. If the oxidizing agent acts in the same way with these substances and they have symmetrical structures, the first would give two molecules of bromacetic acid and the second two molecules of dibromacetic acid. Bromacetic acid was obtained from the former, but the latter yielded penta-

bromacetone. Diacetyl and hydrocyanic acid unite 'quantitatively to form a dicyanohydrin, which is readily converted into
dimethylracemic acid. Attempts to prepare a monocyanohydrin
of diacetyl were unsuccessful, however. When diacetyl and
hydrocyanic acid were brought together in molecular proportions, there was evidence of a reaction, but no compound could
be isolated. The resulting viscous liquid was treated with
hydrochloric acid under different conditions, in order to saponify
it, but no acid was obtained. Dibromdiacetyl and hydrocyanic
acid united, forming a dicyanohydrin. Dichlordiacetyl was prepared by the action of chlorine on diacetyl dissolved in chloroform.

On Diacyl Anilides. By H. L. WHEELER, T. E. SMITH
AND C. H. WARREN. *Am. Chem. J.*, 19, 757-766.—The fact
that formanilide differs in many of its physical properties from
other anilides has been offered as evidence that it is not a true
anilide, and consequently does not contain the formyl group

$-C\overset{\displaystyle O}{\underset{\displaystyle H}{\big\langle}}$. As the formyl group is present in the diacyl ani-

lides, a crystallographic study was made of formylbenzene-
sulphanilide and its homologues containing the acetyl and propionyl groups in order to discover if any marked differences in
physical structure existed. It was found that the anilides containing acetyl and propionyl show close crystallographic analogy,
both being monoclinic, hemimorphic, and pyroelectric, their
axes and the angle β being similar, whereas, formylbenzene-
sulphanilide is orthorhombic. It follows, therefore, that the departure in physical properties of formanilide from those of its
homologues can not be taken as evidence against its being a true
anilide. In addition to the compounds mentioned, the following
are described : *n*-butyrylbenzenesulphanilide, benzoylbenzenesulphanilide, benzoylbenzenesulph-α-(and β)-naphthalide. The
crystallography of benzoylbenzenesulphanilide and formanilide
is also given. The latter crystallizes in the monoclinic system,
whereas acetanilide is orthorhombic.

**Note on the Somewhat Remarkable Case of the Rapid
Polymerization of Chloral.** By J. W. MALLET. *Am. Chem.
J.*, 19, 809-810.—A sample of anhydrous chloral, which had
been kept in a sealed tube for over a year, exploded with considerable violence while undisturbed in a room at about 20° C.
As the compound was found in the form of metachloral, it is
probable that the explosion was due to sudden polymerization.

**Synthesis of Hexamethylene-Glycol Diethyl Ether and
Other Ethers from Trimethylene Glycol.** By ARTHUR A.

NOYES. *Am. Chem. J.*, 19, 766–781.—The results of the investigation are as follows: The monoethyl ether of trimethylene glycol is readily prepared by the action of sodium and ethyl iodide on a considerable excess of the glycol. The diethyl ether was prepared from the monoethyl ether, but not from the glycol itself, by a similar process. By the action of phosphorus trichloride, tribromide, and diiodide on the monoethyl ether, the hydroxyl was replaced by halogen with the production of ethyl chlor-, brom-, and iodpropyl ethers. The yields obtained were about 50, 75, and 30 per cent. respectively. Hydriodic acid gas at 0° caused replacement of the ethoxy as well as the hydroxyl group. The diethyl ether of hexamethylene-glycol was found to be produced in very small quantity by the action of sodium on an ethereal solution of ethyl chlorpropyl ether and in somewhat larger quantity by the similar reaction with the corresponding brom and iodo ether; but the best yields did not exceed 30 per cent. of the theoretical. By heating the ethyl brompropyl ether in a water-bath for several hours, first with a slight excess of concentrated aqueous potassium cyanide solution, and then with concentrated potash and acidifying, γ-ethoxyvalerianic acid was obtained in a 40 per cent. yield. It was shown that by the electrolysis of the salt of this acid a small quantity of hexamethylene-glycol diethyl ether is probably formed. Attempts to prepare a metallic derivative by the action of sodium amalgam and of the zinc-copper couple on the ethyl brompropyl ether, were unsuccessful. The Fittig synthesis was successfully carried out with the bromether and brombenzene, a 54 per cent. yield of ethyl γ-phenylpropyl ether being obtained. The ethoxypropyl radical was also readily introduced into malonic ester by heating its sodium compound with the ethyl brompropyl ether, the yield being 34 per cent. The substituted malonic ester so obtained gave on saponification and subsequent heating to 170° of the acid separated from its salt, δ-ethoxyvalerianic acid in a 37 per cent. yield. Another substance was produced at the same time in considerable quantity, but its nature was not determined.

Chemical Bibliography of Morphine, 1875-1896. BY H. E. BROWN. *Pharm. Rev.*, 15, 204–205.—This important bibliography is being made under the supervision of A. B. Prescott, and when completed will be of great value.

W. O. CROSBY, REVIEWER.

Fractional Crystallization of Rocks. BY GEORGE F. BECKER. *Am. J. Sci.*, 154, 257–261.—This paper supplements the author's vigorous and effective criticism of the theory of magmatic differentiation based upon molecular diffusion in accordance with Soret's principle, as first proposed by Iddings (see this *Rev.*, 3, 67). In the former paper the author was content to demolish the prevalent theory, and confessed himself unable to propose any better alternative hypothesis than that the differences between well-defined rock types are due to original and persistent heterogeneity in the composition of the globe. But now he accepts differentiation as a fact and brings forward the simple and familiar principle of fractional crystallization to account for it. He conceives that during the circulation of the magma, due to inevitable convection currents, convection being the mortal enemy of any process of separation involving molecular flow, the less fusible constituents crystallize on the relatively cool walls, and the residual magma is composed in increasing proportion of the more fusible constituents, until the condition of maximum fusibility or eutexia is reached, after which differentiation necessarily ceases. The limitation imposed upon this process by viscosity is recognized; but in dikes and laccolites of mobile lavas, convection and fractional crystallization are inevitable. And this simple principle, which is the very opposite of magmatic differentiation and almost imseparable from consolidation, unquestionably affords the best explanation yet proposed of the differences observed in many igneous masses; and, as the theory requires, many homogeneous masses are undoubtedly approximately eutectic.

Eopaleozoic Hot Springs and the Origin of the Pennsylvania Siliceous Oolite. BY GEORGE R. WIELAND. *Am. J. Sci.*, 154, 262–264.—The author accepts the view of Bergt and Hovey that the beautiful siliceous oölite occurring in bowlders derived from the calciferous formation near the State College in Center County, Pa., is not pseudomorphic after ordinary or calcareous oölite, but due to direct deposition from silica-laden waters of Hot Springs. Additional confirmation of this view is afforded by the fact that in the limited oölite area and nowhere else in the entire region are found in considerable numbers bowlders of a built-up structure which may have formed the veins of hot springs or geysers. An analysis of the oölite shows 99.1 per cent. of silica. Analyses of two less pure varieties of siliceous oölite are also given.

Pseudomorphs from Northern New York. By C. H. SMYTH, JR. *Am. J. Sci.*, 154, 309–312.—Describes pseudomorphs of pyroxene after Wallastonite from Diana, and of mica after scapolite and pyroxene from near Gouverneur. The chemical relations of the species are considered, but no analyses are given.

On the Chemical Composition of Hamlinite and its Occurrence with Bertrandite at Oxford County, Maine. By S. L. PENFIELD. *Am. J. Sci.*, 154, 313–316.—Only recently has enough of this rare phosphate for a complete analysis been found. The mineral was separated by heavy solutions from every associated species except apatite, and from this by boiling in dilute hydrochloric acid, all possible precautions being taken to obtain the hamlinite in a state of purity. Traces of adhering feldspar and mica, however, could not be wholly avoided; and to this source are referred the small proportions of Fe_2O_3, SiO_2, and of alkalies with their equivalent of the Al_2O_3, which the analysis reveals. Deducting these and calculating the remainder to 100 per cent., we have as the normal composition of hamlinite ; P_2O_5, 30.20 ; Al_2O_3, 32.67 ; SrO, 19.25 ; BaO, 4.18 ; H_2O, 12.53 ; F, 2.01 ; corresponding very closely to the formula $[Al(OH)_2]_3[SrOH]P_2O_8$, the Sr being replaced by Ba in the ratio of 7 : 1, and the OH by F in the ratio of 13 : 1. In its chemical composition hamlinite holds a unique position among minerals as strontium and barium have never before been observed as essential constituents of a phosphate, and this is the first time that a pyrophosphate has been recorded.

Italian Petrological Sketches, V., Summary and Conclusion. By HENRY S. WASHINGTON. *J. Geol.*, 5, 349–377.—The volcanoes of the main Italian line, extending from Tuscany to Naples, embrace two types—great strata volcanoes, characterized by leucitic lavas and a great variety of products, and smaller but more prominent cones, the lavas of which are non-leucitic and consist of but a single extreme (acid or basic) type for each cone, the most of the lavas being referable to the trachy-dolerites, which include ciminite, vulsinite, and toscanite. In the descriptions of these subtypes all known reliable analyses, 41 in number, are quoted and discussed, and the normal proportions of the component minerals computed for each rock. In discussing the correlation of the trachy-dolerites, the author proposes a general classification of the feldspathic effusive rocks based upon the dominant feldspar and the proportion of silica. The classification and nomenclature of the leucitic lavas are regarded as provisional, the leading types being leucitite, leucite-basalt, leucite-basanite, leucite-tephrite, leucite-trachyte, and leucitephonolite. Each type is briefly characterized, and 18 analyses are selected for comparison and discussion. The interesting fact is thus brought out that the silica ratio, the upper limit of which is 60 per cent., does not present a perfect gradation, but the analyses are grouped about 49 and 56 per cent. A similar clustering of the analyses about definite points is noted for the trachy-dolerites, and for other constituents than silica, especially Fe_2O_3, FeO, MgO, and CaO. That is, in neither group do the analyses form a gradual series, from the most basic to the most acid, as Brögger understands a series to be constituted. A similar relation is shown to hold for the lavas of other regions; and an explanation is found in the theory of magmatic differentiation, the process having completed its course in these instances. The author concludes, however, that the two main groups of rocks do not represent two distinct primary differentiation products, but that their differences are due to diverse differences of extrusion and solidification. In other words, primary magmatic differentiation gives the distinction of acid and basic observed in each group, but the varying conditions of extrusion determine the development, now of leucite and again of feldspars. The parent magma is shown to have been rich in potash and lime, and its normal composition is supposed to have been approximately : SiO_2, 57–58 ; Al_2O_3, 17–18 ; total iron oxides as FeO, 6–7 ; MgO, 2–3 ; CaO, 5.0–6.5 ; Na_2O, 2–2.5 ; K_2O, 7–8 ; H_2O, 1–1.5 per cent.

Secondary Occurrences of Magnetite on Islands of British Columbia by Replacement of Limestone and by Weathering of Eruptives. By JAMES P. KIMBALL. *Am. Geol.*, 20, 13–27.—Two separate types of ferriferous deposits, due to hydrochemical replacements, are recognized. One is a morphological replacement of limestone by double decomposition between ferrous salts and calcic carbonate, the former being derived from ferrous silicates; the other type is a partial, and not necessarily pseudomorphic, replacement of ferrous silicates in weathered basic rocks, or more explicitly, a residual concentration or fixation of iron oxides incidental to development of soluble alkaline carbonates from weathering oxidation or splitting up of ferriferous silicates. The common association of the deposits of this type with limestone is regarded as significant, calcic carbonate taking the place, in part, of the alkaline carbonates as an agent for the precipitation of ferric oxide. Among the causes determining the locus of deposition are recognized molecular or concretionary attraction of the mass for homologous matter in passing solutions, or even in suspension, the last being otherwise designated by the author as the extra molecular tendency of ochreous material to form concretions or aggregations. As might be inferred from the above, the two types of deposit, that due to replacement of limestone, and that due to the replacement of basic silicates through the agency of contiguous limestone, are often intimately associated. The probable details of these metasomatic changes are stated ; and the illustrative occurrences on three islands are described, the paper concluding with four analyses of magnetite from these and related deposits.

On the Magnetite Belt at Cranberry, North Carolina, and Notes on the Gensis of this Iron Ore in General in Crystalline Schists. By JAMES P. KIMBALL. *Am. Geol.*, 20, 299–312.—This paper is in line with the preceding. The Cranberry iron ore deposits are described as obscurely defined lenses of magnetite in a persistent belt of pyroxene and amphibole; and the author, rejecting the theory of original magmatic differentiation as an explanation of these and many similar ore bodies, regards the magnetite and its encircling zone of epidote as derivatives through hydrochemical or weathering processes from the enclosing basic silicates. This view is supported by the observation that epidotization of pyroxene from weathering action is practically limited to the superficial zone of rock decay, increasing with proximity to the surface. But no facts are cited to show that the magnetite is also limited in depth ; and the proof that the epidote is coeval with the magnetite or genetically related to it is inconclusive. Three analyses of the Cranberry ore are quoted, showing the composition of the purest magnetite obtainable, of the mixed magnetite and epidote, and of the ore as prepared for market, the percentages of metallic iron being respectively 64.64, 32.37, and 45.93. A notable proportion (seven per cent. in one instance) of the iron obtained in the furnace is attributed to ferrous and ferric oxides in the associated silicates —pyroxene and its derivative (epidote)—an analysis of pure pyroxene giving 24.01 per cent. of metallic iron. In the further discussion of the origin of these and similar ore deposits. the author holds that even, if we grant that magmatic differentiation has played a part, it must have been supplemented by subsequent hydrochemical processes to convert the original ferriferous silicates (pyroxene, etc.) to bodies of highly concentrated magnetite.

The Fisher Meteorite. Chemical and Mineral Composition. By N. H. WINCHELL. (CHEMICAL ANALYSIS, BY C. P. BERKEY.) *Am Geol.*, 20, 316–318.—Prof. Winchell gives the results of microchemical tests for determining a mineral resembling maskelynite. Not enough of the mineral is present to warrant a quantitative analysis ; and the presence of maskelynite is inferred from the feeble polarization, absence of cleavage, presence of lime and soda, and the fact that the glass from which it seems to have crystallized contains soda, and no soda has been detected in the other minerals. Mr. Berkey gives a bulk analysis of the meteorite, showing the presence of Si, Al, Fe, Ni, Ca, Mg, and S. The sulphur is present in small quantity in the mineral troilite and no alkalies were found.

Note on the Hypersthene-Andesite from Mt. Edgecumbe, Alaska. By H. P. CUSHING. *Am. Geol.*, 20, 156–158.—The specimens examined were collected by H. F. Reid in the crater of the volcano, which has probably not been long extinct. They are a quite normal hypersthene-augite-andesite made up of plagioclase, hypersthene, augite, and magnetite, in the order of abundance. It is the only occurrence of andesite reported from Alaska, except the hornblende-andesite of Bogosloff Island, and it is of interest from the fact that similar lava is found throughout the whole extent of the American Sierra belt.

Ferric Sulphate in Mine Waters, and Its Action on Metals. By L. J. W. JONES. *Proc. Col. Sci. Soc.*, June 5, 1897, 1–9.— The corrosive action of the water of the Stanley Mine, Idaho Springs, Col., is so great that ordinary iron pipe cannot long resist it. The substitution of copper and various alloys for iron proved unavailing. On standing, the water yields a muddy brown precipitate which, after drying, gave on analysis: Fe_2O_3, 53.57; Al_2O_3, 2.87; SiO_2, 10.85; SO_3, 11.46; H_2O, 21.14; total, 99.89. It is evidently a hydrated basic sulphate of iron. An analysis of the water filtered off from the precipitate shows, besides silica and sodic chloride, sulphate of Na, K, Al, Zn, Mn, Mg, Ca, Fe (ferrous and ferric), and Cu, the ferric, calcic, magnesic, and sodic sulphates predominating, and the whole slightly exceeding three parts in one thousand. Free sulphuric acid is not present, and the corrosive action of the water is attributed in part to the copper sulphate, but chiefly to the ferric sulphate. After noting the action of the water the lead, copper, and bronze, the conclusion is reached that wooden pipes are the most serviceable.

1. Lead and Zinc Ores of Iowa. By A. G. LEONARD. **2. The Sioux Quartzite and Certain Associated Rocks.** By S. W. BEYER. **3. Artesian Wells of Iowa.** By W. H. NORTON. *Iowa Geol. Survey*, 6, 11–429.—In the first paper the chief sources of the world's supply of lead and zinc are cited, and their mode of occurrence in north-eastern Iowa fully described. The lead occurs mainly as sulphide, and the zinc as carbonate; the two ores being associated, but seldom intermingled. The concentration has been effected by lateral secretion from the limestone by infiltrating surface waters.

2. The Sioux quartzite extends across the north-western corner of Iowa into Minnesota and South Dakota ; but the author considers especially its exposures and its relation to certain associated rocks in the latter state. The quartzite is there cut by a large mass of olivine diabase ; and the alteration products of the feldspar, augite, and olivine of the diabase are described.

3. After a general discussion of the definition and theory of artesian wells and of artesian conditions in Iowa, the author gives a detailed record of the wells in the state and chemical analyses of the waters, many of which are strongly mineral. The contained gases are ammonia, hydrogen sulphide, and carbon dioxide, oxygen not having been observed. The dissolved solids are carbonates of Ca, K, and Fe ; bicarbonates of Mg and Na ; phosphates of Mg and Na ; sulphates of Ca, Mg, Na, and K ; ferric oxide ; chlorides of Mg, Na, and K ; silica and alumina, some of the wells containing a greater proportion and larger variety than others. Their sanitary, therapeutic, and industrial value are discussed, and analyses are also given of the river, surface, and drift waters of the state.

The Marquette Iron-bearing District of Michigan. By CHARLES RICHARD VAN HISE AND WILLIAM SHIRLEY BAYLEY. *Monographs U. S. Geol. Survey*, 28, 1–608.—This elaborate and beautifully illustrated report is designed as a final account of the geological structure and petrography of the oldest iron-producing area of the Lake Superior region. The petrographic descriptions are accompanied now and then by analyses of the rocks—granitite, gneiss, schist, slate, greenstone, peridotite, serpentine, etc. The chief iron-bearing horizon is the Negaunee formation ; and it is here that the chemical interest of the report culminates. Petrographically the iron-bearing formation comprises sideritic slates, which may be grüneritic, magnetitic, hematitic, or limonitic ; grünerite-magnetite-schists ; ferruginous slates ; ferruginous cherts ; gaspilite ; and iron ores. The descriptions of these rocks, with analyses of the sideritic slates, grünerite-magnetite-schists, and jaspilite, are followed by a section on the origin of the ores, which may be summarized as follows : The ores are not eruptive. The original form of the iron is the carbonate-siderite ; and the whole of the iron-bearing formation was probably originally a lean cherty carbonate of iron with some calcium and magnesium. From this rock the ferruginous cherts and jaspers were developed by peroxidation of the iron and solution of the calcareous materials. The ore bodies proper are due to a secondary enrichment by the action of downward percolating water. They invariably rest upon an impervious substratum, usually of soaprock, and the convergence of the downward-flowing waters determined by pitching synclines or the intersection of the impervious soaprock by greenstone dikes has favored the concentration of the ore. The immediate cause of the deposition of the iron at these points is found in the oxidation of the dissolved carbonate by oxygen-bearing waters coming more directly from the surface. It is further essential to the formation of pure ores that the waters, rendered alkaline by the alteration of alkaline silicates to soap rock, should dissolve and remove large amounts of silica (chert, etc.).

Geology of the Denver Basin in Colorado. BY SAMUEL FRANKLIN EMMONS, WHITMAN CROSS, AND GEORGE HOMANS ELDRIDGE. *Monographs U. S. Geol. Survey*, 27, 1–556.—In this very complete study of a limited area of only moderate geologic complexity, we find analyses of the rocks of greatest chemical interest, including Wyoming and Niobrava limestones, the Benton and Laramie iron stones (earthy and concretionary carbonates), löess, dolerite and augite from it, basalt, augite-mica-syenite, tuff, coals, and fire-clays ; and also artesian waters. The löess is distinguished from that of the Mississippi and Rhine valleys by its higher proportion of alkalies, indicating that, owing to the avidity of the climate, the feldsparthic constituent is but little decomposed. The numerous analyses of coal were made partly in Denver, and partly in Washington ; and the curious fact is developed that, although they were all made by the same chemist, in the same manner, with the same care, and at the same temperature, the Washington analyses showed from 3 to 5 per cent. less moisture than the Denver analyses, while the volatile combustible matters are found to vary in the opposite direction, increasing as the water decreases. The specific gravities also vary inversely as the water and directly as the volatile combustible matters. It is a significant fact that the sum of the volatile combustibles and water is approximately the same for both the Washington and Denver analyses. The discussion of these anomalous and seemingly contradictory facts leads to the conclusion that the hygrometric condition of the atmosphere must be the cause of these persistent discrepancies. In the drier atmosphere of Denver, under less barometric pressure, the coals parted with a greater amount of moisture than it is possible to expel under the usual conditions of analysis in the more moist atmosphere and with the higher barometer of the east. Since the water not expelled must be reckoned as volatile combustible matter, the economic value of the coal is involved ; and it is suggested that the limits of the atmospheric influence should be determined and taken into account in all future analyses of coal. The fixed carbon is found to increase with the depth, ranging in the Golden mines from 38.32 per cent. at 130 feet, to 43.42 per cent. at 650 feet. The further discussion includes the ash, fuel ratio, and classification of the coals. A comparison of the fire-clays with those of Eastern and European localities, and with pure kaolin, shows that they are in the main of excellent quality. In connection with an extended account of the artesian wells, relating chiefly to the geological conditions, available supply of water, rate and diminution of flow, four analyses by Prof. Chauvenet are quoted, the total solids ranging from 17.20 to 56.60 parts in 100,000, sodium carbonate largely predominating.

The Gold Quartz Veins of Nevada City and Grass Valley, California. BY WALDEMAR LINDGREN. *U. S. Geol. Survey, Ann. Rep.*, 17, *part 2*, 1–262.—The first essential to understanding the geology of this region is the distinction of the bed-rock series of crystalline schists and ancient igneous rocks and the relatively modern gravels, sands, clays and volcanic rocks which overlie them horizontally. The igneous rocks of the bed-rock series are in the main of mixed granitic and dioritic (granodiorite) character and of complex mineralogical and chemical composition. They also include diabases, gabbro, pyroxenites, peridotites, porphyrite and amphibolite, analyses of which are given, as well as of serpentine derived from the basic eruptives, and the superjacent rhyolite tuffs, and associated clays and sandstones. The sedimentary rocks, including clay slate, siliceous argillite, (of which one complete analysis is given) quartzitic sandstone and chert, are supposed to be altered Carboniferous strata, and are distinctly older than the granodiorite and and diabasic series ; while above them is a series of clay slates of the upper Jurassic. The metamorphic processes to which these rocks have been subjected are : Regional or dynamochemical metamorphism ; contact metamorphism ; hydro, hydrothermal, and solfataric metamorphism ; serpentinization of magnesian silicates ; and surface weathering. Each process is briefly described and its products enumerated. The development of the iron sulphides is specially considered, as having an important bearing upon the genesis of the ore deposits; and for the same reason the section on the mineralogy of the veins is accompanied by analyses of the mine waters and the deposits which they form, and of the liquid inclusions of the quartz. Several analyses of the sulphurets (concentrates), amounting to 2 to 3 per cent. of the ore, show the normal ratios of the base metals, as well as of gold and silver. The changes in the rocks, due to the formation of the veins, are discussed at some length, with numerous analyses. In the discussion of the genesis of the veins the conclusion is reached that they were formed at a great depth by ascending thermal waters, lateral secretion playing but a very minor part ; and in confirmation of this view numerous assays and analyses are cited, showing that the wall rocks rarely contain appreciable traces of gold. The details of the discussion relate to the origin of the metals and gangue, solubility and synthesis of the gangue minerals, sulphides, and gold, relation of solubility to temperature and pressure, precipitation of the gold and mode of deposition of the vein minerals.

Geology of Silver Cliff and the Rosita Hills, Colorado. By WHITMAN CROSS. *U. S. Geol. Survey, Ann. Rep.*, 17, *part 2*, 263–403.—In the geological history of this district an ancient complex of granite and various gneissic rocks was cut by dikes of syenite, diabase and peridotite, and then subjected to prolonged erosion previous to the outbreak of volcanic action. This volcanic center is comparable to the one at Cripple Creek, and from it poured at successive periods varieties of andesite, diorite, dacite, rhyolite, trachyte, and finally the Bassick agglomerate. Nearly all of these types were analyzed, and the rocks of the volcanic series are tabulated for comparison. The decomposition products are of great extent and interest ; and analyses are given of quartz-alunite rock (the first known occurrence in this country of crystalline alunite as a rock constituent), quartz-diaspore rock, pitchstone and the siliceous clay derived from it, and muscovetized rocks derived from several vulcanic types. The sequence of the lavas is discussed and the evidences of differentiation noted.

The Mines of Custer County, Colorado. By SAMUEL FRANKLIN EMMONS. *U. S. Geol. Survey, Ann. Rep.*, 17, *Part 2*, 405–472.—This monograph supplements the preceding, discussing the mining geology of Rosita and Silver Cliff, as that does the general geology. It is in the main an account of four mines, representing four types of deposits. The Pocahontas-Humbolt is a true fissure vein. The Bassick is the most typical example yet described of a neck vein, consisting of a well-defined ore-shoot or chimney formed by fumarole action in the coarse andesitic agglomerate filling an old volcanic neck, and the ore, of which a single analysis is given, occurring as successive coatings on the rounded masses and bowlders of andesite. The Bull Domingo is structurally similar to the Bassick, being a nearly vertical chimney in a conglomeratic mass of the granite and gneiss country rock, with the ores coating the bowlders and fragments ; but the origin of the bowlder zone or stock is found in the intersection of a complicated series of fractures and not in volcanic action.   At Silver Cliff the ore (silver chloride) occupied irregular cracks in the surface flow of rhyolite, but in the deep workings of the Geyser Mine, where the rocks are firmer, the ore (argentiferous sulphides) forms a well-defined but narrow vein.   This mine, now one of the deepest in the West, is described in detail, with complete analyses of the decomposition products of the rhyolite, the ores and gangues, the calcareous and slightly metalliferous sinters deposited by the mine waters, the vadose or surface waters from the upper levels of the mine, and the similar but more concentrated deep waters from the lower levels.   The relations of these waters to the ore deposition are discussed, and as evidence that the ores of this district may have been derived from the enclosing formations.   Ten assays of the country rocks are given, all but four showing appreciable amounts of silver.   In conclusion this district is compared, as regards the character of the ores, with the analogous Cripple Creek District, forty miles north of it.

The Tennessee Phosphates. By CHARLES WILLARD HAYES. *U. S. Geol. Survey, Ann. Rep.*, 17, *Part 2*, 513–550.—A preliminary account of these phosphates appeared in the 16th Annual Report (see *this Rev.*, 2, 57).   They are classed as structural varieties of the two main types—the black and the white phosphates, the former originating in the deposition of a bed of phosphatic organisms in the Devonian Sea, while the latter is a secondary and essentially residuary deposit due to the differential solvent action of meteoric waters on phosphatic limestones. Analyses are given only for the white phosphates, giving from 27.4 to 33.4 per cent. of calcium phosphate ($Ca_3P_2O_8$).

The Underground Waters of the Arkansas Valley in Eastern Colorado. By GROVE KARL GILBERT.   *U. S. Geol. Survey, Ann. Rep.*, 17, *Part 2*, 551–601.—The water-bearing strata of this district are the sandstones of the Dakota group, the lowest member of the Cretaceous formation.   A detailed account of the geological structure of the valley precedes the discussion of the water supply.   The analyses of the artesian waters show a rather complex mineral constitution, with salts of Na, Ca, and Mg predominating, the maximum total solids being over 4,000 parts in 1,000,000.

Preliminary Report on Artesian Waters of a Portion of the Dakotas. By NELSON HORATIO DARTON.   *U. S. Geol. Survey, Ann. Rep.*, 17, *Part 2*, 603–695.—The water-bearing formation, as in the Arkansas Valley, is the Dakota sandstone, which receives its supply where it outcrops to the westward in the foot hills of the Rocky Mountains and the Black Hills, and which underlies nearly the whole of North and South Dakota, rising again to the surface near their eastern boundaries. The water is saline, containing considerable amounts of NaCl, and carbonates and sulphates of Na, Ca, and Mg, but rarely enough to affect its usefulness.

The Water Resources of Illinois. BY FRANK LEVERETT. *U. S. Geol. Survey, Ann. Rep.*, 17, *Part 2*, 695–828.—The rainfall and drainage of the state are discussed and the sources of supply of water for power and for the use of cities, villages, and rural districts, are described. The wells afford chiefly surface waters, but numerous analyses are given for both shallow and artesian wells, showing the proportions of inorganic and organic impurities. Few of the waters are strongly mineral, but many are seriously contaminated by drainage.

GENERAL AND PHYSICAL CHEMISTRY.

A. A. NOYES, REVIEWER.

The Dissociation of Electrolytes as Measured by the Boiling-Point Method. BY HARRY C. JONES AND STEPHEN H. KING. *Am. Chem. J.*, 19, 753–756.—The authors have determined the rise in boiling-point produced by dissolving potassium iodide and sodium acetate in alcohol, their chief object being to show the constancy of the values obtained by Jones' modification of the Beckmann apparatus (*this Rev.*, 3, 183) and the practicability of determining dissociation values by means of it. In the case of potassium iodide, the concentration varied in five experiments between 1.8 and 3.1 per cent., and the dissociation values betweed 25.4 and 27.2 per cent. ; while in the case of sodium acetate for concentrations between 1.0 and 2.1 per cent., the dissociation was only 1.0 to 1.8 per cent. These results are regarded as preliminary.

The Rate of Solution of Solid Substances in their own Solutions. BY ARTHUR A. NOYES AND WILLIS R. WHITNEY. *J. Am. Chem. Soc.*, 19, 930–934.—The authors have found, by agitating sticks of benzoic acid and of lead chloride with water for different lengths of time, that "the rate at which a solid substance dissolves in its own solution is proportional to the difference between the concentration of that solution and the concentration of the saturated solution."

On a Possible Change of Weight in Chemical Reactions. BY FERNANDO SANFORD AND LILIAN E. RAY. *Phys. Rev.*, 5, 247–253.—The authors have carried out a series of experiments similar to those of Landolt (*Ztschr. phys.!Chem.*, 12, 1). They find that, when ammoniacal silver nitrate and grape sugar solutions are caused to react within a closed vessel, the change of weight observed does not exceed the probable errors of the weighings, which is calculated to be about 0.04 milligram or $\frac{1}{3.000.000}$ part of the weight of the reacting substances.

Note on the Rate of Dehydration of Crystallized Salts. BY THEODORE WILLIAM RICHARDS. *Proc. Am. Acad.*, 33, 23-27. —The author has made experiments showing that crystallized barium chloride ($BaCl_2.2H_2O$), when placed in a desiccator over phosphorus pentoxide, loses its water rapidly and fairly uniformly until one molecule of water has escaped, and that then the rate of loss suddenly becomes much less, but again remains quite uniform as the dehydration progresses. The author seems to have overlooked the fact that similar experiments giving the same result have been made with this very salt by Müller-Erzbach (*Wied. Ann.*, 27, 623). Andreae (*Ztschr. phys. Chem.*, 7, 241) has also made accurate and extensive experiments in the same direction. The author has also compiled from existing data a table showing the specific gravity of sulphuric acid solutions which have definite vapor-pressures at 20°, 25°, and 30°.

Two Liquid Phases, II. BY WILDER D. BANCROFT. *J. phys. Chem.*, 1, 647-668.—There is given in this article " a graphical summary of the equilibria in three-component systems when one pair, two pairs, and three pairs of the components can form two liquid phases." In the latter part of the paper the author claims that the chemists who have investigated the principles of solubility-effect have stated that the solubility of any substance can be affected only by those other substances which bring into the solution one of its dissociation-products. As a matter of fact, no such absurd statement has ever been made by the "quantitative physical chemists;" on the contrary, the reviewer has emphasized on different occasions (*Ztschr. phys. Chem.*, 6, 243 ; 9, 603) the fact that the solubility-principles cannot be expected to hold true exactly, when the second substance is of such a nature or is added in so large an amount as to produce an appreciable change in the solvent power of the solvent.

The Electrolysis and Electrical Conductivity of Certain Substances Dissolved in Ammonia. BY HAMILTON P. CADY. *J. phys. Chem.*, 1, 707-713.—The author has made the interesting discovery that many salts dissolved in liquid ammonia at —34° have even a greater electrical conductivity than when dissolved in pure water at +18°. The ammonia employed was the commercial article used in the manufacture of ice ; it had a conductivity of 71×10^{-7}. As it no doubt contained impurities, and as there were difficulties in the determinations, the results are regarded only as preliminary. They serve, however, to give a general idea of the prevailing conditions. The molecular conductivity of salts was found to increase with the dilution as in the case of water. Some salts, especially mercury iodide and cyanide, which have only a very slight conductivity in water, were found to be good conductors, showing that the dissociating power of ammonia is greater than that of water. The rate of motion of ions through it seems also to be greater than through water. The author has also submitted the solutions of a number of salts in ammonia to electrolysis : in the case of lead, mercury, and silver salts, the metals are deposited on the cathode ; in the case of iodides, a dark-colored explosive compound, believed to be HN_4I, separated at the anode ; with potassium iodide, p assium amide was probably formed at the cathode ; etc. The most remarkable results were obtained with a solution of metallic sodium. Although an excellent conductor, there was no indication of electrolysis when a current was passed through : no deposit formed on the electrodes ; no gas was set free ; and no polarization-current could be detected. The article seems to the reviewer to open up a most interesting and important field of investigation ; and it is to be hoped that the author will continue the work under improved conditions as soon as possible.

Precipitation of Salts   By A. ERNEST TAYLOR.   *J. phys. Chem.*, 1, 718–733.—The article describes a continuation of the work previously referred to (*this Rev.*, 3, 75, 122, 152).

A Revision of the Atomic Weight of Nickel. I. The Analysis of Nickelous Bromide. By THEODORE WILLIAM RICHARDS AND ALLERTON SEWARD CUSHMAN. *Proc. Am. Acad.*, 33, 97–111. —Commercial nickel and pure nickel obtained by the carbon monoxide process were subjected to extensive processes of purification. From the former, two samples of nickel bromide were obtained differing from one another in the number of purifying operations to which they had been submitted ; and from the latter two further samples even more carefully purified were prepared. In addition to the usual methods of purification, the ammonia compound $NiBr_2.6NH_3$ was repeatedly crystallized from strong ammonia ; and the nickel was three times fractionally precipitated by electrolysis. All four samples of nickel bromide were sublimed, ignited at $400°$ in a current of hydrobromic acid, and cooled in nitrogen. The exact quantity of silver required for its precipitation was determined ; and the silver bromide formed was weighed, thus giving two independent ratios, 2AgBr : NiBr, and 2Ag : NiBr,. The mean value of the atomic weight obtained from the former ratio in seven experiments made with the two most carefully purified samples was 58.690 (±0.003), while that obtained from the latter ratio was 58.691 (±0.005). Very strong evidence of the purity of the substance analyzed is furnished by the fact that the four samples, which had been submitted successively to more numerous processes of purification, gave nearly the same results; namely, 58.677, 58.683, 58.688, 58.689. The authors adopt 58.69 (O = 16) as the most probable value. That given by Clarke in his last year's report was also 58.69.

A Revision of the Atomic Weight of Cobalt. I. The Analysis of Cobaltous Bromide. BY THEODORE WILLIAM RICHARDS AND GREGORY PAUL BAXTER. *Proc. Am. Acad.*, 33, 115–128. —This investigation is closely analogous to that on the atomic weight of nickel described in the preceding review. The cobaltous bromide used was purified, after removal of the metals of the hydrogen sulphide group, by three processes: first, by precipitation of the potassium cobaltic nitrite and subsequent electrolytic deposition of the metal; second, by precipitation in the form of the purpureochloride and subsequent electrolytic deposition; and third, by a combination of both these processes. The bromide was in each case finally sublimed, ignited in a current of hydrobromic acid, and cooled in one of nitrogen. The mean value of the atomic weight obtained from nine determinations of the ratio 2AgBr:CoBr, was 58.995 (±0.003), while that from eight determinations of the ratio 2Ag : CoBr, was 58.987 (±0.005). The four differently purified samples gave 58.987, 58.992, 58.995, and 59.004. The authors adopt 58.99 as the most probable value, while that given by Clarke in his last year's report was 58.93.

H. M. GOODWIN, REVIEWER.

On Electrosynthesis. BY W. G. MIXTER. *Am. J. Sci.*, 154, 51–62.—Electrosynthesis the author defines to be the chemical union of substances by an electric discharge as distinct from combination effected by the heat of the discharge. A feeble glow discharge from an induction coil driven by one storage cell and giving a spark in air about one centimeter long was used in the experiments for effecting the combination. The amount of combination of oxygen with hydrogen in a discharge tube connected in series with the tube under investigation, served in all cases as an arbitrary standard of reference. A measure of the chemical action was obtained by absorbing the products of the combination directly in the vacuum tube and noting the corresponding diminution of pressure on a manometer. Experiments were made at varying pressures. If the number of hydrogen and oxygen molecules combining be taken as 100, the corresponding values for the mixtures of other gases with oxygen are as follows : Carbonic oxide, 113 ; methane, 149; ethylene, 300 ; acetylene, 320 ; and ethane, 150. From a consideration of these results the author concludes that the glow discharge renders gaseous molecules chemically active, and that the molecular charges involved in "electrosynthesis" are analogous to those often produced by light, or by heat at temperatures below that at which gaseous dissociation is measured.

On the Temperature-Coefficient of the Potential of the Calomel Electrode, with Several Different Supernatant Electrolytes. By THEODORE WILLIAM RICHARDS. *Proc. Am. Acad.*, 33, 1-20.—This is an experimental study of the electromotive force of elements of the type Hg, HgCl, MCl₂, HgCl, Hg, the two electrodes being at different temperatures. A careful preliminary study of the causes affecting the potential difference of calomel electrodes revealed the fact that the depolarizer HgCl is slightly decomposed into mercuric chloride and mercury ; the amount of decomposition being greater, the higher the temperature and the more concentrated the soluble electrolyte, MCl₂, present. To this fact is attributed the slight inconstancy of normal calomel elements. It is suggested that a decinormal calomel electrode would be a more constant standard electrode, on this account. Measurements of the temperature-coefficient between 0° and 30° of calomel electrodes in contact with normal, decinormal, and centinormal solutions of eleven different soluble chlorides showed that it invariably diminishes with increasing concentration. The diminution is shown to be very approximately in the ratio of the logarithm of the concentrations. Hydrochloric acid and ammonium chloride are anomalous in their behavior.

ANALYTICAL CHEMISTRY.

ULTIMATE ANALYSIS.

H. P. TALBOT, REVIEWER.

The Estimation of Phosphorus in Steel. By R. W. MAHON. *J. Am. Chem. Soc.*, 19, 792-795.—The author describes a rapid method, successfully used with steels low in carbon and silicon and free from arsenic, which, he claims, will afford accurate results in about eight minutes. The procedure includes the partial neutralization of the nitric acid and precipitation of the phosphomolybdate at a temperature near the boiling-point, and after shaking for fifteen seconds. The precipitate and filter are placed in a measured excess of caustic alkali ; and the excess is determined by standard acid, with phenolphthalein as an indicator.

The Titration of Stannous Salts with Iodine. By S. W. Young. *J. Am. Chem. Soc.*, 19, 809–812.—The author states that the reaction between stannous chloride and iodine, by which the former is oxidized, takes place smoothly in acid solution, and may be utilized for the determination of tin. The iodine solution is standardized against potassium bichromate with the intervention of a stannous chloride solution of known value. The statements are made that "thiosulphate cannot be satisfactorily used in acid solution to titrate against iodine," and that "the standard of iodine by thiosulphate is likely to be a trifle high;" but no authorities are quoted nor experimental data offered in support of these assertions.

The Electrolytic Determination of Cadmium. By Daniel L. Wallace and Edgar F. Smith. *J. Am. Chem. Soc.*, 19, 870–873.—The paper presents data which refute the statements of Heidenreich (*Ber.*, 29, 1585) and Avery and Dales (*this Rev.*, 3, 123), in which they deny that cadmium can be precipitated electrolytically from solutions of the acetate and sulphate, and that a separation of cadmium and copper can be effected in nitric acid solution. Under the conditions prescribed in this paper the precipitation of the cadmium by electrolysis appears to be complete, as previously affirmed.

The Analysis of Bearing-Metal Alloys, with a New Volumetric Method for Determining Copper. By W. E. Garrigues. *J. Am. Chem. Soc.*, 19, 934–948 ; *Proc. Eng. Soc. W. Penna.*, 13, 415–430. The analysis of bearing-metal alloys is treated in this paper from a general standpoint, their sources of inaccuracy are discussed, and remedies are suggested. For the determination of copper, the precipitation of the metal as cuprous thiocyanate, filtration, and the treatment of the precipitate with an excess of caustic alkali are suggested. The excess of the alkali is then determined by titration.

Henry Fay, Reviewer.

Iodometric Determination of Selenious and Selenic Acids. By James F. Norris and Henry Fay. *Am. Chem. J.*, 18, 703–706.—It is shown that selenious acid can be titrated with sodium thiosulphate, one molecule of the former being equivalent to four of the latter. Definite portions of pure selenious acid, which had been resublimed in the presence of a small quantity of nitric acid, were treated in ice-cold solution with an excess of sodium thiosulphate in the presence of enough hydrochloric acid to set free all of the thiosulphuric acid ; and the excess of sodium thiosulphate was found by titrating back with iodine. An excess of hydrochloric acid may be used, provided the solution is kept cold. For the determination of selenic acid measured portions were treated with 25 cc. concentrated hydrochloric acid, and the solution was diluted to 100 cc. The solution is boiled for one hour, precautions being taken not to allow the volume to fall below 75 cc.; it is then cooled, diluted with ice-water, and the selenious acid determined by titration. The results of both determinations are accurate ; and the method can be used for the determination of a mixture of selenious and selenic acids. The exact nature of the reaction between sodium thiosulphate and selenious acid has not been determined.

W. R. WALKER, REVIEWER.

Separation of Aluminum and Beryllium by the Action of Hydrochloric Acid. By F. S. HAVENS. *Am. J. Sci.*, 154, 111–114. The method of Gooch and Havens (*this Rev.*, 3, 30) for the separation of aluminum from iron is here extended with very little variation to the separation of aluminum from beryllium. After removal of the aluminum, the beryllium is determined by evaporating the ether-acid solution to dryness, moistening with strong nitric acid, igniting, and weighing as beryllium oxide.

J. F. NORRIS. REVIEWER.

Combustion of Organic Substances in the Wet Way. By I. K. PHELPS. *Am. J. Sci.*, 154, 372–383.—In a former paper (*Am. J. Sci.*, 152, 70) the author has shown that carbon dioxide can be estimated iodometrically. He has now applied the method to the estimation of carbon in organic substances, the combustion being effected in solution. The carbon dioxide evolved is passed into a standardized solution of barium hydroxide, the excess of which is determined by heating with iodine and titrating the excess of the latter with arsenious acid. One molecule of barium hydroxide is equivalent to one molecule of iodine. Ammonium oxalate, barium formate, and tartar emetic were successfully analyzed in this way when potassium permanganate was used as the oxidizing agent. The carbon in these substances, as well as in phthalic acid, cane-sugar, and paper, was determined by oxidizing with potassium bichromate. The amount of oxygen necessary to oxidize an organic substance was determined by using a known weight of potassium bichromate and estimating the excess of the latter by digesting with hydrochloric acid, absorbing the evolved chlorine in a standard solution of sodium arsenite, and titrating with iodine. Reference must be made to the original article for a description of the apparatus and details of the method.

PROXIMATE ANALYSIS.

W. R. WHITNEY, REVIEWER.

On the Action of Sulfuric Acid upon Strychnine in the Separation of this Alkaloid from Organic Matter. BY E. H. S. BAILEY AND WM. LANG. *Kan. Univ. Quart.*, 6, 205–207 ; *Am. J. Pharm.*, 70, 18–21.—The authors have found that the limit of the test for strychnine, in which the color effect produced by potassium bichromate and sulphuric acid is employed, is 0.00025 milligram of the alkaloid. They also found that, if the dry strychnine residue be first heated in the water-bath fifteen minutes with the concentrated acid, before the crystal of bichromate is introduced, the delicacy of the test is reduced to 0.0011 milligram ; while in employing a chloroform extraction of the alkaloid from other organic matter, or as nearly as possible, conforming with the test as it is usually applied, a quantity of strychnine less than 0.02 milligram could not be detected. That is, the greatest source of error in the ordinary strychnine determinations lies in incomplete extraction by chloroform.

J. F. NORRIS, REVIEWER.

The Volumetric Determination of the Nitro Group in Organic Compounds. BY S. W. YOUNG AND R. E. SWAIN. *J. Am. Chem. Soc.*, 19, 812–814.—The authors show that a quantitative determination of the nitro groups in dinitrobenzene can be made by reduction with a standardized solution of stannous chloride and titrating the excess of the latter. The method has already been applied to a large number of nitro compounds by Limpricht (*Ber. d. chem. Ges.*, 11, 35, 40).

ASSAYING.

H. O. HOFMAN, REVIEWER.

Fire-Assay for Lead. BY JOHN F. CANNON. *Eng. Min. J.*, 64, 604.—The assay is made as follows : Weigh out 5 grams of ore, place it in a 5-gram crucible containing 20 grams lead flux (16 parts sodium bicarbonate, 16 potassium carbonate, 8 glass borax, 5 flour), mix, tap to settle the mixture, add 20 grams of lead flux, tap, add 4–5 nails, place in a bright-red muffle, fuse in 20–25 minutes, close the muffle when the fusion has become quiet, heat to perfect fluidity, take out the crucible, remove the nails, tapping them to free them from adhering lead, tap the crucible and pour its contents on a level cast-iron plate, stringing out the slag to a thin thread, when even the smallest lead button will be found at the extreme end of the slag. Free the button from slag, hammer out thin on an anvil, and weigh.

The Jones Coupel Moulder. BY THE PARKE AND LACY Co. *Min. Sci. Press*, 75, 357 ; *Eng. Min. J.*, 64, 521.—This is a device by means of which the bone ash in the coupel mould is compacted by continuous and increasing pressure exerted by a system of levers and toggles worked by a foot-lever.

Corrected Assays. BY E. H. MILLER. *School Mines Quart.*, 19, 43-47.—The author determined the losses in silver occurring with the niter and cyanide methods from assaying an argentiferous galena ore and a dry gold-bearing silver ore, containing some malachite. He found that with the cyanide method silver is carried off in the slags, which is not fully recovered by re-treatment.

TECHNICAL CHEMISTRY.

G. W. ROLFE, REVIEWER.

The Principal Amid of Sugar-Cane. BY EDMUND C. SHOREY. *J. Am. Chem. Soc.*, 19, 881–889.—The author has isolated the characteristic amid of the juice from fifteen samples of sugar-cane and demonstrated that this body is glycocoll (glycocin) and not asparagin, as generally supposed. The differentiation has been carefully worked out by an elaborate comparison of the characteristic reaction of the sugar-cane amid with those of pure glycocoll and asparagin. The formation of hippuric acid by synthesis of this amid with benzoic acid as well as its being a derivative of gelatin is especially significant to animal and vegetable physiology. This paper is important as promising to throw light on the constitution of the proteids.

Crystallization in Motion. BY E. P. EASTWICK, JR. *La. Planter and Sugar Mfgr.*, 20, 43–45.—A very clear exposition of the theory and its application in the production of second sugars. The advantages claimed for the new process over the usual method of boiling " blank" and crystallizing in the "hot-room" are : (1) Increased quantity and better quality of sugar ; (2) Decreased quantity and lighter-colored molasses ; (3) Reduced cost of building and machinery ; (4) Reduced expense of labor ; (5) Completion of the production of second sugars by the end of the crop. These advantages are certainly great ones, but it is doubtful whether the statement of the author that the process has gone beyond the experimental stage will be generally accepted.

Crystallization by Motion. BY JOS. E. KOHN. *La. Planter and Sugar Mfgr.*, 20, 45.—The author gives arguments from a theoretical standpoint why crystallization in motion is desirable.

Note on Hydrolysis of Starch by Acids. BY G. W. ROLFE AND GEO. DEFREN. *J. Am. Chem. Soc.*, 19, 679–680.—The authors regret having omitted to refer in their previous article to work on this subject by Prof. H. W. Wiley.

BIOLOGICAL CHEMISTRY.

W. R. WHITNEY, REVIEWER.

Pomegranate Rind. BY HENRY TRIMBLE. *Am. J. Pharm.*, 69, 634–636.—This material has been used to some extent in tanning, and the author finds in the fresh rind 28.38 per cent. tannin. The pure tannin was investigated, and from its composition and chemical properties the author concludes that it is identical with gallotannic acid.

Hemlock Tannin. BY HENRY TRIMBLE. *Leather Manufacturer,* 8, 88–89.—The percentage of tannin in hemlock bark collected at different seasons, in Pennsylvania and Tennessee, are given. The maximum (calculated for dry bark) is 15.45 ; the minimum, 8.22. The results of an investigation of the nature of this tannin lead the author to the conclusion that it is identical with oak tannin. The article contains a description of the method employed in obtaining the pure tannin, a brief discussion of the use of the bark extract in tanning, and an illustration showing the structure of a section of the bark, as seen under the microscope.

A Contribution to the Knowledge of the Gum from the Oil Tree. BY CHARLES W. DIRMITT. *Am. J. Pharm.*, 70, 10–18. —A description of the gum is given (taken from *Bull. Botanical Department, Jamaica,* 4, 77). This is followed by a description of experiments made to fractionate the gum by distillation. Evidently the fractions obtained were the results of decomposition. The analysis of the gum leads the author to the acceptance of the empirical formula $C_{11}H_{10}O_5$. Its iodine number is 46.39 and its saponification equivalent, 15.10.

California and Western Hemlock. BY HENRY TRIMBLE. *Leather Manufacturer,* 9, 11.—This article describes the structure of the bark of the Tsuga Mertensiana and a tannin determination in it. There was found 11.37 per cent. tannin in the dry bark, and the analysis of this tannin led the author to conclude that it is identical with a large number of oak tannins.

The Caffein Compound in Kola. Part II. Kolatannin. BY J. W. T. KNOX AND A. B. PRESCOTT. *J. Am. Chem. Soc.*, 20, 34–75.—A résumé of the literature of oak tannins is given. This is followed by a description of methods employed in ob-

taining kolatannin, and of its properties, and the analyses of
several compounds of it with bromine and acetyl, as well as corresponding anhydrides.   A discussion of assay methods is given
together with comparative results by several of them, which
show that the method of the authors is the best.

SANITARY CHEMISTRY.

E. H. RICHARDS, REVIEWER.

Water Supply and Sewerage. BY THE STATE BOARD OF
HEALTH OF MASSACHUSETTS (1897). *Ann. Rep.*, 28, 1–597.—
The volume contains the results of the usual monthly examinations of the chief water supplies of the State, together with statistics relating to the experimental filtration of water and sewage.

Report on Food and Drug Inspection. BY C. P. WORCESTER
AND C. A. GOESSMANN.   *State Board of Health, Mass., Ann.
Rep.*, 28, 601–645.

The City Water Supply. BY ADOLPH GEHRMANN AND
CASS L. KENNICOTT. *Rep. Dept. of Health of Chicago for 1897,*
175–214.—The averages of the daily examinations during 1896
of samples from the four pumping stations supplying the city
were, in parts per 100,000 : Free ammonia, 0.0016 ; albuminoid
ammonia, 0.0074 ; chlorides (not chlorine), 0.63 ; bacteria,
daily average per cc., 6093 ; per cent. of times pathogenic bacteria were present, 6.30.   On October 23, 1896, a series of samples at one-mile intervals was taken off shore for 12 miles.

	Total solids.	Free ammonia.	Albuminoid ammonia.	Bacteria per cc.
One mile......	14.50	0.0020	0.0140	4000
Twelve miles .	13.00	0.0000	0.0080	520

The highest ammonia was found at three miles; the largest number of bacteria at five miles.   A mineral analysis of the water of
Lake Michigan is also given.

Sanitary Inspection and Analysis of Ice. BY CASS L. KEN-
NICOTT. *Rep. Dept. of Health of Chicago for 1897,* 215–223.—
The method of ice inspection here described must have a marked
effect upon the quality of the ice supplied in future.   Analyses
are given of 61 samples of water from which ice is gathered in
the states of Indiana, Illinois, and Wisconsin.

**Report of the Rockville Center Laboratory of the Depart-
ment of Health, Brooklyn.** BY HIBBERT HILL AND J. W.
ELLMS.—This report gives the results and states some of the
conclusions arrived at during an investigation into the sanitary
condition of the Brooklyn water supply, extending over a period

of twelve months, in order that the fluctuations, seasonal and otherwise, might be studied and their significance might be determined. Although forced by causes beyond individual control to prepare their report in haste, the authors have collected considerable information concerning 16 surface supplies and 13 groups of driven wells. A closely approximate figure for the normal chlorine of this region would seem to be about 0.55, which corresponds with that of other regions near the sea. The identification of a species of myrioplyllum (water milfoil) as a probable cause of a fishy odor, and an indicated relation of the number of bacteria to the degree of pollution of ground waters, are points worthy of note. Certain practical suggestions are made as to the improvement of the quality of the several sources.

Examination of Food Products Collected by the Station. By A. L. WINTON. *Conn. Agr. Expt. Sta. Ann. Rep.,* 21, 15–63. —It appears that oysters are frequently treated with borax, from 5.5 to 38 grains in a pint of oysters having been found. 27 out of 42 samples of sausages examined contained from 8.4 to 50.4 grains of borax. The total of 11 samples of salt cod-fish examined also contained considerable amounts, as did three samples of cream. A chemical analysis of date stone coffee is given and results of the examination of some 200 samples of milk.

METALLURGICAL CHEMISTRY.

H. O. HOFMAN, REVIEWER.

Matte-Smelting at the Hall Mines, British Columbia. By R. R. HEDLEY. *Eng. Min. J.,* 64, 695–696.—The paper is an illustrated description of the blast-furnace work done at this smeltery. The furnace, 144 by 44 inches at the tuyere-level, puts through in 24 hours nearly 250 tons charge, producing matte with 49 per cent. copper.

The Highland-Boy Mine and Mill, Bingham, Utah. *Eng. Min. J.,* 64, 665–666.—The mill which started work in September last, combines pan amalgamating and cyaniding, the coarse gold being recovered from the ore by the former process, the fine gold by the latter.

The Modern Gold Chlorination Process for the Treatment of Gold Ores. By J. E. ROTHWELL. *Min. Sci. Press,* 75, 573.—A short illustrated description of a chlorinating plant, using the Pearce turret furnace for roasting, and the Rothwell barrel, emphasis being laid more on the practical and economic than on the chemical side of the operations.

Chlorination and Cyaniding. By C. C. Burger. *Eng. Min. J.*, 64, 663.—In comparing barrel-chlorination with cyaniding, the author estimates the cost of chemicals per ton of ore to be from 60 to 70 cents for chlorinating and from 30 to 40 cents for cyaniding, but chlorinating has the advantage of requiring less fine-crushing (14 to 16 mesh *vs.* 30 to 40 mesh) and of giving a higher extraction (90–95 *vs.* 80–85 per cent.). Distance from railroad even does not make chlorinating excessively costly; *e. g.*, at Gibbonsville, Ida., the 20 pounds of chemicals required per ton of ore are hauled 100 miles at 1 cent a pound, and make the additional cost of chlorinating only 20 cents.

Cyaniding Sulphide Gold Ores. By R. Recknagel. *Eng. Min. J.*, 64, 580–581.—A paper very general in its character. Among other things the author calls attention to the fact that the gold in some sulphides is soluble, while in others it is insoluble. A very interesting fact is that of iron disulphide; if present as marcasite, it is liable to have a strong decomposing influence upon cyanide, while if present as pyrite this is not likely to be the case, provided, of course, that the sulphide has not been oxidized by atmospheric action. Roasting removes most of the difficulties, if a dead-roast is obtained ; a sulphatizing roast, followed by dissolving out the sulphates with water before using the cyanide solution, gives unsatisfactory results as to consumption of cyanide (some sulphate remaining insoluble in water, but acting upon cyanide) and extraction of gold (some sulphide remaining undecomposed).

Some Products Found in the Hearth of an Old Furnace upon the Dismantling of the Trethellan Tin Works, Cornwall. By W. P. Headden. *Proc. Col. Sci. Soc., Nov. 6, 1897.*—The paper contains a number of elementary chemical and microscopical analyses of the products and deductions as to the rational analyses and the way the products were formed.

Kryolith, Its Mining, Preparation, and Utilization. By W. C. Henderson. *J. Franklin Inst.*, 145, 47–54.—The author discusses briefly the properties of the mineral, the history of its discovery, its occurrence, and the methods of mining, shipping, and conversion into soda and alumina. Two grades are produced, one of which, 99 per cent pure, goes to Copenhagen, and the other, 92 per cent. pure, to Philadelphia, where the Pennsylvania Salt Co. produces, by the Thomsen process, soda and alumina, the latter serving as raw material for the production of alum. Kryolith is also used as a flux in the electrolytic reduction of alumina and in the production of Kryolith glass.

Direct Recarbonizing of Steel from the Blast-furnace. BY
C. KIRCKHOFF. *Iron Age*, 60, No. 23, p. 4.—C. H. Foote, W.
R. Walker, and E. A. S. Clarke, of the Illinois Steel Co., have
succeeded in using spiegeleisen direct from the blast-furnace in
recarbonizing steel, the spiegel being taken from the blast-furnace into a mixer, whence it is poured for neutralizing as needed.

[CONTRIBUTION FROM THE MASSACHUSETTS INSTITUTE OF TECHNOLOGY.]

REVIEW OF AMERICAN CHEMICAL RESEARCH.

VOL. IV. NO. 4.

ARTHUR A. NOYES, Editor; HENRY P. TALBOT, Associate Editor.
REVIEWERS: Analytical Chemistry, H. P. Talbot and W. H. Walker; Biological Chemistry, W. R. Whitney; Carbohydrates, G. W. Rolfe; General Chemistry, A. A. Noyes; Geological and Mineralogical Chemistry, W. O. Crosby; Inorganic Chemistry, Henry Fay; Metallurgical Chemistry and Assaying, H. O. Hofman; Organic Chemistry, J. F. Norris; Physical Chemistry, H. M. Goodwin; Sanitary Chemistry, E. H. Richards; Technical Chemistry, A. H. Gill and F. H. Thorp.

INORGANIC CHEMISTRY.

HENRY FAY, REVIEWER.

On the Cuprosammonium Bromides and the Cuprammonium Sulphocyanates. BY THEODORE WILLIAM RICHARDS AND BENJAMIN SHARES MERIGOLD. *Proc. Am. Acad. Arts Sci.*, 33, 131-138.—Cuprosammonium bromide, $Cu_2Br_2.2NH_3$, was prepared by dissolving cuprous bromide in the least possible amount of ammonium hydroxide and acetic acid, and evaporating over sulphuric acid in a current of hydrogen. It crystallizes in long colorless prisms and is extremely easily oxidized. It is soluble in ammonium hydroxide and nitric acid, but deposits cuprous bromide with other mineral acids. Cuprosammonium sulphocyanate, $Cu_2(SCN)_2.2NH_3$, is precipitated as a white crystalline powder when ammonium sulphocyanate is added to an ammoniacal cuprous bromide solution. It loses ammonia at the ordinary temperature. Tetrammon-cuprosammonium bromiide, $Cu_2Br_2.(NH_3)_4$, is prepared by placing dry finely powdered cuprous bromide in a bulb tube packed in ice, and passing in dry ammonia gas until the mass is saturated. The compound is a black powder, extremely unstable. Triammon-cuprosammonium sulphocyanate, $Cu_2(SCN)_2.(NH_3)_3$, is prepared in the same manner as the last mentioned salt and is very unstable, giving up ammonia on exposure to the air. The diammon-cupriammonium sulphocyanate, $Cu(SCN)_2(NH_3)_2$, was prepared as brilliant blue crystals, by dissolving cupriammonium bromide in a little acetic acid and adding to this a saturated solution of ammonium sulphocyanate, dissolving the precipitate formed in ammonium hydroxide and allowing it to crystallize.

Action of Sulphur on Silicides.   Production of Silicon.   By
G. DE CHALMOT.   *Am. Chem. J.*, 19, 871–877.—Sulphur begins
to be absorbed by pulverized copper silicide, when heated in
closed tubes, at $200°–250°$ C., and the absorption is complete at
$270°–280°$.   If the temperature is allowed to rise above $300°$ C.
the reaction $Si + S_2 = SiS_2$ takes place ; below $300°$ C. the copper and sulphur react with the liberation of free silicon.   Local
action sometimes takes place with production of a small amount
of silicon sulphide, but the principal reaction is according to the
equation $Cu_2Si + S = Cu_2S + Si$.   Free silicon is not liberated
from manganese, iron or chromium silicides by sulphur.

Solubility of Lead in Ammonia.   By H. ENDEMANN.   *Am.
Chem. J.*, 19, 890–893.—Lead, when immersed in strong ammonium hydroxide, gradually goes into solution, and after some
time the lead becomes covered with a colored coating, the ammonium hydroxide remaining clear and holding in solution after
three days' action 0.0139 per cent. of lead.   After standing some
weeks there is a deposit of hydrated oxide, which contains
only a trace of ammonia.   Dilute ammonium hydroxide behaves
similarly but the separation of hydrated oxide is more rapid.
Ammonium bicarbonate does not act upon lead.

A Study of Zinc Hydroxide in Precipitation.   By VERNON
J. HALL.   *Am. Chem. J.*, 19, 901–912.—To determine the action of substances in solution upon the precipitation of zinc hydroxide, the reaction $ZnCl_2 + 2KOH = Zn(OH)_2 + 2KCl$ was
studied, using the same methods as were previously applied by
the author to the study of the precipitation of ferric hydroxide
(*this Rev.*, 3, 135).   Precipitations of zinc hydroxide from solutions, each containing the same amount of zinc, were made with
two, one and one-half, one, and one-half molecules of potassium
hydroxide, and the resulting precipitate and filtrate were analyzed.   No chlorine was carried down when two molecules of
potassium hydroxide were used, but 7.4 per cent. of the total
chlorine was present in the precipitate with one and one-half
molecules, and the amount decreases as the proportion of potassium hydroxide decreases.   With five and eight molecules of
potassium hydroxide no chlorine was carried down.   In the
same manner the reaction $ZnCl_2 + 2KOH + xK_2SO_4 = Zn(OH)_2
+ 2KCl + xK_2SO_4$ was studied.   No chlorine is found in the
precipitate with two or with one and one-half molecules.   Sulphur trioxide was not carried down with two molecules of potassium hydroxide ; but 12 per cent. was carried down when one
and one-half molecules were used.   More zinc oxide is carried
down in the presence of potassium sulphate than when it is absent.   The effect of concentration is to decrease the quantity of

zinc hydroxide precipitated, and to increase the quantity of chlorine carried down. In the case of iron, there is a tendency to decrease the quantity of ferric hydroxide precipitated and the amount of chlorine carried down. Increase of temperature decreases both the amount of metallic oxide and of chlorine which separates from solution.

On the Effect of Light on the Combination of Hydrogen and Bromine at High Temperatures. By J. H. KASTLE AND W. A. BEATTY. *Am. Chem. J.*, 20, 159–163.—Hydrogen and bromine in sealed glass bulbs heated to 196° C. in the vapor of boiling *o*-toluidine were exposed to the sunlight for varying times and the extent of the combination was noted by observing the change of color of the free bromine. Combination of the gases was effected in the sunlight, but apparently no change was produced by heating in the dark. The results given are only qualitative.

The Constitution of Arsenopyrite. By F. W. STARKE, H. L. SHOCK AND EDGAR F. SMITH. *J. Am. Chem. Soc.*, 19, 948–952.—By passing dry hydrogen over arsenopyrite it was shown that all of the sulphur could be removed and only traces of the arsenic. Heating in sealed tubes with copper sulphate showed that arsenious acid was always produced. About 30 per cent. of the iron was shown to be present in the ferrous state and 4 per cent. in the ferric state. The facts established may be represented by the formula $14Fe''As'''S.2Fe'''As'''S$, which the authors regard as tentative only.

Some New Ruthenocyanides and the Double Ferrocyanide of Barium and Potassium. By JAMES LEWIS HOWE AND H. D. CAMPBELL. *J. Am. Chem. Soc.*, 20, 29–33.—The preparation of strontium ruthenocyanide, $Sr_4Ru(CN)_6.15H_2O$, barium potassium ruthenocyanide, $K_2BaRu(CN)_6.3H_2O$, and barium caesium ruthenocyanide, $Cs_2BaRu(CN)_6.3H_2O$, is described, with a crystallographic comparison of barium potassium ferrocyanide and barium potassium ruthenocyanide. Barium potassium ferrocyanide was found to contain three molecules of water and not five, as stated by Wyrouboff. Barium caesium ferrocyanide, $Cs_2BaFe(CN)_6.3H_2O$, is described.

The Action of Sulphuric Acid on Mercury. By J. R. PITMAN. *J. Am. Chem. Soc.*, 20, 100–101.—Sulphuric acid was shaken with mercury at intervals during forty-eight hours, in a nitrometer with and without air, and no decomposition of the sulphuric acid could be detected. Excess of mercury or sulphuric acid was without influence.

Anethol and its Isomers. By W. R. ORNDORFF, G. L. TERRASSE, AND D. A. MORTON. *Am. Chem. J.*, 19, 845–871.— The chemical properties and molecular weights of anethol and its isomers were studied with the following results: Methylchavicol and estragol are identical and are metamers of anethol. Fluid metanethol has the same molecular weight as anethol, and resembles it in physical and chemical properties closely. The two substances are probably stereoisomers, metanethol being

$$HCC.C_4H_4OCH_3(p)$$

the trans compound, $\|$. Anisoïn, the resin-
$$CH_3CH$$

ous polymeric modification of anethol, acts as a colloid towards the solvents acetic ester, acetone, and benzene. Solid metanethol and the liquid isoanethol have the same molecular weight, which is twice that of anethol. They should be called, therefore, solid and fluid dianethol, respectively. As they both act like saturated compounds, it is possible that they are derivatives of tetramethylene. By heating anethol under pressure to 250°–275° it is transformed into isoanethol, the methyl ester of paracresol, and the methyl ester of parapropylphenol. The author suggests for anethol and its isomers names which are in accord with the new system of nomenclature suggested by the Geneva Conference of Chemists.

Acetylene Diiodide. By G. DE CHALMOT. *Am. Chem. J.*, 19, 877–878.—In preparing ethylene tetraiodide by the method of Marquenne (*Bull. Soc. Chim.* [3], 7, 777) using calcium carbide instead of barium carbide, the author obtained acetylene diiodide as a by-product in the alcoholic washings of the crude tetraiodide. When iodine was added to a solution of acetylene in potassium hydroxide, ethylene tetraiodide was not formed, but a white precipitate, which crystallized from benzene in needles having a very pungent odor. The analogous bromine compound is an oil which takes fire spontaneously in the air. No analyses of the compounds are given.

The Action of Sodium upon Methylpropylketone and Acetophenone. By PAUL C. FREER AND ARTHUR LACHMAN. *Am. Chem. J.*, 19, 878–890.—Freer has investigated the action of sodium on acetone (*Am. Chem. J.*, 12, 155; 13, 319; 15, 582; 17, 1) and on acetaldehyde (*Ibid*, 18, 552; *this Rev.*, 3, 8) and has shown that the behavior of the salts formed indicates that they have the structures $CH_3C\begin{smallmatrix}\nearrow ONa\\\searrow CH_3\end{smallmatrix}$ and $CH_2{=}CHONa$, respectively. The present paper contains the results of a study of the action of sodium on methylpropylketone, acetophenone, and mesityl oxide. The method employed was the one introduced during the study of sodium acetaldehyde, namely, the combined action of the ketone and benzoyl chloride upon sodium suspended in ether. When an ethereal solution of methylpropylketone was treated with sodium, hydrogen was evolved, and a white precipitate was formed, which was probably not a definite individual, although the result of an analysis indicated the composition C_4H_9ONa. By the action of a mixture of benzoyl chloride and methylpropylketone on sodium suspended in ether, a number of compounds were formed. These were separated as follows: The ether was shaken with water to remove the yellowish solid formed, and then with dilute alkali. When the combined aqueous extracts were acidified, a large quantity of benzoic acid separated, together with a trace of an undetermined fatty acid, and a solid polyketone. The latter compound is a dibenzoyl derivative, colors ferric chloride intensely red, dissolves in alkalies and carbonates, and has, consequently, the structure $H.C_3H_6.CO.C(COC_6H_5)_2$, or $C_6H_5.CO.C\begin{smallmatrix}C(OH)C_6H_5\\COC_6H_5\end{smallmatrix}$.

When the ethereal residue, which contained the main products of the reaction, was distilled under diminished pressure, a large amount of tar was formed. The distillate was repeatedly refractionated without being separated into any definite compounds. From the products of the saponification of the different fractions it was shown that isoketone benzoate, $C_5H_7\begin{smallmatrix}OCOC_6H_5\\CH_3\end{smallmatrix}$, and methylpropylcarbinolbenzoate were present. All of the fractions contained chlorine, which could not be removed by distillation over sodium or by heating with alcoholic potash. As benzoyl chloride is not an active chlorinating agent, the formation of these halogen compounds is best explained by assuming the addition of the chloride to the sodium derivative of the ketone, and the subsequent elimination of hydrochloric acid, which unites with the unsaturated isoketone benzoate. The experimental results are in accord with this application of Nef's hypothesis. Since from the above it appears that the action of benzoyl chloride upon sodium methylpropylketone is similar to that with sodium acetone, and since the solid dibenzoylketone contains its benzoyl groups attached to methyl, sodium methylpropylketone has the structure $C_4H_7\begin{smallmatrix}ONa\\CH_2\end{smallmatrix}$. The action of a mixture of acetophenone and benzoyl chloride on sodium was studied in the way outlined above. From the alkaline extract tribenzoylmethane was obtained, and from the ethereal solution a thick tarry mass, which was shown to contain the benzoate of tribenzoylmethane. A portion of the tar, when saponified by heating with sulphuric acid, yielded benzoic acid, acetophenone-pinacone, and a ketone containing halogen, which was not investigated. Acetophenone thus shows the same behavior towards sodium in the presence of benzoyl chloride, as is manifested by acetone and methylpropylketone, with the exception that no isoketone benzoate is formed. It seems, therefore, that as the negative character of the ketone increases the tendency to form O-derivatives decreases. Mesityl oxide forms an unstable sodium compound, which reacts with benzoyl chloride forming a tarry mass which could not be investigated.

On Salts of Nitroparaffins and Acylated Derivatives of Hydroxylamine. By LAUDER W. JONES. *Am. Chem. J.*, **20**, 1–51.—When benzoyl chloride reacts with sodium isonitroethane the metal is first replaced by benzoyl, and a compound of the

following structure is formed : $CH_3-CH=N\begin{smallmatrix}O\\ \\O-CO.C_6H_5\end{smallmatrix}$.

This substance cannot be isolated, but is immediately converted by intramolecular oxidation into the benzoyl ester of acethydroxamic acid, which exists in two forms with the probable structures $CH_3C\begin{smallmatrix}OH\\ \\N-OCOC_6H_5\end{smallmatrix}$ and $CH_3C\begin{smallmatrix}O\\ \\NH.COC_6H_5\end{smallmatrix}$. Being a

stronger acid than isonitroethane, the hydroxamic ester immediately reacts with the isoparaffin salt, regenerating nitroethane, and forming the sodium salt of the benzoyl ester of acethydroxamic acid, $CH_3C\begin{smallmatrix}ONa\\ \\NOCOC_6H_5\end{smallmatrix}$. In the presence of benzoyl

chloride, this salt reacts just as the synthetic salt was found to act, namely, in two ways: by direct replacement forming dibenzoylacethydroxamic ester, and by addition and subsequent elimination of sodium chloride forming α-benzoyl-β-acetylbenzoylhydroxylamine, $\begin{smallmatrix}CH_3CO\\ \\C_6H_5CO\end{smallmatrix}\!\!>\!N.OCOC_6H_5$. The formation of

the latter compound throws some light on the triacylated hydroxylamine derivatives in general, and is the first experimental evidence in favor of the suggestion that one of the three isomers corresponds to a true hydroxylamine type. By the action of chlorcarbonic ester on sodium isonitroethane, carbon dioxide and a neutral oil, which could not be purified, were obtained. As it was expected that this reaction would give the carbethoxyl ester of acethydroxamic acid, $CH_3C\begin{smallmatrix}O\\NH.OCO_2C_2H_5\end{smallmatrix}$, and the carbethoxyl ester of ethylnitrolic acid, $CH_3C\begin{smallmatrix}NO_2\\NO.CO_2C_2H_5\end{smallmatrix}$, these substances were prepared and subjected to destructive distillation and the action of a number of reagents. Since the results were identical with those obtained when the above oil was studied in the same way, it is probable that the latter contained the two esters. The action of benzoyl chloride on sodium isonitromethane seems to proceed in two ways, depending on whether the salt is alcohol-free or not. In both cases, however, the chief product is a neutral oil, and the reaction is, in part, analogous to that between benzoyl chloride and sodium isonitroethane. Sodium chloride is eliminated and the resulting product is converted by oxidation into the benzoyl ester of formhydroxamic ester. From this the sodium salt is formed by the decomposition of sodium isonitromethane, and the former then reacts with benzoyl chloride by direct replacement exclusively, giving the benzoyl ester of benzoylformhydroxamic acid, $HC\begin{smallmatrix}OCOC_6H_5\\N.OCOC_6H_5\end{smallmatrix}$. The reaction was shown to proceed as outlined above by effecting the synthesis of the sodium salt of formhydroxamic acid and studying its reaction with benzoyl chloride. By the action of mercuric chloride on sodium isonitromethane a mercuric salt is formed which is converted, probably, into a salt of a hydroximic acid. This salt immediately loses water, giving rise to mercuric fulminate. At the same time by the further action of mercuric chloride on the salt of the hydroximic acid, a basic salt is formed with probable structure $HC\begin{smallmatrix}O-Hg-OH\\N-O-Hg\end{smallmatrix}$. With dilute hydrochloric acid mercuric fulminate is formed. The author comes to the conclusion that the cases of intramolecular oxidation observed in this research, as well as those reported by Nef, cannot be explained on the assumption that the nitroparaffins have the structure proposed by Hantzsch, $\begin{smallmatrix}H\\R\end{smallmatrix}C-N-OH$. The open formula, however,

$$\underset{H}{\overset{R}{\diagdown}}C=N\overset{\diagup O}{\underset{\diagdown OH}{}}$$ is in accord with all the facts.   In the second

part of the paper efforts to prepare derivatives of the oxime of carbon dioxide are described.   For this purpose the salts and esters of carbethoxyhydroxamic acid were prepared and studied. If they have the formulæ usually assigned to them,

$$C_2H_5O-C\overset{\diagup OH}{\underset{\diagdown N.OM}{}}$$, it seemed possible that they would lose

alcohol and thus give rise to the salts of the oxime desired.   By the action of methyl iodide on the potassium salt of carbethoxyhydroxamic acid, the methyl ester of the acid and α-methyl-β-methylcarbethoxyhydroxylamine were formed.   When decomposed by hydrochloric acid, the former gave α-methylhydroxylamine hydrochloride, and the latter α-dimethylamine hydrochloride.   By the action of ethyl iodide on the salt of carbethoxyhydroxamic acid analogous compounds were obtained.   The ethyl ester reacted with phosphorus pentachloride ; hydrochloric acid and ethyl chloride were evolved, and an oily product was formed which decomposed when poured into water, giving α-ethylhydroxylamine hydrochloride.   As it is probable that an ester of the oxime of carbon dioxide was formed in this reaction, it is being studied further.   The benzyl and benzoyl esters of carbethoxyhydroxamic acid were prepared, and from the former α-benzylhydroxylamine hydrochloride was formed by decomposition with hydrochloric acid.

The Action of the Halogens on the Aliphatic Amines and the Preparation of their Perhalides. By JAMES F. NORRIS. *Am. Chem. J.*, 20, 51–64.—The compounds described by Remsen and Norris (*this Rev.*, 2, 12), formed by the action of bromine and iodine on dimethyl and trimethylamine have been further studied, and it is shown by a number of reactions that the bromides are perhalides containing one added halogen, the trimethylamine compound having the formula $(CH_3)_3NHBr.Br$.   The iodine derivative of trimethylamine is a direct addition-product, and has the structure previously assigned to it $(CH_3)_3N.I_2$.   By the action of the chlorides of iodine, iodine bromide, and bromine chloride on trimethylamine, perhalides were formed containing two halogens.   Analogous compounds were prepared from dimethylamine by the action of the halogens on the salts of the amine.   Perbromides of the general structure $R_2NHBr.Br$ were prepared from dimethyl, trimethyl, diethyl, triethyl, dipropyl, tripropyl, and diamylamines.   As the number of carbon atoms increased, the stability of the compound and their crystallizing power decreased. No perbromides of primary amines could be isolated.

On Acylimido Esters. By H. L. WHEELER, P. T. WALDEN, AND H. F. METCALF. *Am. Chem. J.*, **20**, 64–76. In a previous paper (*this Rev.*, **3**, 60) it was shown that imidomethyl and ethyl benzoates give acyl derivatives with acetyl and benzoyl chloride. The authors have since found that the acylimido esters can also be prepared from the silver salts of the diacyl amides. As silver dibenzamide gave with ethyl iodide a product identical with that obtained from benzimidoethyl ester and benzoyl chloride, the reaction takes place as follows :

$$C_6H_5C\diagup\!\!\!\!\diagdown\,^{NCOC_6H_5}_{OAg} + C_2H_5I = C_6H_5C\diagup\!\!\!\!\diagdown\,^{NCOC_6H_5}_{OC_2H_5} + AgI.$$

The isomeric nitrogen ethyl compound, $(C_6H_5CO)_2NC_2H_5$, was prepared from ethyl benzamide and found to have properties different from those of the acylimido ester. It is shown by this reaction that the metal in the silver salts of diacyl amides is joined to oxygen and that they have, therefore, structures analogous to those of the silver salts of monoacylamides and anilides. The acylimido esters are very reactive substances : with water, in the presence of acids, they decompose forming either an alcohol and a diacylamide or a monoacylamide and an ester. Acetyl, propionyl, butyryl, carbethoxyl, and benzoylbenzimido esters form diacylamides, while benzoylphenylacetimido ester and those imido esters containing the group —COCO.OC₂H₅ give monoacylamides when decomposed by water. These reactions show that diacyl amides have both acyl groups joined to nitrogen. When benzoylbenzimidoethyl ester was heated with benzoyl chloride tribenzamide was formed ; when it was heated in a stream of dry ammonia the decomposition-products were phenyl cyanide, phenyl cyanurate, benzamide, and alcohol. Acylamidines were produced by the action of ammonia or bases on the acylimido esters. Acetyl chloride reacts violently with compounds having the grouping $RC\diagup\!\!\!\!\diagdown\,^{NR'}_{OR''}$. As this arrangement exists in the formula usually assigned to the trimethyl ester of normal cyanuric acid, its behavior with acetyl chloride was studied. The ester was crystallized from the chloride without change. The following compounds are described : The picrate of benzimidomethyl ester, benzoyl-, carbethoxyl-, and ethyloxalylbenzimidomethyl esters, benzimidoethyl ester mercuric chloride, acetyl-, propionyl-, normal butyryl-, benzoyl-, and ethyloxalylbenzimido ethyl esters, normal butyrylbenzamide, tribenzamide, ethyldibenzamide, propionylbenzamide, benzimidopropyl ester, acetyl- and benzoylbenzimidopropyl ester, benzoylbenzimidoiso-butyl ester, phenylacetimidomethyl ester, and benzoylphenyl-acetimidoethyl ester.

Note on Double Salts of the Anilides with Cuprous Chloride and Cuprous Bromide. By WILLIAM J. COMSTOCK. *Am. Chem. J.*, 20, 77–79.—A double salt of the formula $(C_6H_5NH.COCH_3)_3HCl.CuCl$ was formed by cooling a hot solution of acetanilide and cuprous chloride in a mixture of acetic and hydrochloric acids. It crystallizes from alcohol in long, colorless prisms, and is stable in dry air. Analogous bromides containing acetanilide and parabromacetanilide were prepared. Formyl compounds give similar double salts which are very unstable. The salt containing formparatoluide, however, was analyzed and had the formula $(CH_3.C_6H_4NHCHO)_3HBr.2CuBr$.

Action of the Anhydride of Orthosulphobenzoic Acid on Dimethyl- and Diethylaniline. By M. D. SOHON. *Am. Chem. J.*, 20, 127–129.—Dimethylanilinesulphonphthaleïn was formed by heating a mixture of dimethylaniline, the anhydride of orthosulphobenzoic acid, and phosphorus oxychloride on the water-bath. It forms a blue-black brittle mass, slightly soluble in hot water. The aqueous solution dyes silk and wool bright blue. Phosphorus pentachloride, reducing agents, acetic anhydride, or bromine in glacial acetic acid solution do not affect the phthaleïn. An analogous compound was prepared from diethylaniline.

The Molecular Weight of Lactimide. By G. W. RICHARDSON AND MAXWELL ADAMS. *Am. Chem. J.*, 20, 129–133.—The molecular weight of lactimide was determined from the lowering of the freezing-point of its solution in acetic acid. The value found was twice that formerly assigned to it and its formula is accordingly $CH_3CH \left< \begin{array}{c} NH-CO \\ CO-NH \end{array} \right> CHCH_3$.

The Action of Sodium Ethylate upon α,β-Dibromhydrocinnamic Ester, Citradibrompyrotartaric Ester, and α,β-Dibrompropionic Ester. By VIRGIL L. LEIGHTON. *Am. Chem. J.*, 20, 133–148.—A 20 per cent. excess (over 2 molecules) of sodium ethylate in alcoholic solution acts upon α, β-dibromhydrocinnamic ester to form mostly β-ethoxycinnamic ester, with a small quantity of unsymmetrical diethoxyphenylpropionic ester. The ethoxy group was shown to be in the β-position in the former ester by the action of hydrochloric acid which gave benzoylacetic acid. When heated to 110°, β-ethoxycinnamic ester decomposed into carbon dioxide and acetophenone. From the acid the silver and calcium salts were prepared. Diethoxyphenylpropionic acid could not be obtained from its ester by saponification as it decomposed, forming benzoylacetic acid. A 20 per cent. excess of sodium ethylate acts upon citradibrompyrotartaric ester forming ethoxycitraconic ester and diethoxypyrotartaric ester, which was shown to be a secondary product of the reaction, formed by the action of sodium ethylate on ethoxycitraconic ester. The former reaction-product gave carbon dioxide and propionylformic acid when boiled with a 10 per cent. solution of sulphuric acid. Ethoxycitraconic acid and a number of its salts were prepared. The reactions of the diethoxypyrotartaric ester obtained towards phenylhydrazine, dilute sulphuric and hydrochloric acids are evidence that it has the symmetrical structure. The silver and lead salts of diethoxypyrotartaric are well characterized. From the experiments with α,β-dibrompropionic ester no satisfactory conclusions could be drawn.

On Some Bromine Derivatives of 2,3-Dimethylbutane. By H. L. WHEELER. *Am. Chem. J.*, **20**, 148–153.—A tribrom derivative of 2,3-dimethylbutane could not be prepared from 2,3-dibrom-2,3-dimethylbutane by the action of one molecular quantity of bromine alone or in the presence of iron. In both cases 2,3,K',K'-tetrabrom-2,3-dimethylbutane resulted. As a small amount of impurity has a marked effect on the melting-point of this substance, it is probable that the tetrabromides of 2,3-dimethylbutane already described are identical with the above. One molecular quantity of alcoholic potassium hydroxide gives with the dibromide a mixture of 2,3-dimethyl-1,3-butadiën, $CH_2=C(CH_3)-C(CH_3)=CH_2$, and unchanged bromide. The unsaturated hydrocarbon unites with bromine forming the tetrabromide. 1,2-dibrom-2-methylpropane was prepared by warming isobutyl bromide with iron and the calculated quantity of bromine.

Alkyl Bismuth Iodides and Bismuth Iodides of Vegetable Bases. By ALBERT B. PRESCOTT. *J. Am. Chem. Soc.*, **20**, 96–100.—The precipitates formed by Dragendorff's reagent for alkaloids (potassium bismuth iodide) in solutions of salts of tetramethylammonium, pyridine, atropine, brucine, and strychnine were analyzed and found to be stable double salts of the general formula $3(Base + HI).Bi_2I_6$, with the exception of the tetramethylammonium compound which had the composition $N_4(CH_3)_{16}HBi_3I_{12}$, and the atropine salt whose formula may be as above or $C_{17}H_{23}NO_3.HI.BiI_3$. It is shown that Dragendorff's reagent has no advantage over that of Mayer in the estimation of alkaloids.

Salts of Dinitro-α-Naphthol with Various Metallic Bases. BY T. H. NORTON AND H. LOEWENSTEIN. *J. Am. Chem. Soc.*, 19, 923–927.—The preparation and properties of the lithium, magnesium, zinc, and copper salts of dinitro-α-naphthol are described. The solubilities of the ammonium and calcium salts in water, alcohol, and ether are also given.

On Certain Amine Derivatives of Dinitro-α-Naphthol and its Chlorination. BY T. H. NORTON AND IRWIN J. SMITH. *J. Am. Chem. Soc.*, 19, 927–930.—The trimethylamine, aniline, orthotoluidine, and dimethylaniline salts of dinitro-α-naphthol are described. By the action of chlorine on the dry phenol a substance was obtained which could not be purified but was free from nitrogen.

Carbon Compounds used in Medicine, Classified According to Chemical Structure. BY MARSTON TAYLOR BOGERT. *School of Mines Quart.*, 19, 47–88.—This is the first paper of a series to be published under the above title. The properties and applications of the methane derivatives used in medicine are described.

MINERALOGICAL AND GEOLOGICAL CHEMISTRY.

W. O. CROSBY, REVIEWER.

Mineralogical Notes on Cyanite, Zircon, and Anorthite from North Carolina. BY J. H. PRATT. *Am. J. Sci.*, 155, 126–128. These notes are chiefly crystallographic; but a single analysis of the feldspar is given, which identifies it as an anorthite.

Four New Australian Meteorites. BY HENRY A. WARD. *Am. J. Sci.*, 155, 135–140.—These are typical siderites; and although representing independent and rather widely separated falls, the analyses disclose a marked uniformity of composition, especially as regards the proportions of iron and nickel.

On Rock Classification. BY J. P. IDDINGS. *J. Geol.*, 6, 92–111.—This is a general discussion of the principles of rock classification, as applied to the igneous rocks, and with special reference to the chemical composition and relations of the rocks. No fewer than 928 chemical analyses of igneous rocks are compared graphically as regards : first, the silica ; second, the ratio between the silica and alkalies ; and third, the ratio between the alkalies, potash and soda. These comparisons lead to various interesting results, one of which is that the variations in all of the chemical constituents, other than silica, must increase in proportion as silica decreases; from which it follows that the number of different kinds of rocks possible for any given percentage of silica is much greater the lower the percentage of silica. The chemical relations of genetically connected series of rocks are next discussed, and various other topics, the general conclusion being that a systematic classification of all kinds of igneous rocks based, as it should be, upon their material characters, cannot properly be expected to take cognizance also of the laws governing their production, eruption, modes of occurrence, and solidification, as well as their subsequent alteration.

Nodular Granite from Pine Lake, Ontario. By FRANK D. ADAMS. *Bull. Geol. Soc. Am.*, 9, 163–172.—The nodules, which the author regards as a product of the primary magmatic differentiation of the granite, are composed chiefly of quartz, muscovite, and sillimanite ; and the chief peculiarity which they present lies in the fact that the portion of the magma which thus separated out was more acid than the magma as a whole, which is very unusual. The analyses show that although the granite is a very acid one, the chief difference between it and the nodules is that the latter are richer in silica and alumina and poorer in alkalies than the granite itself.

The Phosphate Deposits of Arkansas. By JOHN C. BRANNER. *Trans. Am. Inst. Min. Eng.*, 26, 580–598.—This detailed account of these, as yet, unworked deposits is accompanied by numerous analyses of the nodular phosphate rock, the amount of phosphoric acid ranging from 22.62 to 33.86 per cent. Two analyses of the black and green Eureka shales, with which the phosphate rock is invariably associated, are also given.

An Olivinite Dike of the Magnolia District (Colorado) and the Associated Picrotitanite. By MILTON C. WHITAKER. *Proc. Col. Sci. Soc.*, Feb. 5, 1898, 1–14.—Two analyses of this highly altered olivine rock are given, their most striking features being the low percentages of silica (22.24 and 21.90) and the relatively high percentages of magnesia, carbon dioxide, and the alkalies. The secondary and accessory minerals include serpentine, magnetite, picrotitanite, garnet, calcite, chlorite and micas. The picrotitanite or magnesia menaccanite is the most interesting of these, and three analyses of it by different methods are given and discussed.

Phosphatic Chert. By J. H. KASTLE, J. C. W. FRAZER, AND GEO. SULLIVAN. *Am. Chem. J.*, 20, 153–159.—One of the most characteristic formations in central Kentucky is a deposit of chert which marks the upper boundary of the Trenton limestone. The limestone immediately below the chert is characterized by thin layers which are highly phosphatic, containing from 1.46 to 31.815 per cent. of phosphoric acid, P_2O_5, the average being about 15.9 per cent. ; and this rock is probably the source of the remarkable and enduring fertility of the soil of the Blue Grass section. It has more recently been discovered that the chert itself is invariably phosphatic ; and 39 analyses are given, showing from 0.179 to 3.5 per cent. of P_2O_5, with an average of 1.684 per cent. Other analyses are quoted to show that the chert contains four or five times as much P_2O_5 as the normal Trenton limestone or the soils derived from it. Cherts from the Birdseye limestone and from the Permian limestone of Russia were also analyzed and found to contain 0.5 to 3 per cent. of P_2O_5. Other analyses tend to show that the phosphate is an original feature of the chert, while the fact that the porous, weathered cherts are most highly phosphatic suggests that the phosphate may be the insoluble residue of limestone which was once intimately associated with the chert. A chert breccia is also described, the dark-brown cementing substance of which gave in four analyses from 3 to 4.5 per cent. of P_2O_5, which is supposed to exist in the rock as phosphate of iron.

GEOLOGICAL CHEMISTRY.

W. O. CROSBY, REVIEWER.

Geology of Canada. *Ann. Rep. Geol. Survey, Can.*, 7, (1894). —In the section by F. D. Adams on the Laurentian north of the St. Lawrence river, are given the analyses which were made to establish a criterion for the distinction of gneisses of igneous and sedimentary origin. These analyses formed the basis of a separate paper in the *Am. J. Sci.* (July, 1895), which has been reviewed in these pages. Under the head of Economic Resources are also several analyses of iron ores. The section by G. Christian Hoffman, on Chemistry and Mineralogy, consists chiefly of miscellaneous analyses of a large variety of minerals, including coals, iron ores, celestite, graphite, galenite, tetrahedrite, nickeliferous pyrrhotite, marls, mineral waters, and many assays of gold and silver ores.

GENERAL AND PHYSICAL CHEMISTRY.

A. A. NOYES, REVIEWER.

A Revision of the Atomic Weight of Zirconium. BY F. P. VENABLE. *J. Am. Chem. Soc.*, 20, 119–128.—The ratio $ZrOCl_2$. $3H_2O : ZrO_2$ was determined by heating the former substance at 100° in a current of hydrochloric acid to a constant weight and igniting it, first gently and then to the highest heat of the Bunsen burner for three or four days. The mean value of the atomic weight found is 90.78, that given by Clarke in his report of last year being 90.40. Five sources of error possibly affecting the results are considered by the author. The drying of the hydrated salt seems to the reviewer to introduce another element of uncertainty; for it is almost incredible that a salt which is known to lose water at 180°–210° should not undergo an appreciable loss when heated at 100° for 50–100 hours. It would be desirable to show, at least, that the weight on drying not only became constant, but remained so on long-continued heating.

Fifth Annual Report of the Committee on Atomic Weights. Results Published during 1897. By F. W. Clarke. *J. Am. Chem. Soc.*, **20**, 163–173.—The following is a summary of the atomic weight determinations (referred to oxygen as 16.00) published during last year. The values given by the author in his previous report are also appended in parentheses : Carbon, (Scott, *J. Chem. Soc.*, **71**, 550), 12.0008 (12.01) ; Carbon, (Lord Rayleigh, *Chem. News*, **76**, 315), 11.9989 (12.01) ; Nitrogen, (Leduc, *Compt. rend.*, **125**, 299), 14.005 (14.04) ; Aluminum, (Thomsen, *Ztschr. anorg. Chem.*, **15**, 447), 26.992 (27.11) ; Nickel, (Richards and Cushman, *Proc. Amer. Acad.*, **33**, 97), 58.69 (58.69); Cobalt, (Richards and Baxter, *Proc. Amer. Acad.*, **33**, 115), 58.99 (58.93) ; Cerium, (Wyrouboff and Verneuil, *Bull. Soc. Chim.*, **17**, 679), 139.35 (140.2). The article of Scott on carbon related to the application of a correction to the determinations of previous investigators for the change in volume which potash undergoes when it absorbs carbon dioxide. Work has also been done by Hardin (*J. Am. Chem. Soc.*, **19**, 657) on the atomic weight of tungsten, showing the unreliability of the method usually employed.

Investigation of the Theory of Solubility Effect in the Case of Tri-ionic Salts. By Arthur A. Noyes and E. Harold Woodworth. *J. Am. Chem. Soc.*, **20**, 194–201 ; *Tech. Quart.*, **11**, 65–71.—It was found that the solubility of lead iodide is diminished both by potassium iodide and by lead nitrate in such a way that the product of the concentration of the lead ions into the square of the concentration of the iodine ions remains constant, thus confirming the applicability of the mass-action law to the solubility of tri-ionic salts. The concentration of the saturated solutions was determined by conductivity measurements.

The Relation of the Taste of Acids to their Degree of Dissociation. By Theodore William Richards. *Am. Chem. J.*, **20**, 121–126.—The author finds that he can just detect, by tasting, the acidity of a nearly one-thousandth normal hydrochloric

acid solution, and can distinguish weak solutions differing from
one another in concentration by 25 per cent.; and he shows that
tenth-normal acid can be titrated with alkali with an error of
less than one per cent., using the sense of taste as an indicator
of neutrality. The sour taste of different acids was not found
to be proportional to their degree of dissociation ; for example,
a 0.001 normal hydrochloric acid had a taste like that of an acetic
acid solution three times as strong, although the concentration of the hydrogen ions in the former solution is about five
times as great as in the latter solution. It was further found
that, in accordance with the laws of mass-action, sodium acetate
greatly diminished the sour taste of acetic acid, though not as
much as the theory requires, while potassium chloride has no
influence on that of hydrochloric acid.

A Redetermination of the Atomic Weight of Zinc. By H.
N. MORSE AND H. B. ARBUCKLE. *Am. Chem. J.*, 20, 195–202.
—Richards and Rogers (*Proc. Am. Acad.*, 28, 200) having shown
that an error exists in those atomic weight determinations
where zinc oxide has been weighed owing to its occlusion of
nitrogen and oxygen, the authors have repeated the earlier determinations of Morse and Burton, using the same sample of
pure zinc, and determining the quantities of these gases occluded
by dissolving the zinc oxide after weighing, in dilute sulphuric
acid, and measuring and analyzing the gas evolved. The mean
value of the atomic weight found without correcting for the occluded gases is 65.328 ; the corrected value is 65.457. The
value found by Richards by the analysis of zinc bromide was
65.40.

Ternary Mixtures, III. By WILDER D. BANCROFT. *J. Phys.
Chem.*, 1, 760–765.

Molecular Weights of Some Carbon Compounds in Solution.
By CLARENCE L. SPEYERS. *J. Phys. Chem.*, 1, 766–783.—The
article presents the results of a large number of determinations
of the rise in boiling-point produced by various organic substances when dissolved in water, methyl alcohol, ethyl alcohol,
propyl alcohol, chloroform, and toluene. These solvents were
caused to boil at different temperatures by variation of the external pressure, the main object of the investigation being apparently the determination of the effect of temperature on the molecular raising. No discussion of the results is given, however,
nor is any evidence of their accuracy presented, which is particularly unfortunate, since, according to the reviewer's experience,
a Beckmann thermometer which is subjected to considerable
variations of pressure, may give quite unreliable readings, owing
to the imperfect elasticity of the bulb.

Fractional Crystallization. BY C. A. SOCH. *J. Phys. Chem.*, 2, 43-50.—The author has made determinations of the solubility in pure water at 25° and at 80° and in forty per cent. alcohol at 25° of four pairs of salts, the solutions being saturated at the same time with both of the two salts. The salt-pairs investigated were potassium and sodium chlorides, potassium chloride and nitrate, sodium chloride and nitrate, and potassium nitrate and sodium chloride. The article contains in addition readily derived formulæ which express the extent to which a mixture of two salts in a solution saturated with them can be separated by definite processes involving on the one hand crystallization by variation of the temperature and the quantity of solvent present, and on the other, precipitation by alcohol at constant temperature.

Distribution of Mercuric Chlorid between Toluene and Water. BY OLIVER W. BROWN. *J. Phys. Chem.*, 2, 51-52.— It is shown by experiments that mercuric chloride distributes itself between toluene and water in such a manner that the ratio of its concentrations in the two solvents remains nearly, but not completely, constant. The variations from constancy are believed to be greater than would be accounted for by experimental errors.

Solutions of Silicates of the Alkalies. BY LOUIS KAHLENBERG AND AZARIAH T. LINCOLN. *J. Phys. Chem.*, 2, 77-90.— The authors have determined the freezing-points and electrical conductivities of dilute solutions of the neutral and acid silicates of the alkali metals. Their experiments confirm the conclusion previously drawn by Kohlrausch in the case of sodium silicate that these salts are to a great extent broken up in solution into colloidal silicic acid and the alkaline hydroxides, this hydrolysis being complete, according to the freezing-point results, when one mol of the neutral salt is present in 48 liters, and amounting to about 65 per cent. at a dilution of 8 liters. The acid salts investigated, $R'HSiO_4$ and $R'_2Si_2O_{11}$, are somewhat less hydrolyzed, as the Law of Mass Action requires. The authors find that the conductivity of the salts is much less than that of the free bases of corresponding concentration, and that that of the acid salts is less than that of the neutral salts ; and they appear to attribute these facts, and also the decrease of the conductivity of those salts with the concentration, to a retarding influence of the colloidal silicic acid on the movement of the ions. It seems far more probable, however, that these differences are due mainly, if not wholly, to the fact that the hydrolysis is not complete, and varies in the different cases ; for it is not likely that in solutions so dilute as $\frac{1}{14}$ or $\frac{1}{24}$ normal the dissolved substance, whatever its nature, appreciably affects the rate of movement of the ions.

Vapor-tension of Concentrated Hydrochloric Acid Solutions.
By F. B. ALLAN. *J. Phys. Chem.*, 2, 120-124.—The partial
pressure of hydrochloric acid over its solutions containing from
28.1 to 36.4 per cent, acid was determined by drawing measured
quantities of air through them, absorbing the gas in water, and
titrating. Between these comparatively narrow limits of concentration the partial pressure of the acid was found to vary
from 5.5 to 138.1 mm. of mercury, so that the variation cannot
possibly be accounted for by the change in dissociation of the
acid.

H. M. GOODWIN, REVIEWER.

A New Form of Discharger for Spark Spectra of Solutions.
By L. M. DENNIS. *J. Am. Chem. Soc.*, 20, 1-3.—This device
avoids the difficulty usually met with in observing spectra of
this kind, namely, spattering and rapid consumption of the solution. It consists of a U-shaped tube of unequal arms, in the
shorter of which the lower sparking terminal—a cone of
graphite—projects upward, being sealed in from below by means
of a platinum wire. The solution is fed to this electrode from
the longer arm by the following device : a somewhat smaller
straight glass tube which telescopes air-tight through the open
end of the longer arm, by means of a rubber connector, is
pressed down into the solution until its lower end is at a level
with the cone in the shorter arm. As the solution evaporates at
the electrode, air enters the longer arm through this tube until
the same level of liquid is reestablished in both arms.

A Method of Determining the Resistance of Electrolytes.
By PARKER C. McILHINEY. *J. Am. Chem. Soc.*, 20, 206-208.—
The method described lays no claim to the degree of accuracy
attainable with the usual Kohlrausch method and is of advantage mainly when a large number of resistance measurements
are to be made in a short interval of time. It was devised to
measure the resistance of silicates and similar electrolytes at
high temperatures. The method consists in connecting a known
adjustable resistance R, in series with the conductivity cell of
resistance X, with a battery of constant electromotive force. A
rapidly revolving current reverser is inserted in the circuit between the known resistance and the cell, to avoid polarization
and electrolysis in the latter. The fall of potential through the
known resistance R is measured by means of a D'Arsonval galvanometer. If R is made small compared with X, the drop of
potential through R is approximately proportional to the conductivity of the cell. Formula (5) should be corrected as follows :
$f : g = X : R$.

The Jacques Cell. By WM. OSTWALD. *Am. Electrician*, 10, 16–17.—In this letter to the *Electrician* Ostwald has clearly indicated the probable actions taking place in the Jacques carbon cell, by pointing out its analogy to the Lalande cell, the latter consisting of copper oxide and zinc electrodes in sodium hydrate. On closed circuit the zinc dissolves to sodium zincate, while the cuprous oxide is reduced to metallic copper. The continuous action of the cell requires the reoxidation of the reduced copper by air or oxygen. In the Jacques cell the copper oxide is replaced by iron, which in presence of the free oxygen blown in at the high temperature used, is oxidized to ferric oxide, and the zinc by carbon. The action of the cell consists in the oxidation of the carbon anode to carbonic acid and reduction of the ferric oxide. Oxygen must be supplied at the cathode in order to maintain the action of the cell. The chemical potential of free oxygen is therefore not effective in producing electrical energy in this cell, but only that of ferric oxide. The thermoelectric explanation of Reed is regarded as wholly unfounded.

The Surface-tensions of Aqueous Solutions of Oxalic, Tartaric, and Citric Acids. By C. E. LINEBARGER. *J. Am. Chem. Soc.*, 20, 128–130.—The surface-tensions of aqueous solutions of the three above-mentioned acids was determined at 17.5°, 15° and 15°, respectively, for various concentrations, with the apparatus devised by the author and previously reviewed (*this Rev.* 2, 38). It was found that the surface-tension of oxalic and citric acid solutions rapidly diminished with increasing concentration, whereas the values for tartaric acid increased. For solutions of citric acid varying in concentration between thirty-five and sixty-five per cent. the surface-tension was nearly constant. No attempt was made to explain the results obtained other than to ascribe them to complicated relations resulting from polymerization and dissociation phenomena.

The Vapor-pressure Method of Determining Molecular Weights. By W. R. ORNDORFF AND H. G. CARRELL. *J. Phys. Chem.*, 1, 753–760.—This is a preliminary communication on the application of Ostwald's method of determining the vapor-pressure of solutions as worked out by Bredig and Will, to molecular weight determinations. As no essential modifications have been made in the method and as the results published clearly indicate the preliminary nature of the experiments thus far made, a more detailed review may be reserved for a later communication.

Solubility and Boiling-point. By OLIVER W. BROWN. *J. Phys. Chem.*, 1, 784–786.—The addition of potassium chloride to an aqueous alcoholic solution was found to lower its boiling-point, as already shown by Miller, whereas the addition of a substance which is soluble both in alcohol and water, as urea, was found, as expected, to produce a less anomalous result. The boiling-point was in fact found to rise, but less than in simple aqueous solutions. The experiments are to be continued with alcoholic solutions of different percentage composition.

On the General Problem of Chemical Statics. By P. DUHEM. *J. Phys. Chem.*, 2, 1-43; 91-116.—This paper is, as the author states, intended to be a commentary on and complimentary to Gibbs' memoir "On the Equilibrium of Heterogeneous Substances." On account of its mathematical nature it does not admit of a detailed review. The subject is treated with the author's usual elegance and should be read by all interested in the application of thermodynamics to chemical equilibrium. The following subdivisions of the paper will indicate its general scope and contents. The preliminary chapter treats of the thermodynamic potential of an homogeneous mixture under constant pressure and under constant volume. Chapter 1 deals with general theorems of the chemical statics of homogeneous systems, both at constant pressure and at constant volume ; Chapter 2 deals with heterogeneous systems at constant pressures and contains a deduction of the phase rule and an application of it to systems of different variance. The concluding chapter has to do with the general principles of the equilibrium of heterogeneous systems at constant volume, and the more especial discussion of systems of different variance, and under certain prescribed external conditions.

Correction. By WILDER D. BANCROFT. *J. Phys. Chem.*, 1, 786.—The fact that lead iodide crystallizes in the anhydrous form from aqueous solutions, instead of with two molecules of water as assumed by the author, renders valueless his conclusions relative to this salt in his paper on solids and vapors (*J. Phys. Chem.*, 1, 344).

ANALYTICAL CHEMISTRY.

ULTIMATE ANALYSIS.

H. P. TALBOT, REVIEWER.

The Dignity of Analytical Chemistry. By C. B. DUDLEY. *J. Am. Chem. Soc.*, 20, 81–96.—In this paper (his presidential address), Dr. Dudley has clearly and forcibly stated the claims which Analytical Chemistry may justly put forward for a position of equality with other branches of the science. The article does not admit of a brief review.

Sodium Peroxide in Quantitative Analysis. By C. GLASER. *J. Am. Chem. Soc.*, 20, 130–133.—The paper reviews the published statements regarding the quantitative use of sodium peroxide, and the author details a procedure involving its successful use in the determination of sulphur in coal, coke, and asphalt.

The Volumetric Determination of Cobalt. By HARRY B. HARRIS. *J. Am. Chem. Soc.*, 20, 173–185.—A critical examination of the various published volumetric methods for the determination of cobalt has been attempted, from which the author concludes that "after careful repetition of these methods, making varying conditions whenever deemed advisable, one is justified in concluding that none of them possess the degree of accuracy required in any trustworthy determination of cobalt. A good volumetric method for this purpose still remains to be devised." No suggestions are offered to assist the future investigator.

W. H. WALKER, REVIEWER.

The Estimation of Manganese as Sulphate and as the Oxide. By F. A. GOOCH AND MARTHA AUSTIN. *Am. J. Sci.*, 155, 209–214.—The estimation of manganese in its salts when combined with volatile acids, by weighing as the anhydrous sulphate, is found, contrary to previous investigations, to be both rapid and accurate. The difficulty in obtaining a manganese solution of known strength, owing to the varying degrees of hydration of the manganous salts, was overcome by preparing a perfectly neutral solution of manganous chloride, and from the weight of silver chloride obtained from a definite volume, the weight of manganese present was calculated. Portions of this solution were evaporated to dryness in the presence of sulphuric acid, and the excess of acid driven off by heating the residue in a platinum crucible suspended in a larger porcelain crucible used as a radiator. This outer crucible may be heated to dull redness, and the residue thus ignited to constant weight. The results show the process to be both simple and accurate. The estimation of manganese by weighing as the different oxides is described, but found by experiment to be far less satisfactory than the method just outlined.

PROXIMATE ANALYSIS.

A. H. GILL, REVIEWER.

A Delicate Test for the Detection of a Yellow Azo Dye used for the Artificial Coloring of Fats, etc. By J. F. GEISLER. *J. Am. Chem. Soc.*, 20, 110.—The test consists in adding a small quantity of Fuller's earth to the fat, placed in a porcelain dish.

A pink or violet color is produced if the dye is present. The test will detect 14 grains per ton or 1 part per million with ease, and may be used to indicate 0.000000005 mg. of the dye.

Method of Testing Spirits of Turpentine. By C. B. DUDLEY AND F. N. PEASE. *Am. Eng., Car Builder and R. R. J.,* 72, 119. —The four tests employed are : Specific gravity, distillation-point, residue on evaporation, and treatment with oil of vitriol. The gravity is determined by the Westphal balance and varies from 0.862 to 0.872. The distillation-point is determined by boiling 100 cc. in a 500 cc. distillation flask. This point varies from 305° to 308° F. (152°-153° C.) at 29 inches pressure, with the thermometer wholly in the vapor. For the residue on evaporation, 20 grams of the sample are weighed into a 100 cc. platinum dish and evaporated not above 250° F. (121° C.). The residue should not exceed 2 per cent. and usually does not exceed one. The evaporation should take place at 100° C. in cases of dispute. The treatment with oil of vitriol is based upon the fact that pure oil of turpentine is almost wholly polymerized and dissolved by sulphuric acid. Six cc. of the sample are placed in a 30 cc. tube graduated to tenths, held under a cold water faucet and slowly filled with C. P. sulphuric acid. It is allowed to cool, the tube corked, and the contents mixed five or six times, cooling if necessary. The tube is placed vertically and allowed to stand half an hour. The material unaffected by the acid is the adulterant and its volume is measured. It is not usually more than 3 per cent.

G. W. ROLFE, REVIEWER.

The Lecithins of Sugar-cane. By EDMUND C. SHORKY. *J. Am. Chem. Soc.,* 20, 113-118.—The paper presents a discussion of the determination of lecithins in sugar-cane juice as well as the limitations of the present analytical methods for separating these bodies from other nitrogen compounds present. The author suggests that the greasy deposits in the higher vacuum cells of multiple effects of sugar houses result in part from the decomposition of lecithins.

Comparison of the Standard Methods for the Estimation of Starch. By H. W. WILEY AND W. H. KRUG. *J. Am. Chem. Soc.,* 20, 253-266.—A large amount of experimental data is given which proves that all the existing polarimetric methods for determining starch are unreliable. The Lindet method is only approximate. The diastase method gives satisfactory results providing precautions are taken to pulverize the sample to extreme fineness and allow sufficient time for the action of malt, preferably containing pepsin. It is recommended to remove the fat first and also to repeat the malt treatment after boiling and cooling. While noting that the results of cereal analyses are at best but approximate, the authors do not believe that any constituents are present that are unaccounted for unless amounts of complex carbohydrates so small as to be reasonably negligible.

The Solubility of the Pentoses in the Reagents Employed in the Estimation of Starch. By W. H. KRUG AND H. W. WILEY. *J. Am. Chem. Soc.*, 20, 266–268.—The authors show by analyses that the solution of the pentosans by malt extract is inappreciable. Incidentally, they point out the important fact that hexose carbohydrates give a noticeable amount of furfurol when distilled with 12 per cent. hydrochloric acid. This, it would seem, might have an important bearing on many "pentosan" determinations.

Note on Fehling's Solution. By J. B. TINGLE. *Am. Chem. J.*, 20, 126–127.—The author recommends the Pavy solution, modified by Purdy, in which the tartrate is replaced by glycerol. The composition is given of the solution which is especially designed for urine analysis.

W. R. WHITNEY, REVIEWER.

A Simple and Accurate Method of Testing Diastatic Substances. By JOKICHI TAKAMINE. *Am. J. Pharm.*, 70, 141–143.—This is apparently a rapid method of procedure, based on the assumption that taka-diastase does not lose its diastatic power on standing, as do other forms of this ferment. The diastatic action of the material to be measured is compared with the action of the taka-diastase, the actual comparison being made between the color produced by iodine in starch paste hydrolyzed by the sample ferment and shades, or colors, produced when taka-diastase is used.

Vinegar Analysis and Some Characteristics of Pure Cider Vinegar. By ALBERT W. SMITH. *J. Am. Chem. Soc.*, 20, 3–9.—Beside the methods of analysis, this article also states the results of analyses of over fifty different samples of vinegar, the acid, total solids, ash, alkalinity of ash, and the soluble and insoluble P_2O_5 of the ash being given. From these results interesting conclusions are drawn as to differences between pure cider vinegar and other varieties.

Method of Analysis of Licorice Mass. By ALFRED MELLOR. *Am. J. Pharm.*, 70, 136–137.—This is an outline of the methods adopted by consumers and manufacturers in the United States. Moisture, mineral matter, insoluble matter (in cold water), gummy matter, glycyrrhizin, saccharine matter, and extractive substances are determined.

Glucose.in Butter. By C. A. CRAMPTON. *J. Am. Chem. Soc.*, 20, 201–207.—In preparing butter for transportation into warm climates it is sometimes the practice to mix glucose with the salted butter. The author has determined the quantities of glucose in different samples "by difference" and by direct methods; *i. e.*, by the use of Fehling's solution and by the optical method. The percentage of glucose found in the samples varies from 5 to 13 per cent. calculated as confectioner's glucose (16.5 per cent. water).

Methods and Solvents for Estimating the Elements of Plant Food Probably Available in Soils. By WALTER MAXWELL. *J. Am. Chem. Soc.*, 20, 107–110.—A preliminary publication, to be followed later by descriptions of experiments carried on in Hawaii.

The Determination of Small Quantities of Alcohol. By FRANCIS G. BENEDICT AND R. S. NORRIS. *J. Am. Chem. Soc.*, 20, 293–302.—The method is based on the oxidation of alcohol by chromic acid, and presupposes the absence of other reducing agents in the alcoholic solution. The result of an analysis of a solution containing 0.00494 per cent. of alcohol varied by but 0.03 of this amount. With solutions ten times as concentrated the agreement was exact. The paper contains interesting information concerning the absorption of alcohol from air by sulphuric acid.

On the Assay of Belladonna Plasters and the Alkaloidal Strength of the Belladonna Plasters of the Market. By CARL E. SMITH. *Am. J. Pharm.*, 70, 182–189.—This is a report of the Research Committee D, Section II of the Committee of Revision of the U. S. Pharmacopœia. It contains a detailed method of analysis, as well as the results of analyses of ten plasters of American and one of foreign make.

Analysis of the Rhizome and Rootlets of Plantago Major, Linne. By J. FRANK STRAWINSKI. *Am. J. Pharm.*, 70, 189–191.—The drug was successively extracted with petroleum ether, ethyl ether, absolute alcohol, and water, and these extracts examined. A crystallizable material was found in the petroleum ether extract, which the author hopes to study further. No glucosides or alkaloids were found in the residue resulting from the ether extract. Glucose and saccharose were found in the alcohol extract. Starch was found, but no tannin. The ash, amounting to 24.7 per cent. of the weight of the drug, was found to contain the ingredients usually found in plant ashes.

Criticism of a Proposed Method for the Assay of Senega. By EDWARD KREMERS AND MARTHA M. JAMES. *Pharm. Rev.*, 16, 45-49.—The authors show that the fact that methyl salicylate may be obtained from most samples of senega root by distillation with steam, is not a satisfactory criterion in testing for this drug. Samples known to have been false senega yielded the salicylate, while some true senegas did not. The authors point out the fact that addition of acid in the distillation increases the yield of methyl salicylate, which is evidence that this ester is a product of hydrolysis, perhaps of a glucoside.

Test for Hydrocyanic Acid in Mitchella Repens. BY RICHARD FISCHER. *Pharm. Rev.*, 16, 98-99.—The author failed to find this acid in samples of the partridge berry or squaw vine examined by him. The search was prompted by the report that hydrocyanic acid occurs in this plant.

F. H. THORP, REVIEWER.

A Method of Estimating Tannin. BY J. N. HURTY. *Leather Manufacturer*, 9, 10.—A ten per cent. solution of the extract is made by dissolving the requisite quantity in 500 cc. boiling water, and, after cooling, diluting with the proper amount of water. After mixing, two samples of 100 cc. each are evaporated on a steam-bath in aluminum dishes, weighed, and the average of weight taken as the total solids. The remainder of the solution is filtered on S. & S. No. 597 paper and the first 100 cc. of the filtrate rejected. Two more samples, each of 100 cc. are evaporated as before and the average of the weighings subtracted from the total solids. The difference is the insoluble matter. Two more 100 cc. samples are run upon two 10 gram portions of hide powder in 100 cc. beakers, well stirred and allowed to stand five minutes; they are then put into glass percolators stoppered with absorbent cotton. The turbid and colored percolates are thrown away and the wet hide powder packed by pressing with a large glass rod. New portions of the filtered extract are then poured through the percolators and 5 cc. portions of the clear liquid are collected and tested with the original filtered extract for soluble hide. If no soluble hide is found, 105 cc. of each percolate is reserved and then 5 cc. more is collected and tested for tannin by adding a few drops of soluble hide solution. If both tannin and soluble hide are absent in the percolates, 100 cc. of the reserved portions are evaporated as before and the average of the weights taken as non-tannins. This weight is subtracted from the total solids from the filtered extract solution and the difference called tannin. The method is claimed to be more rapid than the official "shake process." Several misprints make the directions somewhat obscure.

Testing of Formaldehyde. By CARL E. SMITH. *Am. J. Pharm.*, 70, 86.—The author has studied numerous methods of assaying formaldehyde with the view of finding a simple, rapid, and reasonably accurate process for use by manufacturers and pharmacists. It is necessary that common impurities, such as acetone and methyl alcohol shall not affect the accuracy of the process. The hydroxylamine method of Brochet and Cambier (*Compt. rend.*, 120, 489) is quick and accurate with pure solutions, but other aldehydes and acetones interfere with its exactness. The iodine method (Romijn, *Ztsch. f. anal. Chem.*, 36, 18) is unsuitable if acetone is present. The cyanide method of Romijn (*ibid.*) requires much care and attention and will only give satisfactory results in practiced hands. The free alkali method consists in heating formaldehyde with sodium or potassium hydrate solution under pressure, in a manner similar to that in the saponification of esters, converting the formaldehyde into methyl alcohol and formic acid. The results are reasonably exact in all but one case, but methyl alcohol and acetone interfere with the result, and there is some danger of an explosion. One anomalous case was observed, but could not be explained. The ammonia method (Legler, *Ber.*, 16, 1333) was found reliable if care was taken to standardize the ammonia solution frequently and to titrate the contents of the flask soon after admixture with the ammonia. Methyl alcohol and acetone have practically no influence. The inaccuracy caused by the loss of ammonia led the author to modify the process as follows: Dissolve two grams of pure neutral ammonium chloride in 25 cc. of water and put into a flask provided with a well-fitting stopper. Add 2.25 grams of the sample and run in from a burette 25 cc. of $\frac{n}{7}$ potassium (or sodium) hydrate. Stopper the flask at once and let stand one-half hour. Then add a few drops of rosolic acid solution and determine the excess of ammonia with $\frac{n}{7}$ sulphuric acid, each cc. $\frac{n}{7}$ potassium hydrate consumed indicating 0.5 per cent. of formaldehyde. A series of tests on commercial samples is tabulated and a list of tests to be followed in the examination of them, together with the precautions to be observed, is given.

TECHNICAL CHEMISTRY.

A. H. GILL, REVIEWER.

Alabama Coal in By-product Ovens. By W. B. PHILLIPS. *Am. Manuf.*, 17, 446.—The paper gives the results of coking washed coal from the Pratt seam in Otto–Hoffman ovens at Jefferson Co., Alabama. Four different runs of about 7 tons each were made. The results are as follows:

| | Analysis. | |
	Coal.	Coke.
Moisture	5.95	dry
Volatile matter	32.69	0.98
Fixed carbon	54.33	90.22
Ash	7.03	8.80
Sulphur	0.94	1.28

Yield of gas 9600 cu. ft. per ton; heating power 630 B.T.U. per cu. ft.; candle power 11.4; coke 70.6 per cent.; tar 90 lbs. per ton; sulphate of ammonia 23.6 lbs. per ton; about 3000 cu. ft. of gas per ton are available.

The Explosive Properties of Acetylene. By F. C. PHILLIPS. *Proc. Eng. Soc. Western Pa.*, 13, 299.

G. W. ROLFE, REVIEWER.

Additional Notes on the Sugar-cane Amid. By EDMUND C. SHOREY. *J. Am. Chem. Soc.*, 20, 133, 137.—The author gives further experimental evidence that the amid described in a previous paper (*this Rev.*, 4, 34) is glycocoll containing a trace of leucine.

Making Concrete in Multiple Effects. By L. M. SONIAT. *La. Planter and Sugar Mfr.*, 20, 248.—This is an account of experiments with the Lillie triple-effect in boiling clarified juice directly to concrete. In order to give the requisite hardening temperature the juice was run "backwards;" *i. e.*, from the highest to the lowest vacuum-cell. The author believes that this process will prove a distinct economy over working centrifugal sugars especially where transportation rates are high. The steam saving is also thought to be considerable. An analysis of the concrete is given.

The Manufacture of Concrete in the Multiple Effect. By S. M. LILLIE. *La. Planter and Sugar Mfr.*, 20, 248–251.—This paper offers a further treatment of the subject on similar lines. A short history of the manufacture of concrete, and statistics as to the profits in this work as carried on twenty years ago are given, as well as analyses. The vital point of the probable influence of this industry on the sugar market is also discussed.

The Cambray Process of Double Sulphuring. By G. CAMBRAY. *La. Planter and Sugar Mfr.*, 20, 203.—A series of analyses are given of mill juices and clarified syrups of El Puente Sugar House, showing an average increase of quotient of purity from 82.87 to 91.20 by this method of clarification.

Sugar Beet Investigations in Wisconsin during 1897. *Bull. Agr. Sta. Univ. of Wis.*, 64, 3–104. **Sugar Beets in South Da-**

kota. *Bull. U. S. Expt. Sta., So. Dakota*, 56, 3-32. **Sugar Beet Investigations.** *Bull. Cornell Univ. Agr. Expt. Sta.*, 143, 493-574. **The Sugar Beet in Illinois.** *Univ. of Ill. Agr. Expt. Sta.*, 49, 1-52. **The Composition and Production of Sugar-Beets.** By L. L. VAN SLYKE, W. H. JORDAN AND G. W. CHURCHILL. *Bull. N. Y. Agr. Expt. Sta.*, 135, 543-572.— These bulletins give evidence of the continually growing interest in this subject. In the main they deal with experiments in cultivation. Besides giving the results of analytical examinations of experimental crops, they are excellent manuals of beet culture.

F. H. THORP, REVIEWER.

Paints, Painting Materials and Miscellaneous Analyses. By H. H. HARRINGTON AND P. S. TILSON. *Bull. No. 44, Texas Agr. Expt. Sta., 1898.*—The chief matter of interest in this bulletin is an investigation of the value of cottonseed oil to be used for paint as a cheap preservative for outside work. Tests of the drying properties of the oil were made with a view to its use instead of linseed oil. Two methods of testing were used: (A) Boiling the oil with different drying agents in order to increase its drying properties. (B) Treating the oil with gasoline or turpentine. Seventeen experiments were made under (A) using borate of manganese, chloride of lime, black oxide of manganese (MnO_2?), nitrous acid, caustic soda, lead acetate and various mixtures of these substances. The time and temperature of the heating were varied. The prepared oils were mixed and ground with a good quality of "white lead," and the paint thus made applied to pine boards and dried in the sun. The best result was obtained with an oil which had been heated to 170° C. for one hour, with 0.3 per cent. of manganese borate. This paint dried in six and a half hours. The next best paint was made with an oil which was heated forty-five minutes to 70° C. with 1 per cent. of sodium hydroxide and then treated with 2.5 per cent. of lead oxide, and heated several (?) hours from 130° to 170° C. This sample, as well as several others, also dried in six and a half hours. But in all cases, the films obtained were less hard and firm than those yielded by linseed oil. Many of the samples required two, six, and even eight days for drying, while others were entirely useless. The oils tested with turpentine and gasoline were less satisfactory. One sample, mixed with 10 parts turpentine, dried in eight and a half hours, giving a fair paint. As a rule, cottonseed oil affords less gloss and hardness in the film, but somewhat more tenacity than linseed oil. The conclusion reached is that a properly made cottonseed oil may well replace linseed paints for cheap outdoor work. It is also suggested that the boiling of the oil might be done by "anyone;" but in the opinion of the reviewer, this statement might well have been omitted, since if left to the care of an inexperienced person, the kettle is almost certain to boil over. The authors also devote some space to discussion of the tabulated analyses of thirteen commercial paints. One "pure white lead" consisted of 66.79 per cent. barytes ($BaSO_4$), 15.59 per cent. zinc oxide and 9.81 per cent. "white lead." Another contained 74.5 per cent. barytes. They also find that red and yellow ochres are most suitable for general use with crude cottonseed oil, but with well prepared "boiled" oil, it may pay to use the best white lead or white zinc. The remainder of the bulletin is devoted to tabulated analyses of mineral waters (36 samples) cottonseed meal, coals, clays, fertilizers, asphalts, ashes, ores, and a lignite tar distillate.

Estimation of Mineral Matter in Rubber Goods. By L. DE KONINGH. *J. Am. Chem. Soc.*, 19, 952.—This is a method designed to remove all the soluble mineral matter by treating the rubber with fuming hydrochloric acid. The rubber plus the insoluble matter is weighed, and the ash determined by ignition as usual. The acid filtrate may then be analyzed separately.

The Semet-Solvay Coke Oven and its Products. By WILLIAM H. BLAUVELT. *Proc. of the Ala. Industrial and Scientific Soc.*, 7, 33.—This is a general description of the Semet-Solvay oven as erected at Ensley, Ala., and a discussion of the properties of the coke, gas, ammonia and tar obtained from it. The usual charge is 4½ tons, and the time of coking is about 24 hours; the time of discharging and charging is about 15 minutes. The amount of gas varies from eight to ten thousand cubic feet per ton of 2000 pounds. The heat utilization is very perfect. The gases from the retort are cooled and passed through scrubbers and condensers to remove ammonia and tar; a part of the gas is then burned with air to heat the retorts and the boilers, supplying steam for operating the plant; the remainder is available for lighting, or other heating purposes. In the retort oven the coking takes place without access of air, and a certain amount of the hydrocarbons distil off, are broken down, and some graphitic carbon is deposited on the coke. This increases the yield of coke over that obtained from the bee-hive oven, into which some air enters, so that the hydrocarbons are largely burned in the oven itself, together with a portion of the coke. In a bee-hive oven with Connellsville coal, yields of 65 per cent. coke are considered good; but from the retort oven 75 per cent. coke is obtained. In the bee-hive the coal lies in a layer about 24 inches deep, over a broad surface ; the bottom of the oven being cold, from the quenching of the previous charge and contact with new coal, the coking begins at the top and extends downward, the coke swells and develops the characteristic cellular structure. This texture is thus entirely beyond control and dependent upon the character of the coal; hence many coals do not yield a good coke in the bee-hive oven. In the retort oven, the charge lies in a tall narrow mass, about 5 feet high by 20 inches wide, while the retort, having been quickly emptied by mechanical means and recharged while very hot, the coking and distillation begins at once and the gases formed at each side, penetrate to the center of the mass, where they meet and rise to the top; thus a cleavage plane is formed midway between the walls. The coke in contact with the hot wall is denser, and that in the center more spongy than the main mass. The cellular structure is more compressed, since the narrow retort allows no expansion in the direction of the flow of the gases and because of the depth of the charge. The cellular structure depends somewhat upon the size of the oven, the temperature, and the time of coking. Owing to the rapid heating of the charge, coals may be coked in these ovens which cannot be used in the bee-hive form, since the bituminous matters are decomposed before they can escape. Tests have been made with retort and with bee-hive cokes, both produced from Connelsville coal, during a year's running of a blast-furnace, the fuel charge being changed at times from all retort coke to all bee-hive coke, or to a mixture of both, and yet no indications of difference in the fuel was observed in the working of the furnace. It is not claimed that the retort product is superior to the Connelsville bee-hive coke, but that certain other coals, which are unfit for bee-hive coking, will yield a fair product when coked in retorts; even from impure coal a coke good enough for brewers, malsters or domestic uses, may be made. The retort coke is harder and is thought to withstand the exposure to the furnace gases in the upper part of the furnace better than bee-hive coke; but it has not the silvery luster or glaze of the latter, and is sometimes too wet, owing to improper quenching after drawing. The effect of moisture in the coke is not decided, but tests have been made indicating that it protects the coke from the furnace gases by its cooling action, and by thus allowing more fuel to reach the zone of fusion, should decrease the fuel consumption. It is shown that chemical analysis is of little value in indicating the coking power of coal. But a method devised by Louis Campredon and used in the Vignac Works, in France, is explained. Weighed portions of the finely powdered coal are mixed with varying quantities of sand and the mixtures heated to red heat in closed crucibles until the coal is carbonized. When cold, a powdery or more or less coked mass is left ; from the test portion showing the maximum quantity of sand that the coal can bind together, the coking qualities are estimated ; the weight of the coal being unity, the binding power is shown by the weight of agglomerated sand. The most binding coal tried was found equal to 17, while pitch showed a power of 20.—The ammonia is given off most rapidly about ten hours after charging but ceases entirely before the coking is complete, leaving in the coke about one-quarter of the nitrogen originally in the coal. The yield of ammonia is calculated as sulphate and from Pittsburg coal averages about 16 to 22 pounds per ton of coal. The tar is of good quality and owing to the increased manufacture of water gas, it now finds a ready market. The yield of tar from Pittsburg coal varies from 70 to 80 pounds per ton. The by-product oven gas will probably have to be used as fuel, since it contains less illuminants than that from gas-house retorts ; but from good gas coal it would furnish a fairly good illuminant.

Formaldehyde Tannin. *U. S. Pat. No. 598,914. Abstract in Leather Manufacturer*, 9, 39.—This is a new substance patented by H. C. Durkopf, and called methylene digallotannic acid. It is made by the condensation of tannin with formaldehyde in the presence of hydrochloric acid. The formula assigned to it is $CH_2 \Big\langle \begin{smallmatrix} \text{Tannin radical.} \\ \text{Tannin radical.} \end{smallmatrix}$

BIOLOGICAL CHEMISTRY.

W. R. WHITNEY, REVIEWER.

An Exudation from Larix Occidentalis. BY HENRY TRIMBLE. *Am. J. Pharm.*, 70, 152-153.—This is not a resinous exudation but of a saccharine nature, and is said to be used as a food by Indians. It was found to contain 19.38 per cent. reducing sugar, 68.69 per cent. non-reducing sugar, 5.02 per cent. moisture, 0.41 per cent. ash, and 6.47 per cent. fibre.

Diastatic Fungi and Their Utilization. BY JOKICHI TAKAMINE. *Am. J. Pharm.*, 70, 137-141.—The author describes the manufacture of taka-moyashi, taka-koji, and taka-diastase. A fungus, the Japanese moyashi, is employed. This is cultivated in a starchy material as ground corn, which has been steamed to gelatinize the starch. The spores resulting from the growth of the fungus are separated by sifting. This product is known as taka-moyashi. Taka-koji results from mixing the spores with steamed wheat bran and allowing the culture to grow until a maximum diastatic value is reached. Taka-diastase is precipitated from the aqueous extraction of taka-koji by addition of alcohol.

Preliminary Observations on a Case of Physiological Albuminuria. By TORALD SOLLMANN AND E. C. McCOMB. *J. Expt. Medicine*, 3, 138–145.—This is the study of a case of occurrence of coagulable proteids in human urine which lasted for a considerable period and was unaccompanied by symptoms of disease. During four months of investigation the quantity of proteid present varied but little from day to day. The average daily quantity was 0.5317 gram. It was found to vary directly as the urea and inversely as external temperature. The influence of sleep, of dietary changes, of diuretics, and of drugs acting on the circulation, were also observed.

On the Occurrence of Methyl Salicylate. By EDWARD KREMERS AND MARTHA M. JAMES. *Pharm. Rev.*, 16, 100–108. —This contains a collection of abstracts of the literature on the occurrence of this ester in plants.

The Present Condition of Formaldehyde as a Disinfectant. By C. O. PROBST. *Ohio Sanitary Bull.*, 2, 36–42.—While the author's paper is only a summary of recent work on the subject, the discussion which follows it shows how generally this disinfectant is being used and also that its superiority over burning sulphur is held in doubt. The author evidently believes that failures in the case of the formalin may be attributed to causes which will soon be removed by a clearer understanding of the conditions necessary and quantities of the gas requisite for proper fumigation.

Alkaloidal Constituents of Cascarilla Bark. By W. A. H. NAYLOR. *Am. J. Pharm.*, 70, 237–239.—The author believes that from this bark he has extracted two alkaloids, betaine and cascarilline.

E. H. RICHARDS, REVIEWER.

Dietary Studies in New York City. By W. O. ATWATER AND CHARLES D. WOODS. *Expt. Sta. Bull. U. S. Dept. Agr.*, 46, 1–117.—In many respects this is the most satisfactory of the dietary studies thus far published by the U. S. Dept. of Agriculture, partly because a summary only of the results is given in the text and the details reserved for the appendix, and partly because of the variety of occupation and condition of the 23 families. These studies were made among the so-called poor in the worst congested districts of New York, and yet 11 of the 23 families paid daily 20 cents and over per person for food, and 2 of them over 40 cents, while only 3 spent as little as 16 cents. The calculation of the cost of each 1,000 calories in the different dietaries brings out more clearly the lack of economy in buying. A well-selected dietary may be made up so as to cost from 5 to 5.5 cents per 1,000 calories ; only 4 of the list given come within this limit, while 8 cost over 7 cents per 1,000 calories, and one of these high cost dietaries furnished the lowest total food value found.

A Digest of Metabolism Experiments. By W. O. ATWATER AND C. F. LANGWORTHY. *Expt. Sta. Bull. U. S. Dept. Agr.*, 45, 1–434.—The Department of Agriculture has done great service to all students of problems relating to food and nutrition, in bringing together a summary of the results hitherto obtained, however contradictory they may be.

G. W. ROLFE, REVIEWER.

Observations on Some of the Chemical Substances in the Trunks of Trees. By F. H. STORER. *Bull. Bussey Inst., Harvard Univ.*, 35, 386–408.—This is an able and suggestive paper on the nature and origin of the carbohydrates stored in the wood of trees. It also discusses analytical methods for the differentiation of starches and of pentosans and their limitations, the distribution of starch and pentosans in the tree, and their comparative food values interpreted by the habits of rodents and certain economic uses. Many analytical data are given and many authorities cited.

SANITARY CHEMISTRY.

E. H. RICHARDS, REVIEWER.

Recent Work in England on the Purification of Sewage. By LEONARD P. KINNICUTT. *J. Am. Chem. Soc.*, 20, 185–194.—The author writes from personal observation of the favorable results of experiments made in the more efficient utilization of those bacterial agencies which, in absence of air, accomplish the quick solution of a large part of the organic substances met with in ordinary town sewage.

Municipal and Other Water Supplies. *North Carolina Board of Health Sixth Biennial Rep.*, *1897*, 56–96.—Analyses of twenty samples of water from various towns in North Carolina are given, with reports and correspondence in regard to bacteriological examinations. The whole statement is instructive as to the status of sanitary knowledge and control in many of the states.

The Interpretation of Sanitary Water Analysis. By
FLOYD DAVIS. *Eng. Mag.*, *1898*, 68.

Engineering Chemistry of Boiler Water. By H. HEFFE-
MANN. *Railway and Eng. Rev.*, *1898*, 203.

The Purification of River Water Supplies. By A. HAZEN.
Eng. Mag., *May*, *1898*.

AGRICULTURAL CHEMISTRY.

W. R. WHITNEY, REVIEWER.

The Relative Sensibility of Plants to Acidity in Soils. By
WALTER MAXWELL. *J. Am. Chem. Soc.*, 20, 102–107.—Of a
number of vegetables, cereals, etc., the greater part were found
to be exceedingly sensitive to the presence of acid. Millet and
maize were least affected by acid.

Milk Fat from Fat-free Food. By F. H. HALL. *N. Y.
Agr. Expt. Sta. Bull.*, *Popular Edition*, 132.—An experiment is
described in which materials which were practically fat-free, were
fed to a cow during a period of three months. The fat which
the cow produced in the milk during this time was not derived,
the author believes, either from fat in the food, from stored fat
of the animal's body, or from proteids in its food, nor was it
formed at the expense of two or more of these. A part at least
must have come from carbohydrates of the food. ' That fat may
be formed from carbohydrates has been pretty generally recognized and was discussed in the *U. S. Expt. Sta. Rec.*, 7,
538, and 8, 179. That protein does not produce fat is also there
stated as the result of Pfluger's work.

Cotton Culture. Fertilizer Formulas. By. R. J. RED-
DING. *Georgia Expt. Sta. Bull.*, 39, and **Experiments with
Cotton.** By J. F. DUGGAR. *Ala. Agr. Expt. Sta. Bull.*, 89 and
91.—These contain accounts of experiments with different varieties of cotton plants under the influence of different fertilizing
mixtures.

Analyses of Commercial Fertilizers. *Kentucky Agr. Expt.
Sta. Bull.*, 71.

**The Chemical Composition of Utah Soils, Cache and San-
pete Counties.** *Utah. Agr. Coll. Bull.*, 52.

**Cooperative Experiments made by the Ohio Students'
Union.** *Ohio Agr. Expt. Sta. Bull.*, 88.

Cooperative Fertilizer Experiments with Cotton in 1897. *Ala. Agr. Expt. Sta. Bull.*, 91.

Experiments with Corn. *Ala. Agr. Expt. Sta. Bull.*, 88.

Experiments with Cotton. *Ala. Agr. Expt. Sta. Bull.*, 89.

Concentrated Feed-Stuffs. *Hatch Expt. Sta. of Mass. Agr. Coll. Bull.*, 53.

Calories of Combustion in Oxygen of Cereals and Cereal Products, Calculated from Analytical Data. By H. W. WILEY AND W. D. BIGELOW. *J. Am. Chem. Soc.*, 20, 304–316.

F. H. THORP, REVIEWER.

Corrosive Sublimate and Flour of Sulphur for Potato Scab. Experiments made in 1896. By H. GARMAN. The Use of Corrosive Sublimate for Potato Scab in 1897. By H. GAR-MAN. *Ky. Agr. Expt. Sta. Bull. No.* 72, Feb., 1898.— The first of the above investigations was made to deter-mine whether the scab fungus was introduced into the soil on the seed potatoes. Before planting, the potatoes were rolled in flowers of sulphur or soaked for an hour in a solution of mer-curic chloride (4½ ounces in 30 gallons of water). The test showed that sulphur had little effect in preventing scab but that mercuric chloride did check the disease very greatly. The second investigation was to determine the strength of the mer-curic chloride-solution that could be safely used in treating the seed potatoes. The results showed that treatment of the seed reduced the proportion of scabby potatoes ; that the percentage of scab diminishes with the increase of the strength of the solu-tion used ; that short soaking of the seed in strong solution was as effective as long exposure to weaker solutions ; and that very strong solutions caused some reduction in the yield, if the soak-ing was prolonged. Treatment for one hour with 4½ ounces of corrosive sublimate per gallon of water was found most suita-ble. Stronger solutions may be used for short soaking.

ASSAYING.

H. O. HOFMAN, REVIEWER.

A Cement for the Assay Office. By R. MARSH. *Eng. Min. J.*, 65, 37.—The author recommends for repairs of cracks in muffles from 1 to 2 parts litharge and 10 parts bone ash mixed dry and then moistened with water to the consistency of paste. This cement will stand fire, be tight, hard and strong, and will not crack.

The Assay of Copper Bullion. By A CORRESPONDENT. *Eng. Min. J.*, 65, 223.—The method recommended is as follows: Weigh 1 assay ton into a No. 5 beaker, add 120 cc. cold water, then 100 cc. nitric acid, boil when violent action ceases until red fumes disappear, dilute to 300 cc., filter through 11 cm. paper, sprinkle 2¼ grams test-lead over it and, having gathered together the edges, press into a 2¼ inch scorifier charged with 5 grams test-lead so as to flatten out the point of the filter. Add to the hot filtrate a slight excess of brine, stir vigorously (by hand, by compressed air, or other mechanical device), when silver chloride will settle out in about 30 minutes with 20 oz. copper. Filter through a double filter, sprinkle with test-lead as before, and transfer to scorifier containing gold filter, place in front of muffle, dry paper and burn. Then place scorifier in the opening of muffle to finish incineration, cover sintered mass with 15 grams test-lead and ½ gram borax glass and scorify. This gives a button weighing about 4 grams, which is cupelled in a small cupel and parted.

The Assaying of Gold Bullion. By C. WHITEHEAD AND T. ULKE. *Eng. Min. J.*, 65, 189.—This paper is a short description of the method employed at the mint. This is discussed under the heads, melting and sampling, gold assay, determination of base metals, determination of silver.

The Assaying of Silver Bullion. By C. WHITEHEAD AND T. ULKE. *Eng. Min. J.*, 65, 250–251.—This paper discusses the assay of silver bullion at the mint in the same manner as a previous one did the gold assay. It covers grading and sampling of bullion, standardizing of the sodium chloride solution, determination of the silver, influence of alloyed metals and organic matter, and accuracy and rapidity of the assay.

METALLURGICAL CHEMISTRY.

H. O. HOFMAN, REVIEWER.

The Influence of Altitude on Smelting. By H. LANG. *Eng. Min. J.*, 65, 131–132.—The author attributes the greater consumption of fuel at a high altitude than at a low one to the fact that as the intensity of combustion depends upon the atmospheric pressure, at a high altitude the gases leave a furnace in a more attenuated condition than at a low one and consequently absorb and carry away a greater amount of heat which has to be made up by a greater amount of fuel. Referring to the above paper, J. T. Smith (*Ibid.* p. 277) states that by arching over the top of the cupola in an iron foundry and having the side-exits closed by flap-doors, a greater pressure was maintained than that of the air outside and a considerable saving of fuel effected, the percentage of fuel being 7 on the pig charged. With a cupola 36″ in diameter, narrowing down to 22″ at the crucible, 18000 lbs. were melted down in one hour and thirty-five minutes from the time the blast was started.

Recent Smelting Practice In Colorado. By L. S. Austin. *Eng. Min. J.*, 65, 282–283.—The practice of lead and copper smelting in Colorado and other parts of the country has undergone important change which is briefly reviewed in this paper. The subjects discussed are : Sampling ores and base bullion, size of furnaces and management, recovery of flue-dust, automatic charging, roasting of ores, the separation and disposal of slag, handling of materials, reverberatory smelting, blast-furnace fuel and changes in character of ore smelted. The conclusion that the writer arrives at is that the changes have been mainly in the mechanical details for the saving in labor and the better separation of waste from main and intermediary products.

The Garretson Copper Smelting Furnace. By O. S. Garretson. *Eng. Min. J.*, 65, 160–161.—The aim with this blast-furnace is to carry on in one continuous operation the three necessary processes for treating a sulphide copper ore, *viz.*, roasting, smelting, and converting. While it has the high endorsement of Dr. E. D. Peters, Jr., the reviewer cannot but believe that this furnace will fail to do satisfactory work more often than it will prove a success, because the three processes require for the best work conditions differing too much to be obtainable in a single furnace. Should this great difficulty be temporarily overcome, it is almost inevitable that the slightest changes will prove detrimental to one or the other process.

A Copper-Arsenic Compound. By H. C. Hahn. *Eng. Min. J.*, 65, 401.—The author found in a salamander of a lead blast-furnace dark green hexagonal crystals which upon analysis showed the following composition in percentages : Cu. 61.03, Fe. 0.97, Pb. 0.44, As. 35.50, Sb. 1.08, S. 0.89, Total 99.91, corresponding to the formula Cu_3As. The sp. gr. at 10° C. was 7.976, at 20° C. 8.008, at 30° C. 7.578.

The Electrolytic Production of Zinc. By Editor. *Eng. Min. J.*, 65, 336.—The paper briefly reviews the leading processes for the electrolytic reduction of zinc; *viz.*, the processes of Ashcroft, Siemens-Halske, Dieffenbach and Hoephner, and concludes that at present zinc can be electrolytically produced with profit only when the zinc is removed from ores containing more valuable metals so as to permit them to be satisfactorily treated.

Notes on Aluminum. By J. M. SMITH. *Assoc. Eng. Soc.*, 20, 1–19; *Discussion*, 19–23.—The paper offers little that is new, being a brief review of the properties of aluminum and its alloys, of the ores, of the manner of preparing the alloys and of the Hall, the Héroult and Minet processes for producing the metal. The discussion brings out the proposed use of aluminum wire as electric conductor and the use of the metal in the manufacture of semi-steel.

The Dry Separation of Gold and Copper. By F. R. CARPENTER. *Eng. Min. J.*, 65, 193.—After a general discussion of methods of separating gold and copper, the writer outlines his recently patented process (No. 897,139, January 11, 1898), which consists in exposing metallic copper containing silver and gold to an oxidizing fusion when cuprous oxide and, in the presence of silica, cuprous silicate will form. Both are readily fusible and take up little silver and hardly any gold. By drawing off the oxidized copper from the surface of the metal bath, as with litharge in cupelling, a satisfactory separation of copper from precious metal can be effected. An addition of lead toward the end of the process protects the silver so that a larger percentage remains with the gold than would be otherwise the case. The oxide or silicate of copper is sold to copper works or used again for collecting precious metal in smelting pyritic ores free from or low in copper.

Temperature in Amalgamation. By F. R. CARPENTER. *Eng. Min. J.*, 65, 126–127.—At the Homestake gold mines, Lead City, S. D., two amalgamating batteries were run by T. J. Grier upon the same ore, in one the battery water had a temperature of 50° F., in the other, of 60° F. and over. The yield in gold with the former was found to be decidedly greater than with the latter. The explanation of this appears to be that warm water, making the low-grade amalgam from the fine gold fluid, causes it to run off more readily from the plates than cold water which leaves the amalgam more pasty and hard and more firmly adhering to the plate.

Experiments on Roasting Telluride Ores. By E. D. SKEWES. *Eng. Min. J.*, 65, 488.—This paper is a record of experiments made in Western Australia. In dry-crushing a silver-bearing telluride gold ore with 3.66 per cent. sulphur and 0.33 per cent. copper and then roasting, the sulphur being eliminated to 1 per cent., it was found that the loss in precious metal was very much lower than when it had been wet-crushed. The cause of this was discovered to be in the water used which contained 8.5 per cent. solid matter; *viz.*, $MgCl_2$ and NaCl, 6.95, CaO, 0.86,

Al,O, and Fe,O, 0.49, S 0.12.   The chlorides of the water remained with the ore, being decomposed in the roast and volatilizing the precious metals.

Ore Treatment in Boulder County, Colo. BY C. C. BURGER. *Eng. Min. J.*, 65, 129–130.—The gold ores of Boulder County are sulphides and tellurides, the former are treated by the Colorado system of battery amalgamation, the latter by leaching with potassium cyanide and chlorine water.   The paper describes the technical details of the barrel-chlorinating plant of the Delano Mining and Milling Co., some of the novel points of which may be recorded.   As to general arrangement, the crushing, roasting, and chlorinating houses are on level ground instead of on a hill-side, as is, or rather was, usual, the ore being handled by belt elevators and belt and screw conveyors.   The ore is roasted at a high temperature in a Pearce turret furnace which has water-cooled rabble-arms and plow-shaped rabbles of forged steel.   The loss of gold by volatilization is very slight.   The roasted ore is automatically cooled and delivered into the hoppers of the chlorinating barrels.   The raw ore contains 2.5 per cent. sulphur; in the roasted ore the sulphur must be reduced a few hundredths of one per cent. if an extraction of 95 per cent. of the gold is to be attained.   The barrels used hold 5-ton charges.   They have the Rothwell quartz filter, the quartz having been crushed to pass a 2-mesh sieve.   This allows the slimes to pass through the filter, and these have to be settled in clarifying vats before the gold can be precipitated.   For solution there is required per ton of ore 10 lbs. bleaching powder, 15 lbs. sulphuric acid of 66° B., and for precipitation ¼ lb. sulphur, ½ lb. iron sulphide, and 1¼ lbs. sulphuric acid.

Chlorinating and Cyaniding. BY C. C. BURGER. *Eng. Min. J.*, 65, 427.—A continued discussion (*this Rev.*, 4, 36).

Absorption of Gold by Wooden Leaching Vats. BY F. L. BOSQUI. *Eng. Min. J.*, 65, 248.—Many losses of gold in lixiviation plants are attributed, for want of a better explanation, to absorption by the wooden leaching vats.   An investigation of the subject gave the author results which prove that the amount of gold taken up by the vats is relatively small.   California redwood absorbed 22 per cent. of its weight of solution.   Assuming it to be 50 per cent., a vat weighing 6 tons would absorb 3 tons of solution worth $5.00 in gold per ton.   With plants treating from 3000 to 4000 tons tailings per month, the loss would not be very much noticed.   But the absorption might be cumulative, the leaching vats being part of the time exposed to air, when absorbed solution being partly evaporated, gold might be deposited. The author made a series of experiments with different kinds of woods and with intermittent and continuous contact of wood and solution. He found the amounts of gold absorbed to be exceedingly small by either method of contact.

Influence of the Anodes in Depositing Gold from Its Cyanide Solutions. By E. ANDREOLI. *Eng. Min. J.*, 65, 100–101.—In the Siemens and Halske process, carried out successfully in the Transvaal, gold is recovered from dilute cyanide solutions which would be lost if it were precipitated by zinc shavings. Iron anodes and sheet lead cathodes are used. The iron anode is attacked, forming Prussian blue, which contaminates the electrolyte and hinders a perfect precipitation of the gold, only from 60 to 70 per cent. being recovered. The lead has to be replaced when the gold is separated by cupellation. The author found that an anode of peroxidized lead was absolutely insoluble and lasted any length of time ; he recommends the use of iron cathodes from which the deposited gold is to be separated by immersion in melted lead.

Notes on the Moeblus Process for Parting Gold and Silver as Carried on at the Guggenheim Smelting Works, Perth Amboy, N. J. By P. BUTLER. *Can. Min. Rev.*, 17, 81–88.— The paper describes in a brief elementary way the three departments of the works : electrolytic copper refining by the multiple process, desilverizing of base bullion by the Parkes process, and electrolytic parting by the Moebius process. The many carefully made, but badly printed drawings accompanying the paper, lose some of their value by not being either dimensioned or having the scale given to which they were made.

Molding Sand. By D. H. TRUESDALE. *J. Am. Foundrymen's Assoc.*, 3, 159–176.—The paper treats in detail of the four leading properties of molding sand; *viz.*, refractoriness, porosity, fineness, and bond.

Comparative Fusibility of Foundry Irons. By Th. D. WEST. *J. Am. Foundrymen's Assoc.*, 3, 127–158.—The author shows that chilled iron will melt faster than gray iron of the same composition and advocates the use of sandless pig for foundry purposes. He further discusses the changes in composition due to re-melting, the production of semi-steel castings, and the shrinkage and contraction of cast iron.

Effects of Phosphorus on the Strength and Fusibility of Iron. By TH. D. WEST. *J. Am. Foundrymen's Assoc.*, 4, 123–128.—The author carried on a series of experiments to show the strength and fluidity of foundry iron was increased by the addition of phosphorus.

Carbon and Strength of Iron. By G. R. JOHNSON. *Iron Trade Rev.*, 21, 8. The author contends that in the analyses according to which pig iron is sold, both carbons should be included, not alone silicon, sulphur, phosphorus, and occasionally manganese, as the strength of the pig increases with a diminishing percentage of carbon.

Value of Metalloids in Cast Iron. By M. McDOWELL. *J. Am. Foundrymen's Assoc.*, 4, 97–114. The author gives the subjoined table :

No.	Kind.	Price per ton.	Total C.	Si.	P.	S.	Mn.
1	Foundry	$12.00	3.68	2.90	0.75	0.01	0.30
2	Foundry	11.50	3.30	2.25	0.70	0.02	0.40
3	Foundry	11.00	3.25	1.50	0.30	0.03	0.50
4	Gray forge	10.50	3.80	1.00	0.65	0.04	0.62
5	Mottled forge	10.00	3.75	0.70	0.50	0.05	0.05(?)
6	White forge	9.50	3.65	0.40	0.35	0.10	0.96

to show how the value of pig iron increases with the silicon it contains and what harm is done by the sulphur, each unit of which is said to neutralize 10 units of silicon. The table also demonstrates the necessity of the presence of manganese and the value of phosphorus. In a second table, representing six different heats, the author brings out the changes foundry iron may undergo in composition and physical properties.

Oxidation of Foundry Metals. By T. D. WEST. *J. Am. Foundrymen's Assoc.*, 4, 115–123. The paper is an experimental study into the loss of metal incurred in foundries from oxidation which is claimed to be greater when pig iron is cast in chills instead of in sand.

Annealing of Malleable Cast Iron. By G. C. DAVIS. *J. Am. Foundrymen's Assoc.*, 4, 86–91.—This paper is mainly a review of one read by G. P. Roylston before the English Iron and Steel Institute. The author, however, brings out some of the differences between English and American practice. For instance, in England the pig iron used contains 0.112 per cent. manganese giving a casting with 0.043 per cent. manganese ; in this country the mixture contains 0.40–0.80 per cent. and the casting 0.20–0.40 per cent. manganese. The silicon in the product from a Western foundry ranged from 0.75 to 0.98 per cent. Sulphur must be low, under 0.11 per cent., as, if present in larger amounts, it interferes with annealing. Thus, a ½-inch casting with 0.133 per cent. S. showed after the first annealing G. C. 0.547, C. C. 0.515 ; after the second annealing G. C. 0.505, C. C. 0.380. Roll-scale is generally used here instead of iron ore as in England, the scale being reoxidized at intervals by spreading on the floor and sprinkling with ammonium chloride. The annealing pots are cast from cupola metal and last for five or six heats.

What is Semi-steel? By A. E. OUTERBRIDGE. *Digest of Phys. Test,* 3, 99–103.—The object of this paper is to show that calling the product obtained by charging a cupola with pig iron and steel and wrought iron scrap by the new name, semi-steel, is not justified, as chemically it shows no resemblance to steel or malleable iron, and physical tests show an equally great divergence ; in fact, the so-called semi-steel is simply a strong, close-grained cast iron.

APPARATUS.

A. H. GILL, REVIEWER.

A Constant Temperature Device. By H. P. CADY. *J. Phys. Chem.,* 2, 242.—It comprises an apparatus in which a circulation of water at a constant temperature is attained by means of a siphon emptying into a rotating funnel. The rotation causes the water in the center of the funnel to become depressed which sets the siphon in action.

A New Form of Hydrogen Generator. By E. W. MAGRUDER. *Am. Chem. J.,* 19, 810.—The apparatus is designed to take the place of the Bunsen electrolytic generator, and obviates the difficulties of that form by having the electrodes in separate tubes. The apparatus is said to yield pure hydrogen and may be obtained of Greiner, of New York.

A New Form of Discharger for Spark Spectra of Solutions. By L. M. DENNIS. *J. Am. Chem. Soc.,* 20, 1.

A New Electrolytic Stand. By G. T. HOUGH. *J. Am. Chem. Soc.,* 20, 268.

A Modified Air-bath. By F. P. VENABLE. *J. Am. Chem. Soc.,* 20, 271.

A Collector for Distillation of Ammonia from Water. By F. P. DUNNINGTON. *J. Am. Chem. Soc.,* 20, 286.

A Device to Prevent Loss by Spattering. By A. H. LOW. *J. Am. Chem. Soc.,* 20, 233.

REVIEW OF AMERICAN CHEMICAL RESEARCH.

VOL. IV. No. 7.

ARTHUR A. NOYES, Editor ; HENRY P. TALBOT, Associate Editor.
REVIEWERS: Analytical Chemistry, H. P. Talbot and W. H. Walker;
Biological Chemistry, W. R. Whitney ; Carbohydrates, G. W. Rolfe;
General Chemistry, A. A. Noyes ; Geological and Mineralogical Chemistry, W. O. Crosby ; Inorganic Chemistry, Henry Fay ; Metallurgical Chemistry and Assaying, H. O. Hofman ; Organic Chemistry, J. F. Norris ; Physical Chemistry, H. M. Goodwin ; Sanitary Chemistry, E. H. Richards ; Technical Chemistry, A. H. Gill and F. H. Thorp.

INORGANIC CHEMISTRY.

HENRY FAY, REVIEWER.

On the Action of Hydrogen Sulphide upon Vanadates. BY
JAMES LOCKE. *Am. Chem. J.*, 20, 373-376. — Hydrogen
sulphide reacts with sodium vanadate in the cold, but owing to
the formation of sulphovanadate on the surface the reaction is
not complete unless the liquid is heated to 500°-700°. At this
temperature the author finds that the absorption of hydrogen sulphide is almost violent. The increase of weight in the sodium
vanadate corresponds to that required by the reaction :

$$Na_3VO_4 + 3H_2S = Na_3VOS_3 + 3H_2O.$$

The sulphovanadate formed in this way corresponds to the salt
described by Krüss, which was made in the wet way. Sodium
pyrovanadate reacts with hydrogen sulphide with the formation
of the sulphovanadate, $Na_4V_2O_3S_4$, which resembles potassium
permanganate in appearance. The salt is exceedingly hygroscopic, dissolves in water to a deep reddish purple solution,
which gradually loses hydrogen sulphide, changing to the colorless sodium pyrovanadate. This change of color is also characteristic of the free sulphovanadic acids, but it was found impossible to isolate a pure compound. The vanadates of the heavy
metals are not so readily changed to the sulphovanadates. Lead
pyrovanadate, however, passes over to the sulphovanadate,
$Pb_2V_2O_3S_4$, on heating in a current of hydrogen sulphide at a
full red heat.

Some New Lime Salts. BY H. C. HAHN. *Eng. Min. J.*,
65, 404.—By evaporating slowly a solution of calcium hydroxide
which had been saturated with hydrogen sulphide crystals of
the composition $CaS.3H_2O$ were obtained ; by heating, filtering,

and again saturating with hydrogen sulphide and evaporating, crystals having the composition CaS.H$_2$O were deposited.

The Solubility of Silver Chloride. By H. C. HAHN. *Eng. Min. J.*, 65, 434.—A solution containing 475.3 grams of CaCl$_2$ in one liter dissolves 2.835 grams silver chloride at 0°, and 8.147 grams at 100°. It is stated that the solubility is greater in saturated solutions of magnesium chloride.

Upon the Salts of Hydronitric Acid. By L. M. DENNIS AND C. H. BENEDICT. **Crystallographic Notes.** By A. C. GILL. *J. Am. Chem. Soc.*, 20, 225–232.—The lithium, sodium, potassium, cesium, and rubidium salts were prepared by neutralizing the corresponding hydroxides with the free acid, and the calcium, strontium, and barium salts by dissolving the oxides in the free acid. The lithium and barium salts both crystallize with one molecule of water, and their crystalline forms cannot be compared with the other members of their series. With the exception of the sodium salt which crystallizes in the hexagonal system, the salts of the other alkalies are tetragonal and isomorphous. The anhydrous salts of the alkaline earth group are probably orthorhombic.

Introductory Note on the Reduction of Metallic Oxides at High Temperatures. By FANNY R. M. HITCHCOCK. *J. Am. Chem. Soc.*, 20, 232–233.—This note states that nitrogen is given off from tungstic and molybdic oxides when reduced in a current of hydrogen.

Some Properties of Zirconium Dioxide. By F. P. VENABLE AND A. W. BELDEN. *J. Am. Chem. Soc.*, 20, 273–276.—Ignited zirconia is practically insoluble in all acids except hydrofluoric, and is unattacked by heating with sodium carbonate. The specific gravity of an ignited sample was found to be 5.409. Determinations of water in zirconium hydroxide, which had been washed free from ammonia and with alcohol and ether, or with petroleum ether, gave figures which do not correspond to any definite degree of hydration, although when washed with petroleum ether approximately 26.00 per cent. of water remained. The hydroxide precipitated in the cold is insoluble in water, but readily soluble in dilute and strong hydrofluoric, hydrochloric, and hydrobromic acids. Oxalic acid dissolves it readily. When precipitated from a hot solution the solubility in acids is much decreased. Ammonia is without action on the hydroxide, but a saturated ammonium carbonate solution dissolves one part in a hundred. Varying quantities of carbon dioxide are taken up in the dry and moist conditions.

Ammonium Selenide. By VICTOR LENHER AND EDGAR F. SMITH. *J. Am. Chem. Soc.*, **20**, 277–278.—A solution of five grams of ammonium molybdate in 50 cc. of water to which 20 cc. of strong ammonia had been added, was saturated with hydrogen sulphide; and the resulting deep red solution was concentrated in a vacuum over sulphuric acid. Black, anhydrous, orthorhombic prisms of ammonium selenide separated. The crystals, after they are freed from metallic selenium by extraction with carbon bisulphide, dissolve in water to a deep red solution, which precipitates selenides from neutral or alkaline solutions of metallic salts.

Action of Sulphur Monochloride upon Minerals. By EDGAR F. SMITH. *J. Am. Chem. Soc.*, **20**, 289–293.—Sulphur monochloride reacts with many minerals with the formation of the chlorides of the metals present; with sulphides the sulphur of the mineral remains dissolved in the sulphur monochloride. The reaction is both oxidizing and substituting, and takes place with some minerals in the cold, while with others complete decomposition is accomplished only by heating.

Note on Liquid Phosphorus. By F. P. VENABLE AND A. W. BELDEN. *J. Am. Chem. Soc.*, **20**, 303–304.—The authors have failed to obtain the liquid modification of phosphorus, previously described by Houston and Thompson.

The Oxyhalides of Zirconium. By F. P. VENABLE AND CHARLES BASKERVILLE. *J. Am. Chem. Soc.*, **20**, 321–329.— The oxychloride, $ZrOCl_2.8H_2O$, crystallizes from water; and from its solution hydrochloric acid precipitates the salt $ZrOCl_2.6H_2O$. Either of these salts dried at $100°-125°$ in hydrochloric acid gas, forms the oxychloride, $ZrOCl_2.3H_2O$, which may also be formed by crystallizing a solution of zirconium hydroxide in hydrochloric acid. Oxybromides prepared by dissolving zirconium hydroxide in hydrobromic acid are described; but much difficulty was experienced in obtaining definite compounds on account of the tenacity with which the hydrobromic acid was held by the crystals. Salts represented by the formulas $ZrOBr_2.13H_2O$, $ZrOBr_2.14H_2O$, $ZrOBr_2.4H_2O$, $ZrBr(OH)_3.2H_2O$, and $ZrBr(OH)_3.H_2O$ are described, but the analytical data would indicate that some of them at least are not definite compounds, but merely transition forms. Gelatinous compounds were formed in several cases, but these were not studied.

ORGANIC CHEMISTRY.

J. F. Norris, Reviewer.

On the Action of Acetic Anhydride on Phenylpropiolic Acid.

By Arthur Michael and John E. Bucher. *Am. Chem. J.*, 20, 89–120.—It has been shown by the authors that acetic anhydride reacts with acetylenedicarboxylic acid to form the anhydride of acetoxylmaleic acid, which gives oxalacetic acid on treatment with water. If acetic anhydride should react with phenylpropiolic acid in an analogous manner, the anhydride of β-acetoxylcinnamic acid would be formed, and this compound on decomposing with water would yield a β-hydroxycinnamic acid. As it seemed of theoretical interest to ascertain whether a product obtained in such a manner is identical with the benzoylacetic acid got by saponification of its esters, the above reaction was studied. The reaction took place in an entirely different way, however, a derivative of naphthalene being formed. When phenylpropiolic acid was heated with acetic anhydride, the anhydride of 1-phenyl-2,3-naphthalenedicarboxylic acid was formed, together with a small quantity of the mixed anhydride of acetic and phenylpropiolic acids. From its solution in alkalies the former is precipitated as the anhydride by acids. The sodium, barium, calcium, and silver salts, and the methyl ester of the acid were analyzed. By reduction of the anhydride, $C_{18}H_{10}O_3$, in alkaline solution with sodium amalgam, 1-phenyl-1,2,3,4-tetrahydro-2,3-naphthalenedicarboxylic acid was formed. Neutralized with ammonia its solution gave white crystalline precipitates with solutions of barium, calcium, and mercuric chlorides. Silver nitrate precipitated the salt $C_{18}H_{14}O_4Ag_2$. The anhydride of the acid was also prepared. When reduced in acetic acid solution the anhydride $C_{18}H_{12}O_3$ gave a lactone of the formula $C_{18}H_{14}O_3$, and when oxidized by potassium permanganate, 1-phenyl-2,3,5,6-benzenetetracarboxylic acid was formed. From the latter the silver, barium, calcium, and lead salts, the methyl and benzyl esters, and anhydride were prepared. When the barium salt of the acid was distilled with barium hydroxide diphenyl was obtained. A small amount of an acid isomeric with the above, which also yielded diphenyl when distilled with barium hydroxide, was obtained in the oxidation. The hydrocarbon obtained from the anhydride $C_{18}H_{12}O_3$ was shown to be α-phenylnaphthalene by oxidizing it to o-benzoylbenzoic acid. The authors point out the analogy between the formation of phenylnaphthalenedicarboxylic anhydride from phenylpropiolic acid and the formation of the isatropic acids from atropic acid (Fittig, *Ann. Chem.*, 206, 34). They also offer explanations of the steps involved in the condensation of acetic anhydride and phenylpropiolic acid.

⌐ **On the Conversion of Methylpyromucic Acid into Aldehydopyromucic and Dehydromucic Acids.** By H. B. HILL AND H. E. SAWYER. *Am. Chem. J.*, 20, 169–179.—In order to establish experimentally the structure of dehydromucic acid, its synthesis from methylpyromucic acid was effected. Methylpyromucic acid was converted into ω-brommethylpyromucic acid, and the latter was changed into the dibrom derivative by the action of bromine in the sunlight. As the yield was small, methylpyromucyl chloride was made, treated with bromine, and the resulting ω-dibrompyromucyl bromide heated with water, when aldehydopyromucic acid was obtained. The phenylhydrazone and oxime of the latter acid were prepared. Aldehydopyromucic acid was readily converted by oxidation with silver oxide into dehydromucic acid.

On the 3,4,5-Tribromaniline and Some Derivatives of Unsymmetrical Tribrombenzol. By C. LORING JACKSON AND F. B. GALLIVAN. *Am. Chem. J.*, 20, 179–189.—In the course of an attempt to prepare the vicinal tetrabrombenzene, which did not lead to the desired result, 3,4,5-tribromaniline was prepared. As its properties did not agree with those which had been assigned previously to a compound of this structure, the substance and a number of its derivatives were carefully studied. The structure of the compound was established by its preparation from *p*-nitraniline. The dibrom substitution-product was first prepared, the amido group was next replaced by bromine, and finally the nitro group was reduced. 3,4,5-tribromaniline melts at 118°–119° and forms a chloride, bromide, and sulphate which are more stable than the analogous salts of *s*-tribromaniline. The following derivatives were prepared : tribromphenylurethane, tribromacetanilide, tribromnitraniline $NH_2 1, Br_2 3,-4,5,NO_2 2$, and the acetyl derivative of the last compound. In order to characterize more fully the 2,4,5-tribromaniline previously described by the authors (*Am. Chem. J.*, 18, 247), derivatives similar to the above were studied. In their previous paper the authors state that tribromresorcine is the product of the action of sodium ethylate on tribromdinitrobenzene, $Br_2 1,2,4; NO_2 1,3,5$. The substance has been further studied and found to be tribromnitrophenetol.

Direct Nitration of the Paraffins. By R. A. WORSTALL. *Am. Chem. J.*, 20, 202–217.—When normal hexane is boiled in an open flask with nitric acid (sp. gr. 1.42), or with a mixture of nitric and sulphuric acids, it is converted into nitro- and dinitrohexane. The best yield is obtained, however, when fuming acid is used. To separate the nitro-products, the reaction mixture is dried, and the unchanged hydrocarbon is distilled off. This is treated with more acid. By several days' successive treatment hexane was completely converted into nitro- and oxidation-products, the yield of crude nitro-compounds amounting to about 60 per cent. of the theoretical. Nitrohexane is best separated from the dinitro compound of distillation with steam, with which it is volatile. It was shown to be a primary compound by the application of the nitrolic acid test and was reduced to a primary amine. The dinitrohexane formed has probably the structure $CH_3(CH_2)_3CH(NO_2)_2$. It carbonizes on distillation, cannot be solidified by a mixture of ice and salt, and dissolves slowly in concentrated aqueous solutions of potassium and sodium hydroxide forming a deep red solution. Normal heptane is capable of more rapid nitration than hexane. Twenty hours' treatment with nitric acid (sp. gr. 1.42) gave about 20 per cent. of the theoretical yield. The action of a mixture of sulphuric and nitric acids yields a considerable amount of dinitroheptane which forms a layer between the nitroheptane and the acids. As the dinitro-compound yielded on reduction hydroxylamine and ammonia, probably it has the structure $CH_3(CH_2)_3CH(NO_2)_2$. Octane was studied in the same way and gave analogous results. All of the hydrocarbons yielded the same oxidation products ; viz., carbon dioxide, acetic, oxalic, and succinic acids. Carbon dioxide is the most abundant oxidation product. The quantity of other substances formed by boiling the hydrocarbons two days with nitric acid did not exceed two per cent.

On the Silver Salts of 4-Nitro-2-Aminobenzoic Acid and Its Behavior with Alkyl and Acyl Halides. By H. L. WHEELER AND BAYARD BARNES. *Am. Chem. J.*, **20**, 217–223.—The alkali salts of the aminobenzoic acids and the sodium salts of the acyl amides do not give esters directly with alkyl halides, as the alkyl groups attach themselves entirely or partially to nitrogen. As a direct replacement of the metal takes place with the silver salts of the amides, oxygen esters being formed, the action of acyl and alkyl halides on the silver salt of an aminobenzoic acid was studied to determine whether or not oxygen esters were formed in this case. 4-nitro-2-amino-benzoic acid was used, as its esters are solid. The silver salt of the acid gave, as the chief product with ethyl iodide in molecular proportions, 4-nitro-2-ethylaminobenzoic acid. Some 4-nitro-2-ethylaminobenzoic ethyl ester was formed, but no 4-nitro-2-aminobenzoic ethyl ester was observed. The behavior of the silver salt is therefore analogous to that of the sodium salts of the amine acids. Acetyl chloride reacted with the silver salts of anthranilic and 4-nitro-2 aminobenzoic acids forming compounds in which the acetyl group was joined to nitrogen. Since the above silver salts, in which the metal is joined to oxygen, give derivatives with acyl and alkyl halides containing the substituting group joined to nitrogen, it follows that the position of the metal in the salts or the amides cannot be determined by the structure of the reaction products, and that the sodium salts of the amides may have the metal joined to oxygen. The following substances are described : 4-nitro-2-acetaminobenzoic acid, and its sodium and silver salts ; ethyl 4-nitro-2-acetaminobenzoate ; 4-nitro-2-aminobenzoic acid, and its ammonium, sodium, and silver salts ; methyl and ethyl 4-nitro-2-aminobenzoate ; and 4-nitro-2-ethyl-aminobenzoic acid and its ethyl ester.

Formamide and Its Sodium and Silver Salts. By PAUL C. FREER AND P. L. SHERMAN, JR. *Am. Chem. J.*, 20, 221–228. —The authors show that pure formamide cannot be prepared by Hofmann's method. The products of distillation of ammonium formate under ordinary pressure are formamide, water, ammonia, formic acid, ammonium formate, carbon monoxide, hydrocyanic acid, and ammonium cyanide. When distilled in a vacuum ammonium formate readily dissociates, and ammonia and formic acid pass into the distillate. In order to prevent this dissociation ammonium formate was heated to 180° in a current of dry ammonia. The resulting formamide was fractionated in an atmosphere of ammonia under a pressure of about one-half mm. The distillate, which boiled at 85°–95°, was colorless, oily, possessed a high index of refraction, reacted neutral towards litmus, and, when kept free from moisture, was stable. At —1° it solidified to a white crystalline mass of long, irregular needles. Its specific gravity at 4° was 1.16. Sodium formamide did not react with organic halogen compounds ; and when dissolved in alcohol it gave with silver nitrate a silver derivative as a white, curdy precipitate. The orange-red silver salt of formamide described by Titherly (*J. Chem. Soc.*, 72, 460) was shown to be impure, the color being due to a trace of sodamide in the sodium salt from which it was prepared.

On the Decomposition of Diazo Compounds. XIII.—A Study of the Reaction of the Diazophenols and of the Salts of Chlor- and Bromdiazobenzene with Ethyl and Methyl Alcohol. By FRANK KENNETH CAMERON. *Am. Chem. J.*, 20, 229–251.—This paper, the thirteenth from the laboratory of Johns Hopkins University on the diazo compounds, gives the results of an investigation of the influence of the hydroxyl group and the elements chlorine and bromine, when substituted in the benzene nucleus, in determining the decomposition of diazo compounds with ethyl and methyl alcohols. Paradiazophenol chloride was decomposed by methyl and ethyl alcohols in the presence of sulphuric and hydrochloric acids, potassium hydroxide, ammonia, and sulphur dioxide. In all cases the hydrogen reaction took place, phenol being formed. Meta- and orthodiazophenol chlorides gave the same result. Orthochlordiazobenzene salts with methyl and ethyl alcohols gave the hydrogen reaction alone. Metachlordiazobenzene salts with ethyl alcohol gave the hydrogen reaction, but with methyl alcohol the methoxy reaction predominated, metachloranisol being the principal product. Parachlordiazobenzene nitrate and ethyl alcohol gave principally the hydrogen reaction, but at the same time, to no inconsiderable extent, the ethoxy reaction, accompanied by the formation of chlornitrophenol. With methyl alcohol the same diazo compound gave the methoxy reaction and some chlornitrophenol. Parachlordiazobenzene sulphate with ethyl alcohol gave the hydrogen and with methyl alcohol the methoxy reaction. The bromdiazobenzene salts were studied in the same way and gave analogous results. From the above and other work on the diazo compounds the following conclusions are drawn : 1. The alkoxy reaction is normal ; and the hydrogen reaction, when it takes place, is a modification induced by special conditions. 2. Regarding water as the first member of the series, the more complex the alcohol, the greater the tendency towards the hydrogen reaction. 3. Acid radicals, COOH, Cl, Br, NO_2, etc., induce the hydrogen reaction, and their influence is probably in the above order. 4. The presence of the substituting radical in the ortho position is most favorable to the hydrogen reaction, and in the meta position is more favorable than in the para.

An Investigation of Some Derivatives of Orthosulphobenzoic Anhydride. By MICHAEL DRUCK SOHON. *Am. Chem. J.*, **20**, 257–278.—The anhydride of *o*-sulphobenzoic acid was prepared by heating molecular quantities of the acid potassium salt of the acid and phosphorus pentachloride. It is a well-characterized stable substance and crystallizes from benzene. The anhydride dissolves in methyl and ethyl alcohols forming acid esters from which the silver and potassium salts were prepared. By heating the anhydride with the following phenols at above 135° sulphonphthaleïns were formed : phenol, *o*-cresol, *p*-cresol, resorcinol, orcinol, hydroquinol, pyrogallol, *m*-amidophenol, and *p*-amidophenol. By the action of dry ammonia on the anhydride dissolved in ether, the ammonium salt of benzamidesulphonic acid, $C_6H_4{<}^{CONH_2}_{SO_2NH_4}$, was formed. From the latter the potassium and barium salts were prepared, but it was impossible to obtain the free acid in crystalline condition. The anhydride reacts with aniline as it does with ammonia. The resulting anilido acid, $C_6H_4{<}{CONHC_6H_5 \atop SO_3H}$, is a very soluble oil,

whereas the isomeric sulphanilido acid, $C_6H_4{<}{COOH \atop SO_2NHC_6H_5}$, is readily obtained in crystalline condition. The following salts of benzanilidosulphonic acid were prepared and analyzed : aniline, barium, ammonium, potassium, and cadmium. The sodium, copper, and lead salts evaporated to hard glasses and the silver salt decomposed on evaporation. By the action of phosphorus pentachloride on the potassium salt, he formed the anil,

$$C_6H_4{<}{CO \atop SO_2}{>}NC_6H_5.$$

p- and *o-*toluidine reacted with the anhydride giving toluidine salts of benztoluidosulphonic acids, $C_6H_4{<}{CONHC_7H_7 \atop SO_2NH_2C_7H_7}$.

From these the barium and potassium salts were prepared. Benzamide and acetamide were converted into nitriles by the anhydride. Phosphorus pentachloride acting on the anhydride produced both chlorides of *o-*sulphobenzoic acid, the formation of the unsymmetrical chloride being favored by high temperature, continued action, and an excess of phosphorus pentachloride.

The Relation of Trivalent to Pentavalent Nitrogen. By ARTHUR LACHMAN. *Am. Chem. J.,* **20,** 283–288—The author has undertaken a study of the conditions necessary for the existence of trivalent and pentavalent nitrogen and for the passage of one form into the other. In this preliminary communication the action of hydrochloric acid, zinc ethyl, and hydroxylamine on nitrosamines, $R_2{=}N{—}N{=}O$, is described. Dimethyl and diethylnitrosamines form unstable hydrochlorides which are decomposed when heated at 60° in a sealed tube. If a rapid stream of hydrochloric acid gas is passed into diethylnitrosamine, nitrosyl chloride passes off and diethylamine hydrochloride is formed. The analogous reaction between diphenylnitrosamine and hydrochloric acid gas (both carefully dried) was found to take place quantitatively according to the reaction

$$(C_6H_5)_2NNO + HCl = (C_6H_5)_2NH + NOCl.$$

It is probable that an unstable addition-product containing pentavalent nitrogen of the formula $(C_6H_5)_2{>}N{<}{H \atop Cl} \atop ON{\nearrow}$ was first formed. Diphenylnitrosamine and zinc ethyl form a stable addition-product, which is decomposed by water or alcohol without evolution of a gas, the products being zinc hydroxide, diphenyl-amine, and an unstable base which reduces Fehling's solution. When diphenylnitrosamine and hydroxylamine, in molecular quantities, were boiled in methyl alcohol solution, either in the presence or absence of free alkali, nitrous oxide was given off in theoretical quantity.   As no hyponitrites were formed the reaction can be explained only by assuming the direct decomposition of an addition-product of the following formula :

$$(C_6H_5)\diagdown N\diagdown{\overset{NO}{\underset{NH.OH}{}}}$$

From the above it is seen what limits exist for the stability of pentavalent nitrogen.   A marked contrast in the chemical nature of a part of the five substituting groups is necessary to insure stability.

On Paramethoxyorthosulphobenzoic Acid and Some of Its Derivatives. By P. R. MOALE.   *Am. Chem. J.*, 20, 288–298.— *p*-Methoxy-*o*-sulphobenzoic acid was prepared by the method of Parks (*Am. Chem. J.*, 15, 320).   In the treatment of *p*-methoxy-benzoic sulphinide with dilute hydrochloric acid the free acid was formed and not the ammonium salt, which is the chief product of the reaction when benzoic sulphinide is decomposed by dilute acids.   From the acid the neutral and acid calcium, magnesium, and lead salts and the acid potassium salt were prepared.   Sulphonfluoresceïns were made by the action of resorcinol, orcinol, and phenol on the acid, but were not obtained in pure condition.

Decomposition of Paradiazoorthotoluenesulphonic Acid with Absolute Methyl Alcohol in the Presence of Certain Substances. By P. R. MOALE.   *Am. Chem. J.*, 20, 298–302.— When *p*-diazo-*o*-toluenesulphonic acid was decomposed by methyl alcohol in the presence of sodium methylate, the hydrogen reaction took place to a small extent, whereas with the alcohol alone only the methoxy product is formed.   When the decomposition of the diazo compound was effected in the presence of sodium ethylate or potassium hydroxide, tarry masses were obtained from which no pure compounds could be isolated. With dry ammonia gas in methyl alcohol *p*-toluidine-*o*-sulphonic acid was obtained.

Parabenzoyldiphenylsulphone and Related Compounds.   By LYMAN C. NEWELL.   *Am. Chem. J.*, 20, 302–318.—*p*-Toluene-sulphone chloride was converted by means of the Friedel-Crafts reaction into *p*-tolylphenylsulphone, which was oxidized by chromic acid to *p*-phenylsulphonebenzoic acid. This acid forms well-characterized calcium, barium, and sodium salts and an acid chloride from which an amide and an anilide were prepared. *p*-Phenylsulphonebenzoyl chloride condenses with benzene in the presence of aluminium chloride forming *p*-benzoyldiphenylsulphone, a compound analogous to the one obtained in the same way by Remsen and Saunders (*Am. Chem. J.*, 17, 362) from the chlorides of *o*-sulphobenzoic acid. The para sulphone, unlike the corresponding ortho compound, is not decomposed into benzoic acid and diphenylsulphone when fused with potassium hydroxide. On the other hand it does yield a hydrazone and an oxime.

Derivatives of Silicon Tetrachloride. BY JOSEPH F. X. HAROLD. *J. Am. Chem. Soc.*, 20, 13-29.—With a view to developing some new analogies between the behavior of silicon tetrachloride and that of the tetrachlorides of the other elements of the fourth group in the periodic classification of the elements, its action on nitriles and aromatic amines was studied. It was found, however, that silicon tetrachloride did not react with hydrocyanic acid, benzonitrile, acetonitrile, tolunitrile, succinonitrile, chlorcyanogen, sulphur dichloride, nitrogen dioxide, or the chlorides of phosphorus, whereas the tetrachlorides of tin and titanium enter into reaction with these substances. Silicon tetrachloride and aniline gave a substance of the composition $SiCl_2(C_6H_5NH)_2$. With *o*-toluidine an analogous compound was formed. The chloride with benzamide gave benzaldehyde and with acetamide acetonitrile.

Preliminary Note on Some New Derivatives of Vanillin. BY A. E. MENKE AND W. B. BENTLEY. *J. Am. Chem. Soc.*, 20, 316-317.—By the action of sodium amalgam on chlorvanillin chlorvanilloin was formed. Nitric acid and vanillin yield three products : dinitroguiacol ; a substance, which by further treatment with acid gives dinitroguiacol, and by oxidation with potassium permanganate nitrovanillic acid ; and an unknown substance which was not identified. Tetrachlorpyrocatechin was formed when chlorine was passed into an alcoholic solution of pyrocatechuic acid.

Atropine Periodides and Iodomercurates. BY H. M. GORDIN AND A. B. PRESCOTT. *J. Am. Chem. Soc.*, 20, 329-338.—The conditions necessary for the formation of atropine enneaiodide were determined and applied to the volumetric estimation of the alkaloid. Atropine penta- and triiodide were also obtained. Double salts of the formulæ $C_{17}H_{23}NO_3HI.HgI_2$ and $(C_{17}H_{23}NO_3HI)_2.HgI_2$ were prepared and analyzed.

92   *Review of American Chemical Research.*

Chemical Bibliography of Morphine, 1875-1896. By H. E. BROWN. *Pharm. Archives*, 1, 49–68.—This is a continuation of a bibliography of morphine, the appearance of which has already been noticed (*this Rev.*, 4, 13).

Alkyl Bismuth Iodides. By A. B. PRESCOTT. *Pharm. Rev.*, 15, 219–220.—See *this Rev.*, 4, 49.

Formaldehyde. By GEORGE L. TAYLOR. *Am. J. Pharm.*, 70, 195–201.—The author gives a brief review of the methods of preparation and analysis and of the uses of formaldehyde.

GENERAL AND PHYSICAL CHEMISTRY.

A. A. NOYES, REVIEWER.

Electrical Disturbance in Weighing. By H. K. MILLER. *J. Am. Chem. Soc.*, 20, 428–429.—The author finds that a flask wiped with a dry cloth shows an apparent increase in weight (in one case as much as 0.08 gram), and attributes this to the production of an electrical charge upon it, which induces an opposite charge in the floor of the balance case.

On the Speed of Coagulation of Colloid Solutions. By C. E. LINEBARGER. *J. Am. Chem. Soc.*, 20, 375–380.—The author concludes from his experiments with ferric hydroxide, silicic acid, and egg albumen, that coagulation started in one part of a colloid solution does not necessarily spread through the whole body of it, and that therefore it is very improbable that colloid solutions are comparable with supersaturated solutions of crystalline substances.

The Decomposition of Sulphonic Ethers by Water, Acids and Salts. By J. H. KASTLE, PAUL MURRILL AND JOS. C. FRAZER. *Am. Chem. J.*, 19, 894–901.—The ethyl ether of *p*-brombenzenesulphonic acid was dissolved in an excess of acetone, and 50 times the theoretical quantity of water in one case, and of alcohol in another, required for the saponification, was added to it; portions of these mixtures were heated in sealed tubes at 98° for different lengths of time (from 20 to 120 minutes), and the percentage decomposed was determined by titration with potash. Good velocity-constants of the first order were obtained, that for water being 3.5 times that for alcohol. It was found that acids and neutral salts, instead of merely acting catalytically, entered into a metathesis, the ethyl radical of the ester uniting with the acid radical of the acid or salt. The saponification and metathesis therefore occur simultaneously; and the authors have determined, at the ordinary and at the boiling temperature, the amount of each taking place in the presence of the three halogen acids, of potassium iodide, and of calcium and magnesium chlorides, 1 cc. of a normal or a half-normal solution of these substances being added to 4 cc. of an acetone solution of the ester. In almost all cases, the amount undergoing metathesis greatly exceeded that undergoing saponification. The authors regard the occurrence of the metathetical reaction as a new characteristic distinguishing the esters of sulphonic acids from those of the carbonic acids, in which latter, mineral acids in aqueous solution simply accelerate saponification. It seems quite probable, however, that the difference in behavior is due to the difference in the solvents, acetone being used for the sulphonic esters and water for the carbonic esters.

A Determination of the Atomic Weight of Praseodymium and of Neodymium. By HARRY C. JONES. *Am. Chem. J.*, 20, 345–358.—The material for the determination of the atomic weight of praseodymium was prepared by crystallizing 1½ kilos of fairly pure ammonium praseodymium nitrate presented by the Welsbach Light Company twenty-one times from a nitric acid solution, only the middle portion from each crystallization being taken. The metal was then precipitated by adding oxalic acid to the hot solution, in order to remove traces of iron, calcium, etc. The oxalate so obtained was found spectroscopically to contain only 0.06 per cent. of neodymium; but as traces of cerium and lanthanum were still present, the oxalate was ignited, the oxide dissolved in nitric acid, and the solution evaporated and poured into hot water, when a cloud of basic cerium nitrate separated. To remove the lanthanum, the double nitrate was again formed and recrystallized repeatedly from very strong nitric acid. Finally the metal was again precipitated as the oxalate, which was ignited forming the peroxide, and this was heated in hydrogen to ,reduce it to sesquioxide. The atomic weight determinations were made by adding concentrated sulphuric acid to the oxide in a platinum crucible, and heating it first gently and then above the boiling-point of the acid in an air-bath to a constant weight. In the case of neodymium, the methods used for purifying the material (which was obtained in the form of the double nitrate of ammonium and neodymium from the same source) and for determining the atomic weight of the element were almost identical with those just described. The average of twelve closely agreeing determinations was 140.45 in the case of praseodymium, and 143.60 in that of neodymium. It is remarkable that the values given by von Welsbach both differ from these by about three units and would agree well with them if interchanged, his values being 143.6 for praseodymium and 140.8 for neodymium.

On the Taste and Affinity of Acids. BY J. H. KASTLE. *Am. Chem. J.*, 20, 466–471.—The author found by experiments with nineteen different acids in $\frac{1}{10}$ normal solution on seventeen persons that in the great majority of cases one acid was pronounced more or less sour than another according as the affinity constant of the one was greater or less than that of the other.

The Equilibria of Stereoisomers, I and II. BY WILDER D. BANCROFT. *J. Phys. Chem.*, 2, 143–158; 245–255. — These papers discuss from a qualitative standpoint the application of the Law of Mass Action and the principle of freezing-point lowering to the equilibrium existing between two stereoisomers which undergo transformation into one another. In the first paper, the effect on the freezing-point of heating each of the stereoisomers by itself is considered; it is evident that a short heating will cause a lowering of the freezing-point of both of the pure compounds, owing to the formation of some of the other stereoisomeric modification; longer heating, near the melting temperature, will cause a further lowering, increasing in the case of one modification continuously until the liquid has the composition corresponding to the equilibrium-condition of the two stereoisomers, after which no further change in freezing-point occurs; while, in the case of the other modification, the freezing-point will steadily decrease on continued heating as long as it separates as a solid from the solution, but will suddenly begin to rise as the solution becomes so concentrated in the first modification as to cause *it* to separate, and will continue to rise until the former constant freezing-point corresponding to equilibrium is attained. In the second paper the effect is considered of the addition of a third substance not reacting chemically 'with the stereoisomers, for example, of the addition of a solvent. The author cites from the work of previous investigators numerous examples of some of the cases considered, showing them to be in accordance with his conclusions.

Acetaldoxime. BY HECTOR R. CARVETH. *J. Phys. Chem.*, 2, 159–167. **Benzilorthocarboxylic Acid.** BY C. A. SOCH. *J. Phys. Chem.*, 2, 364–370. **The Benzoyl Ester of Acethydroxamic Acid.** BY FRANK K. CAMERON *J. Phys. Chem.*, 2, 376–381.—These articles are descriptions of experimental investigations of the stereoisomeric transformation of the substances mentioned in the titles along the lines indicated in the preceding review.

Naphthalene and Aqueous Acetone. BY HAMILTON P. CADY. *J. Phys. Chem.*, 2, 168–170.—Naphthalene in excess was added to seven different mixtures of acetone and water, and the composition of the solution was determined at the temperature at which it resolved itself into two liquid layers, the temperature being also noted.

Indicators. BY JOHN WADDELL. *J. Phys. Chem.*, 2, 171–184.—The author shows that the addition of ether, benzene, or chloroform, and in some, but not in all cases, also of alcohol or acetone, to aqueous or alcoholic solutions in which the indicators fluorescein, phenacetolin, phenolphthalein, paranitrophenol, cyanine, methyl orange, turmeric, lacmoid, and corallin are present in the form of their salts causes a change in color to that of the free acid or base. This change of color takes place much more readily when the indicator is present in the form of its acetate than when in that of its hydrochloride, and when in the form of its ammonium salt than when in that of its potassium salt. The author apparently considers that the change of color is most probably explained by the assumption that the electrolytic dissociation of the salt to whose ion the color is due is greatly reduced by the organic solvents ; but at the close of the article he raises two objections to this hypothesis ; namely, first, that it is then necessary to assume that the undissociated indicator salt and acid or base of the indicator have just the same color in all the cases investigated ; and, second, that the hypothesis does not offer any explanation of the fact that the salts of weak acids (acetic) or bases (ammonia) undergo the change so much more readily than those of strong acids and bases (like hydrochloric acid and potash). These objections seem to the reviewer to warrant completely the rejection of the above explanation; and it appears to him that the following one, which is in part suggested by the author himself, is far more probable : the very weak acids and bases of which indicators consist, are so much less dissociated in organic solvents than in water, that their salts become nearly completely hydrolyzed when these solvents are added to the aqueous solutions of the salts, in spite of the fact that the concentration of the ions of the water is also reduced somewhat ; hence the addition of the solvents liberates the free indicator acid or base, which exhibits its usual color, and it will, of course, liberate it much more readily if the other component of the salt is also a weak base or acid.—By noting whether the change of color occurred on adding the solvents to the acid solution, or to the alkaline solution of the indicator, or to both solutions, the author was able to determine whether it acts as a base or as an acid, or as both. He concludes from his experiments that phenolphthalein, paranitrophenol, and turmeric are acids ; that cyanine, methyl orange and lacmoid are bases ; and that fluorescein, phenacetolin and corallin act both as acids and bases.

Benzene, Acetic Acid and Water. By JOHN WADDELL. *J. Phys. Chem.*, 2, 233–241.—This investigation is a continuation of the so-called "Mass Law Studies." See *this Rev.*, 3, 75, 152. No general conclusions are reached.

Boiling-Point Curve for Benzene and Alcohol. By E. F. THAYER. *J. Phys. Chem.*, 2, 382–384.—The author has determined by means of a molecular weight apparatus the boiling-point of various mixtures of benzene and alcohol. The boiling-point of benzene, 79°.5, was found to be lowered very rapidly by the first additions of alcohol, reaching 69° when seven per cent. of it was added. The boiling-points of the mixtures then remained fairly constant—between 69° and 67°—until about 70 per cent. of alcohol was present, after which they rose gradually up to that of pure alcohol, 78°. No reference is made to the extended investigations of Konowalow (*Wied. Ann.*, 14, 34, 219) in a similar direction.

Molecular Weights of Liquids. By CLARENCE L. SPEYERS. *J. Phys. Chem.*, 2, 347–361 ; 362–363.—The case of two miscible liquids, both of which have appreciable partial vapor-pressures, is considered with reference to the relation between the change in partial pressure of one liquid and the concentration of the other (expressed as the ratio of the number of mols of it to the total number present). Where proportionality exists between these two quantities, the author concludes that no association of molecules occurs. Where it does not exist, he thinks that there are errors in the observations or that association takes place.

H. M. GOODWIN, REVIEWER.

Spectroscope without Prisms or Gratings. By A. A. MICHELSON. *Am. J. Sci.*, 155, 215–217.—As is well known, the resolving power of a grating is proportional to the product of the total number of lines by the order of the spectrum observed. The great efficiency of the best modern gratings is due to the former factor almost exclusively, spectra of a higher order than the second or third being rarely observed. In the very beautiful device described by the author, the second factor is made of paramount importance, the spectra observed being of the order of several thousand. This is effected by observing the transmission spectrum produced by passing a beam of light from a collimater through a series of overlapping optically plane rectangular blocks of glass of equal thickness. These blocks must all be so nearly optically perfect that the accumulated error in the difference of path of light passing through the whole system shall not be a half wave-length. A system consisting of twenty elements 5 mm. thick would have a resolving power of 100,000, about equal to that of the best gratings. The great advantage of this arrangement over ordinary gratings is that all spectra are superimposed in a very high order, consequently the maximum possible brilliancy is obtained. The Zeeman effect was easily observed with an instrument constructed with but seven elements.

On the Surface Tension of Liquids under the Influence of Electrostatic Induction. By SAMUEL J. BARNETT. *Phys. Rev.*, 6, 257–285.—The author has employed the beautiful method of ripples as perfected recently by Dorsey for investigating the change of surface tension of liquids with their degree of electrification, with the object of ascertaining whether the phenomenon is better explained by Helmholtz and Lippmann's or by Warburg's theory. The apparatus was modified in some details. Contrary to expectation, it was found that measurable effects could be produced only by very high electrification. This was affected by making the liquid surface one plate of an air condenser, the other being of glass coated with tin-foil and grounded. A Holtz machine served to charge the system. The potential was measured by applying the whole potential difference to the terminals of two condensers of known capacity, connected in series, and measuring the potential difference between the plates of one of them by means of a quadrant electrometer. Mercury and water were the two liquids investigated. It was found in all cases that the effect of electrification was to diminish the surface tension. The diminution was much greater in the case of water than in the case of mercury, but in both cases it was independent of the sign of the charge. A similar diminution in the apparent surface tension of water was also noted when alternating charges from the secondary of an induction coil was used. The author regards these results as being in accord with Helmholtz "charge" theory, but in disagreement with any electrolytic hypothesis. He points out, however, that the effect may possibly be wholly due to electrostatic actions in the liquid and not to "forces tangential to the surface acting between the elementary electric charges."

Normal Elements. By D. MCINTOSH. *J. Phys. Chem.*, 2, 185–193.—Experiments were made with numerous cells, mostly of the Clark type, for the purpose of obtaining one with an electromotive force of about 0.5 volt and suitable for a standard of reference. The most promising consisted of $Pb/PbCl_2$, $ZnCl_2/Zn$. This cell has an electromotive force of 0.5 volt at 20° when the zinc chloride solution has a gravity of 1.23. Its temperature coefficient between 0° and 20° was found to be 0.0001 volt per degree. The electromotive force E in volts of the cell

Cu/saturated CuSO$_4$, Hg$_2$SO$_4$/Hg was found to be as follows :
$E = 0.3613 + 0.0006 [16.5° — t°]$.   A Pb/PbCl$_2$, HgCl/Hg
cell failed to give satisfactory results.   The attempt to reduce
the effect of mechanical disturbances by the use of the agar-agar
jelly proved unsuccessful with concentrated copper sulphate.

The Transference Number of Hydrogen. By DOUGLAS MC-
INTOSH.   *J. Phys. Chem.*, 2, 273–289.—The method chosen for
determining this quantity was that suggested by von Helmholtz
and based on his theory of concentration cells.   If the electromotive force of two cells of the type Zn/ZnSO$_4$, Hg$_2$SO$_4$/Hg
of different concentrations, connected in opposition, that is,
mercury to mercury, be compared with the electromotive force
of the corresponding ordinary diffusion cell, Zn/ZnSO$_4$ concentrated, ZnSO$_4$ dilute/Zn, it is easy to show that the ratio of the
former to the latter is equal to $\dfrac{v}{u + v}$, Hittorf's transference ratio.
In order to determine the ratio for hydrogen in various acids,
the electromotive force of gas batteries, with hydrogen
electrodes, both with diffusion and without, was determined.
Poggendorff's compensation method with the Lippmann electrometer was used for making the measurements.   The mean value
of the ratio $\dfrac{v}{u + v}$ was for hydrochloric acid 0.159, hydrobromic
acid 0.158, hydriodic acid 0.161, sulphuric acid 0.174, and oxalic
acid 0.163.   The observed values of which the above are the
mean were nearly independent of the dilution which varied from
1 to 1000 liters.   This, together with the fact that the above
values are much lower than Hittorf's values, leads the author to
conclude that the method is inapplicable to gas cells, owing possibly to the solubility of hydrogen in the electrolyte.   Certain
other measurements on concentration cells with chlorine, bromine
and iodine electrodes, led to results of no particular importance.

Single Differences of Potential. By HECTOR R. CARVETH.
J. Phys. Chem., 2, 289–322.—The object of this research " is to
examine the theories of to-day with respect to single differences
of potential ......, and to attempt to answer the question, Do
these methods give the true potential difference ?''   A consideration of the theories of Helmholtz and of Warburg leads to the
conclusion that an ideal dropping electrode is not realized in
practice, owing to the fact that the dropping mercury does not
assume in any case its maximum surface tension when falling
into an electrolyte.   Certain apparently discordant results of
Paschen lead the author to think that no potential difference
between metal and liquid is accurately known.   Notwithstanding his criticism of the dropping electrode method, the author

makes use of it for investigating further the potential difference between a number of metals in solutions of their salts. It was found that fairly constant measurements could be obtained when the mercury broke below the surface of the liquid, but that the actual values varied widely with the adjustment of the electrode. A steady variation of potential with the dilution was observed in certain cases, but a comparison with the values computed from Nernst's theory was not satisfactory. Measurements of the potential difference between $Hg/HgCl$, KCl and between Hg/KCl agree well, if in the latter case the electrode was allowed to stand some time in contact with the electrolyte. In certain cases, contrary to Paschen's results, the potential difference was found to vary with the nature of the cation. Experiments on irreversible electrodes confirmed the results of Warburg and Paschen as to the effect of oxygen on the electrodes. Rothmund's results on the potential difference of amalgams are criticised, and the general conclusion is drawn .from the whole investigation that ''neither on the ground of Helmholtz's theory, nor that of Warburg, nor that of Nernst is there reason for regarding one single potential difference known.''

Study of a Three-Component System. BY HECTOR R. CARVETH. *J. Phys. Chem.*, 2, 209-229.—The author has determined the freezing-points of mixtures of lithium, sodium, and potassium nitrates, taken in pairs, and also when all three were simultaneously present. The measurements were made with a mercury thermometer, and an accuracy greater than one or two degrees was not attempted. The results are discussed graphically by the triangular diagram of Roozeboom, and the regions representing systems of different variance are determined. An approximate computation of the molecular weight of one inorganic salt dissolved in another, from the lowering of the melting-point produced, shows it to be much lower than ordinarily assumed, and to be largely dependent on the concentration.

Note on Thermal Equilibrium in Electrolysis. BY D. TOMMASI. *J. Phys. Chem.*, 2, 229-232. From the decomposition-products observed in the electrolyses of solutions of several nitrates, nitrites, chlorates, sulphates, arsenates and arsenites between platinum electrodes placed very close together, the author concludes : " First, when a substance is submitted to two equal and contrary chemical actions, the reaction which involves the most heat will take place in preference, provided always that it can begin. Second, of two chemical reactions, that one which requires less heat to start it will always take place in preference, even though it evolves less heat than the other reaction."

The Specific Heat of Anhydrous Liquid Ammonia. By
LOUIS A. ELLEAU AND WILLIAM D. ENNIS. */. Franklin Inst.,*
145, 189–214; 280–294.—The authors obtain the value 1.0206
for the mean specific heat of liquid ammonia between 0° and 20°.
This is the mean of nine determinations made by the method
of mixtures. The liquid ammonia, about 12 grams, was
contained in a closed steel cylinder, and was initially cooled in a
chamber surrounded by ice to nearly 0° before immersion in the
calorimeter. The ammonia used was once distilled from the
commercial product.

ANALYTICAL CHEMISTRY.

ULTIMATE ANALYSIS.

H. P. TALBOT, REVIEWER.

The Moist Combustion Method of Determining Carbon. By
GEORGE AUCHY. */. Am. Chem. Soc.,* 20, 243–253.—The data
given in this article indicate that, in the moist combustion
method, a single potash bulb containing a potash solution of
1.27 sp. gr. is insufficient to insure the complete retention of the
carbon dioxide evolved, and the substitution of a solution of
1.40 sp. gr. does not always obviate the necessity for a second
potash bulb. It is also shown that a change in weight of the
bulbs, due to loss or gain of moisture, must be carefully guarded
against. The author finds the occasional absorption by the
potash bulbs of some chloro-chromic compound which passes the
purifying train, the most serious source of error in this procedure.

Electrolytic Determinations. By EDGAR F. SMITH AND
DANIEL L. WALLACE. */. Am. Chem. Soc.,* 20, 279–281.—The
authors confirm statements previously made regarding the electrolytic determination of uranium (*Am. Chem. /.,* 1, 329), and
cadmium (*this Rev.,* 4, 30). They also controvert the statements of Heidenreich (*Ber.,* 29, 1587), who claims to have
found their previous directions to be inaccurate.

Boric Acid Determination. By THOMAS S. GLADDING. */.
Am. Chem. Soc.,* 20, 288–289.—The procedure includes the distillation of the boric acid in the presence of sirupy phosphoric
acid and methyl alcohol. The acid is titrated in the distillate
after the addition of glycerine. A cut shows the details of the
apparatus employed.

**On the Determination of Potash without the Previous Re-
moval of Iron, Cadmium, Etc.** By C. C. MOORE. */. Am.*

Chem. Soc., 20, 340–343.—The usual procedure is much shortened by the addition of the hydrochloroplatinic acid to the acid solution of the substance, after filtration from any insoluble residue. The platinum compound is added only in sufficient quantity to combine with the potash, and, after evaporation almost to dryness, the double potassium-platinum salt is washed with acidulated alcohol (made by passing dry hydrochloric acid gas into cool ninety per cent. alcohol), then with ammonium chloride solution and finally with eighty-five per cent. alcohol. The double salt is then dried and weighed. Ammoniacal salts must first be destroyed.

Analytical Notes upon the Estimation of Phosphorus in Steel. By R. W. Mahon. *J. Am. Chem. Soc.*, 20, 429–453.— The molybdate-magnesia method for the determination of phosphorus in steels is critically examined by the author, and experimental data are given in profusion. A table of contents at the opening of the paper shows the division and scope of the subject-matter.

The Colorimetric Estimation of Small Amounts of Chromium, with Special Reference to the Analysis of Rocks and Ores. By W. F. Hillebrand. *J. Am. Chem. Soc.*, 20, 454– 460.—The rock or ore is fused with sodium carbonate and nitrate, the manganese reduced by methyl or ethyl alcohol, the silica and aluminum removed (and re-examined for chromium), and if the amount of chromium is very small, mercurous nitrate added to throw down phosphoric, vanadic, chromic and carbonic acids in combination with mercury. This precipitate is ignited and the residue fused with sodium carbonate and the color of the solution compared with that of a standard chromate solution, made alkaline with sodium carbonate. Details as to apparatus are given in the original paper. The procedure exceeds in accuracy the usual gravimetric process.

Volumetric Estimation of Vanadium in Presence of Small Amounts of Chromium, with Special Reference to the Analysis of Rocks and Ores. By W. F. Hillebrand. *J. Am. Chem. Soc.*, 20, 461–465.—The rock or ore is treated exactly as described in the preceding review for the colorimetric estimation of chromium. The chromate solution may then be made acid with sulphuric acid and the chromic and vanadic acids reduced by sulphurous acid gas, the excess being expelled by boiling. The vanadium is then determined by titration with potassium permanganate. A considerable quantity of chromium, or the presence of arsenic or molybdenum, interferes with the titration.

F. H. THORP, REVIEWER.

The Protection of Steam-Heated Surfaces. BY C. L. NOR-TON: *Engineer, No.* 459, 165.—This investigation was made in order to show the relative efficiency of various kinds of steam-pipe covering now on the market, to ascertain the fire risk attendant upon the use of certain methods and materials for insulating steam-pipes, and to show the gain in economy attendant upon increase in thickness of coverings. A piece of steam-pipe was electrically heated from the inside, the electrical energy supplied was measured, and the amount of heat furnished was calculated. By keeping the pipe at a constant temperature, the amount of heat supplied is obviously equal to the heat lost by radiation, convection, and conduction. The pipe, closed at one end by a welded-in plate and at the other by a tight cover, was filled with heavy cylinder oil, into which the heating coil, an effective stirrer, and a thermometer, were introduced. The apparatus was suspended in the middle of a room, by non-conducting cords, and the thermometer was read by a telescope, the observer being at a distance to avoid the production of air currents and the addition of heat from the person. Fourteen different samples were tested, the best results being obtained with a covering composed of granulated cork pressed in a mould and rendered non-combustible by a fire-proofing process. Nearly as good was a covering composed of 90 per cent. magnesium carbonate. A cover made of thin sheets of asbestos paper, separated from one another so as to enclose a considerable amount of air, was also very effective. Asbestos, when not porous and not containing entrapped air, is a good conductor of heat. Carbonate and sulphate of calcium are not good heat insulators. Coverings containing wool, hair-felt, or wool-felt, have high efficiency as non-conductors, but are considered dangerous on account of fire risks. Computations of the money saving effected by covers, show that in general a cover pays for itself in a little less than a year of 310 ten-hour days; or in four months, working 365 days of twenty-four hours each. As a protection to wooden surfaces, an asbestos board, made of corrugated asbestos paper sheets cemented with sodium silicate, was found very effective, though not remarkable as a non-conductor of heat.

Sadtler's Reducing Agent. *Leather Manufacturer,* 9, 105.—These are merely general remarks upon the use of hydrogen peroxide as a reducing agent for chrome tanned skins (see *this Rev.,* 2, 106). A slightly acid bath is used, the peroxide being added in successive small portions.

Commercial "Moellon" Degras. *Leather Manufacturer,* 9,

110.—Moellon is a degras made by the oxidation of cod or fish oil in the pores of skins from which the degras is then pressed. Abstracts of the various methods for preparing moellon, with a few remarks upon its use in tanning, make up the larger part of this article.

The Liming of Soils. By H. J. WHEELER. *U. S. Dept. Agr., Farmers' Bull. No. 77.*—This bulletin gives a popular exposition of the theory and practice of using lime as a fertilizer. The various materials employed are described and their relative values compared. Many of the conclusions are drawn from the results of experiments at the Rhode Island Experiment Station where the author is the chemist.

Commercial Fertilizers. *N. Y. State Sta. Bull.,* 129, 351–421 ; *Miss. Sta. Special Bull.,* 42 and 45 ; *La. Sta. Bull.,* 49, 163–198 ; *R. I. Sta. Rep., 1896,* 211–220 ; *N. J. Bull.,* 124, 48 ; *Ark. Sta. Rep., 1897,* 101–118 ; *Md. Sta. Bull.,* 49, 105–160 ; *Pa. Dept. Agr. Bull.,* 33 ; *Vt. Sta. Rep., 1896–97,* 28–30 ; *Ga. Dept. Agr. Bull.,* 33, 116 ; *Me. Sta. Bull.,* 38.

The Control of the Temperature in Wine Fermentation. By A. P. HAYNE. *Cal. Sta. Bull.,* 117, 19.—The causes of the spontaneous heating of fermenting must and methods of controlling it are considered in this bulletin. If the temperature exceeds 100° F. the wine ferment becomes inactive ; but injurious bacteria continues to thrive. A method of cooling must, employed in France, is described, in which the must is pumped through several hundred feet of 1½ inch piping arranged in columns and cooled externally by water dripping upon the pipes from a tank above them. At the California Experiment Station a column composed of flat tubes 4 by 1½ inches is used, the total length of pipe being only 43 feet. The pipes are cooled by a spray of water blown against them by a strong blast of air, the evaporation adding greatly to the cooling effect.

Petroleum Briquettes in Germany. By MAX BOUCHSEIN. *U. S. Consular Rep.,* 57, 254.—This is a description of a process patented by J. Kohlendorfer, by which petroleum refuse is worked into a solid substance for a cheap fuel. Soda lye and a saponifiable fat (such as tallow) are heated with superheated steam with exclusion of air until the saponification begins and then a four-fold quantity of petroleum refuse is added, and the mixture heated about an hour, with constant stirring, but keeping the temperature below the boiling-point of the petroleum. The mixture is then run into molds and solidifies on cooling. Coal-dust, saw-dust, or other combustibles may be added during the process if desired. The grease may be replaced by resin or resin acid if a harder product is desired. The product contains about 80 per cent. of petroleum oil and only 5 per cent. of non-combustible matter.

GEOLOGICAL AND MINERALOGICAL CHEMISTRY.

W. O. CROSBY, REVIEWER.

Auriferous Conglomerate of the Transvaal. BY GEO. F. BECKER. *Am. J. Sci.*, 155, 193–208.—The auriferous conglomerate of the Witwatersrand gold field constitutes the most important gold deposit ever known. A strip of country a couple of miles in width and about 30 miles in length has yielded since 1887 about $240,000,000, and only about one-fifteenth of the accessible gold has been extracted. The nature of this wonderful deposit is a subject of manifest interest to geologists and mining engineers, who, however, have arrived at various conclusions. No considerable doubt exists that the conglomerate is a marine littoral deposit, but some observers have held that the gold is detrital, being an original part of the conglomerate; others that it is a chemical precipitate from the ocean in which the pebbly beds were laid down ; and still others that the precious metal reached the uplifted but uncemented gravel in solution, so that the ore-bearing strata are allied to ordinary veins. After presenting all the more striking facts which it is necessary to consider in testing the theories propounded to account for the deposition of gold on the Rand, the author rigidly tests each theory by the facts and reaches the conclusion that the deposition of the gold was in no sense chemical, that there are no valid objections to the theory of marine placer origin, and that no noteworthy features are left unexplained by this theory. The argument is so clear and cogent that the detrital theory of the Transvaal gold may be regarded as firmly established.

San Angelo Meteorite. BY H. L. PRESTON. *Am. J. Sci.*, 155, 269–272.—This describes a metallic meteorite found in 1897, in Tom Green County, Texas. It weighed 194 pounds, and is distinguished by unusually perfect octahedral structure, the comparative scarcity of troilite, and the sharpness and regularity of the Widmanstätten figures. Analysis shows: Fe, 91.958 ; Ni, 7.860; and traces of Co, Cu, P, S, Si, and C.

On Clinohedrite, a New Mineral from Franklin, N. J. BY S. L. PENFIELD AND H. W. FOOTE. *Am. J. Sci.*, 155, 289–293.—This is a monoclinic hydrous silicate of zinc and calcium. The crystals are especially interesting, as they belong to that division of the monoclinic system characterized by a plane of symmetry but not an axis of symmetry. The theoretical composition as deduced from two analyses is : SiO_2, 27.92 ; ZnO, 37.67 ; CaO, 26.04 ; H_2O, 8.37 ; total, 100.00. This corresponds very nearly to the ratio $1:1:1:1$, giving the formula $H_4ZnCaSiO_4$, analogous to that of calamine. The formula may also be written $(ZnOH)(CaOH).SiO_2$; and that hydroxyl is present is proved by the fact that water is not expelled much below a faint redness.

On Rhodolite, a New Variety of Garnet. By W. E. HIDDEN AND J. H. PRATT. *Am. J. Sci.*, 155, 294–296.—This garnet, found in a very limited area in Macon County, North Carolina, is distinguished by its strikingly beautiful rose tints; its surprisingly small amount of coloring-matter; its gem-like transparency; its freedom from internal imperfections; and its remarkable brilliancy when cut as a gem. As yet it is known only as rolled pebbles and fragments in gravel. Analysis gives : SiO_2, 41.59; Al_2O_3, 23.13; Fe_2O_3, 1.90: FeO, 15.55; MgO, 17.23; CaO, 0.92; total, 100.32. The ratio of protoxide, sesquioxide and silica is nearly $3:1:3$, which classifies the mineral as one of the garnets. The ratio of MgO to FeO is almost exactly $2:1$, which would indicate that the mineral is composed of two molecules of a magnesium-aluminum garnet (pyrope) and one molecule of a ferrous iron-aluminum garnet (almandite). The theoretical composition is calculated for this ratio and the formula deduced.

Some Lava Flows of the Western Slope of the Sierra Nevada, California. By F. LESLIE RANSOME. *Am. J. Sci.*, 155, 355–375.—The lavas here described occur in the valley of the Stanislaus river and are of later date than the rhyolite and the andesitic breccias and tuffs of the region. The distinctive chemical feature of these rocks is a rather high percentage of total alkalies, with the potash somewhat in excess of the soda. Chemically they stand between typical andesites and typical trachytes, and belong to a group which it seems necessary to classify under a new name, and they are here called *latite*. Three successive flows are recognized, the first of which, the Table Mountain flow, extended a distance of at least twenty miles. The first and third flows are augite-latite, and the second, biotite-augite-latite. Seven analyses are given, and many others of allied rocks are quoted in tabular form for comparison. The potash approximates 5 per cent., and yet the majority of the thin sections contain no recognizable potash-bearing mineral. Hence the author concludes that where biotite is absent the potash must exist in the residual glass of the ground mass. The classification of the latites is discussed at some length and their lithologic individuality definitely established.

On Krennerite, from Cripple Creek, Colorado. By ALBERT H. CHESTER. *Am. J. Sci.*, 155, 375–377.—Krennerite has not previously been identified from this country. The mineral from

Cripple Creek occurs in brilliant, orthorhombic crystals of a pale yellowish-bronze color, but tin-white on cleavage faces, and up to 2 mm. in length. An analysis by Prof. W. S. Myers gave, after deducting 1.21 per cent. of insoluble matter and calculating to one hundred : Au, 43.86 ; Ag, 0.46 ; Te, 55.68 ; corresponding to the formula $AuTe_2$, and agreeing very closely with sylvanite (monoclinic) and calaverite (triclinic).

Geology of the St. Croix Dalles. Part II.—Mineralogy. By CHAS. P. BERKEY. *Am. Geol.*, 21, 139–155.—The article describes (p. 153) a greenish-yellow sulphate occurring as an efflorescence on the pyritiferous Dresbach shales. Analysis gave : SiO_2, 12.946; Fe_2O_3, 22.828 ; Al_2O_3, 4.141 ; K_2O, 1.844 ; Na_2O, 4.659 ; CaO, 2.210; SO_3, 32.500 ; H_2O, 17.840 ; organic matter, traces; total, 98.968 per cent. Although probably a mixed substance, as indicated by the high proportion of silica, it is similar in complexity and general range to voltaite.

Residual Concentration by Weathering as a Mode of Genesis of Iron Ores. By JAMES P. KIMBALL. *Am. Geol.*, 21, 155–163.—After citing his previous papers on this subject, two of which have been noticed in this *Review*, the author describes a remarkable differential development of ferric and magnetic oxides from an amorphous basic aggregate on a tributary of the Yakima River, in the state of Washington. The ferruginous beds have a thickness of from six to eighteen feet ; the chief occurrence of iron ores is at the base, where they overlie crystalline pyroxene; and here its development is confined to wet places and exposed ledges. In circumstances thus favorable to atmospheric oxidation and percolation of water, magnetite, martite, hematite, and limonite have been exfoliated as an insoluble residuum from decomposition of the basic aggregate, with interlaminated siliceous residuums, to a thickness of from two to eighteen inches ; while immediately beneath the basal pyroxene is to the depth of a few inches, commonly decomposed into a soft chloritic clay. The ferriferous zone fades out gradually upward, and the whole occurrence is economically insignificant and worthless. Seventeen partial analyses by Prof. James A. Dodge show a range in metallic iron from 15.36 to 63.05 per cent. The fixation of ferric oxide is believed to occur through reaction with calcic carbonate, which is also a product of the splitting up of basic silicates. The magnetite is not regarded as wholly residual, but due in part to stoichiometrical transformation of ferric hydrate at ordinary temperatures.

Brazilian Evidence on the Genesis of the Diamond. By ORVILLE A. DERBY. *J. Geol.*, 6, 121–146.—This is a very complete summary of the facts afforded by the diamond washings of Brazil that have a bearing upon the origin of this gem; and a comparison with the evidence from the "dry diggings" of South Africa. The general conclusion reached is that as the case now stands the indications support the view of the formation of the Brazilian diamonds in the phyllites or associated sedimentary deposits through the active agency of neighboring eruptives which are in part pegmatitic, the carbon having been supplied by the phyllites. In order to bring the Kimberley and Brazilian modes of occurrence into line as phases of a single mode of genesis, it seems necessary to put aside the idea that the recent interesting experiments on the artificial production of the diamond afford a solution of its terrestrial origin, and that the Kimberley type of rock and mode of occurrence are essential features. Presumably, also, the genesis must be sought in the rocks affected by the eruptive masses rather than in those masses themselves; and hence the diamond is to be regarded as of metamorphic, and not of truly igneous origin.

[CONTRIBUTION FROM THE MASSACHUSETTS INSTITUTE OF TECHNOLOGY.]

REVIEW OF AMERICAN CHEMICAL RESEARCH.

VOL. IV. NO. 10.

ARTHUR A. NOYES, Editor ; HENRY P. TALBOT, Associate Editor.
REVIEWERS: Analytical Chemistry, H. P. Talbot and W. H. Walker;
Biological Chemistry, W. R. Whitney ; Carbohydrates, G. W. Rolfe ;
General Chemistry, A. A. Noyes ; Geological and Mineralogical Chem-
istry, W. O. Crosby ; Inorganic Chemistry, Henry Fay ; Metallurgical
Chemistry and Assaying, H. O. Hofman ; Organic Chemistry, J. F. Nor-
ris ; Physical Chemistry, H. M. Goodwin ; Sanitary Chemistry, E. H.
Richards ; Technical Chemistry, A. H. Gill and F. H. Thorp.

INORGANIC CHEMISTRY.

HENRY FAY, REVIEWER.

The Action of Zinc on Copper Silicide. BY G. DE CHALMOT. *Am. Chem. J.*, 20, 437-444.—In a previous paper (*this Rev.*, 4, 40) it was shown that sulphur decomposes copper silicide with the liberation of free silicon. It is now shown that zinc acts in a similar manner, setting free crystalline silicon. With the addition of small quantities of zinc, there is no liberation of silicon, owing to the fact that the zinc combines with the free copper in . the silicide. By varying the amounts of zinc it has been estimated that the combination of the zinc and copper ceases when the two elements are present in the proportion to form the compound $ZnCu_2$, so that for the same quantities of zinc there is more silicon liberated the smaller is the quantity of free copper in the silicide.

The Action of Sulphur upon Metallic Sodium. BY JAMES LOCKE. *Am. Chem. J.*, 20, 592-594.—Sulphur in varying proportions dissolved in hot toluene, when brought into contact with sodium under boiling toluene, reacts to form a sulphide which, after washing with hot toluene, has approximately the composition expressed by the formula Na_2S_2. In two experiments where there was some residual sodium, it was bright and clear, showing that it was without action at this temperature on the polysulphide formed. At the temperature of melted naphthalene there was likewise combination ; but there was no evidence in either case of the formation of sodium monosulphide.

On Some Compounds of Trivalent Vanadium. BY JAMES LOCKE AND GASTON H. EDWARDS. *Am. Chem. J.*, 20, 594–

606.—The vanadates of sodium and ammonium were reduced by alcohol and hydrochloric acid to vanadyl dichloride, and this was further reduced to vanadium chloride by means of mercury amalgam in a special apparatus arranged so that all of the work could be carried out in a current of hydrogen. From vanadium hydroxide new salts were prepared so as to study the influence which the atomic weight exerts upon the development of the properties common to aluminum, chromium, manganese, iron, and cobalt. By evaporating *in vacuo* the green solution obtained by dissolving vanadium hydroxide in hydrochloric acid, crystals of the compound $VCl_3.6H_2O$ were obtained. This salt had previously been obtained by Piccini, but was not described by him. It is similar in composition and properties to the chloride of iron, aluminum, and chromium. Double salts could not be obtained from vanadium chloride and the alkali chlorides. The bromide, $VBr_3.6H_2O$, prepared in a similar manner, is described ; but the iodide could not be obtained. Potassium vanadicyanide, $K_4V(CN)_6$, was prepared by dissolving the anhydrous chloride in as small a quantity of water as possible, and by adding to this an excess of a concentrated solution of potassium cyanide. From this solution the new compound was precipitated by the addition of alcohol. This salt forms a member of the series $K_4M(CN)_6$ and gives colored precipitates with solutions of inorganic salts. It was impossible to prepare the sodium and ammonium salts. Potassium vanadisulphocyanide, $K_3V(SCN)_6.4H_2O$, was made by bringing together vanadium trichloride and an alcoholic solution of potassium sulphocyanide and crystallizing in a vacuum. It crystallizes in dark-red crystals, which are readily soluble in water and alcohol, the solution, however, being extremely unstable.

On the Decomposition of Concentrated Sulphuric Acid by Mercury at Ordinary Temperatures. By CHARLES BASKERVILLE AND F. W. MILLER. *J. Am. Chem. Soc.*, **20**, 515–517.— The authors state that the acid used by them in previous experiments contained 99.65 per cent. H_2SO_4, and that it does react with mercury, notwithstanding the evidence to the contrary cited by Pitman. (*This Rev.* **4**, 41.)

BIOLOGICAL CHEMISTRY.

W. R. WHITNEY, REVIEWER.

Proteids of the Pea. By THOMAS B. OSBORNE AND GEO. F. CAMPBELL. *J. Am. Chem. Soc.*, **20**, 348–362.—The authors have discovered that the legumin of the pea, previously described by them, was contaminated with another proteid, to which they have given the name vicilin; and that the purified legumin of the pea and of the vetch, lentil, and horse bean are probably identical. In the pea they recognize the following proteids: legumin, which is the non-coagulating globulin; vicilin, coagulable and relatively soluble; a globulin or albumin to which the name legumelin is given; protoproteose; and deuteroproteose. The results of analyses of these constituents and their properties are given.

Proteids of the Lentil. Proteids of the Horse Bean. Proteids of the Vetch. Proteids of the Soy Bean. By T. B. Osborne and G. F. Campbell. *J. Am. Chem. Soc.*, 20, 362–375; 393–405; 406–410; 419–427.—In these four investigations the authors have followed the methods employed in the preceding one, and have arrived at conclusions concerning the composition of the nitrogen-containing components of these vegetables. They find legumin, legumelin, and proteose in the lentil, horse bean, and vetch, and vicilin in the lentil and horse bean. The soy bean contains as its chief proteid a globulin somewhat similar to the legumin found in the other vegetables examined, but of somewhat different composition. The soy bean also contains some legumelin and proteose.

The Proteids of the Pea, Lentil, Horse Bean, and Vetch. By T. B. Osborne and G. F. Campbell. *J. Am. Chem. Soc.*, 20, 410–419.—The authors compare the compositions and properties of the legumin, vicilin, legumelin, and proteose which they have isolated from the vegetables mentioned in the title, and give the results of the application of the proteid tests upon these different constituents. The analyses of the four substances which they have obtained agree exceedingly well for the same substance from the different sources, while the differences between the compositions of the different proteids are very marked.

A. G. WOODMAN, REVIEWER.

The Chemistry of Cascara Sagrada. By ALFRED R. L. DOHME AND HERMANN ENGELHARDT. *J. Am. Chem. Soc.*, 20, 534–546.—The authors have made a systematic analysis of a typical sample of cascara sagrada bark. They have obtained the volatile oil which gives to the drug its characteristic odor, and have also isolated a fixed oil which they regard as a mixture of dodecyl palmitate and dodecyl stearate. From the extract with eighty per cent. alcohol, was obtained the glucoside of cascara sagrada, which the authors have named purshianin. This is a dark brown-red crystalline substance which melts at 237° C., and yields emodin when saponified.

Nitrogenous Feeding Stuffs. By C. S. Phelps. *Storrs Agr. Expt. Sta. Bull.*, 18, 1–16.—This bulletin describes a number of nitrogenous feeding stuffs, and contains tables showing their composition and nutritive value.

Mushrooms as Food. *U. S. Dept. of Agr., Farmers' Bull.*, 79, 18-20—Tables are given comparing the composition of mushrooms with that of other articles of food. These show that edible fungi do not possess a high food value.

Evaporation and Plant Transpiration. By Walter Maxwell. *J. Am. Chem. Soc.*, 20, 469–483.—The object of this investigation was to determine, first, the loss of moisture due to direct evaporation from the soil ; and secondly, the relative proportion that escapes by transpiration from the sugar-cane (*saccharum officinarum*) during different periods of growth. The observations were made on considerable quantities of soil, 125 pounds in each case, and covered a period of eight months. The author considers that it is shown by the results that nitrogen, which was applied in the form of sodium nitrate, stimulates growth and causes increased transpiration, and also that nitrogen is the vital element in the growth of plants. The data obtained were found to be of value in practical field irrigation.

Some Spraying Mixtures. *Cornell Univ. Agr. Expt. Sta. Bull.*, 149.—This bulletin contains analyses of some mixtures sold as insecticides, most of them being found to contain arsenic and copper.

The Value of Experiments on the Metabolism of Matter and Energy. By C. F. Langworthy. *U. S. Dept. Agr. Expt. Sta. Record*, 9, 1003–1019.—The author discusses the value of metabolism experiments in comparison with ordinary feeding tests, and advocates a more general trial of such experiments at experiment stations.

The Mineral Constituents of the Tubercle Bacilli. By E. A. de Schweinitz and Marion Dorset. *J. Am. Chem. Soc.*, 20, 618–620.—In continuation of their work upon the composition of the tubercle bacilli, the authors have made careful analysis of the ash. Quite noticeable were the high percentage of phosphorus pentoxide and the absence of other acid radicals. The high percentage of fat in the body of the tubercle bacilli, as previously observed, taken in connection with the high phosphate content of the ash, lead the authors to raise the interesting query as to whether, in prescribing phosphates and cod-liver oil in cases of tuberculosis, we are not supplying nourishment to the bacilli rather than to the individual.

A Volumetric Assay of Opium. By H. M. GORDIN AND A. B. PRESCOTT. *Pharm. Archives*, 1, 121-126.—The method is based on the following plan : The opium alkaloids are set free by ammonia with alcohol, ether, and chloroform. The other alkaloids are removed by benzene, after which the morphine is taken out by percolation with acetone. After evaporation of the acetone the residue is taken up with lime-water, and the morphine is estimated by titration as periodide, according to the method of the authors (*J. Am. Chem. Soc.*, 20, 334). The procedure is described in considerable detail.

Note upon the Volumetric Assay of Opium. By H. M. GORDIN AND A. B. PRESCOTT. *Pharm. Rev.*, 16, 303.—The authors have found that in the estimation of morphine as a periodide by their method, as described in the preceding review, the character of the precipitate obtained varies under certain conditions, which are now being studied.

Factory Tests for Milk. By S. M. BABCOCK, H. L. RUSSELL, and J. W. DECKER. *Wis. Agr. Expt. Sta. Bull.*, 67, 1-20.—This bulletin describes a number of simple tests that can be applied to determine the quality of the milk with regard to its use in the cheese industry. It gives a quite thorough description of the Wisconsin Curd Test as used for this purpose.

The Chemistry of Aloes. By ALFRED R. L. DOHME. *Am. J. Pharm.*, 70, 398-402.—This paper is a summary of recent work on the subject, showing (1) that the resin of aloes is as ester varying with the kind of aloes ; (2) that aloin contain emodin, to which its laxative property is probably due; and (3) that many drugs owe their laxative property to the substance emodin, which is probably a derivative of hydroquinone.

Analyses of Commercial Fertilizers. *Agr. Expt. Sta. Bull.*, R. I., No. 48 ; S. C., No. 35 ; Vt., Nos. 63, 64, and 65 ; Ky., No. 75 ; Me., No. 43.

Valuation of Crude Carbolic Acid. By CARL E. SMITH. *Am. J. Pharm.*, 70, 369-378.—As the result of a number of experiments, the author concludes that a modification of Koppeschaar's bromine test, and a test for alkali, are all that is required to determine the value of a sample for disinfecting purposes.

On the Determination of Fat and Casein in Feces. By HERMAN POOLE. *J. Am. Chem. Soc.*, 19, 877-881.—The author considers methods of fat determination by extraction with ether faulty, because no attempt is made to separate the fat from the cholesterol, bile, and other products extracted. He effects this separation by saponifying the residue from the evaporation of the ether extract with alcoholic potash, and extracting the aqueous solution of the resulting soap with ether to remove cholesterol. The remaining liquid is used for the determination of fat acids. In the determination of casein the chief difficulty lies in its separation from the epithelium cells and foreign matter. This the author effects by extracting the feces in succession with ether, water, and alcohol, and then drying. The residue is digested with dilute hydrochloric acid (1 : 2½) at 50°. This dissolves the casein and leaves the epithelium debris. The casein is calculated from the nitrogen determined by the Kjeldahl method. The results obtained are lower than those given by the old methods ; but the author claims that they are nearer the truth.

Salt River Valley Soils. BY ROBERT H. FORBES. *Ariz. Agr. Expt. Sta. Bull.*, 28, 66–99. — This bulletin contains the results of analyses made of the alkaline soils of Arizona, taken entirely from the surface, and in nearly all cases at a dry season of the year. A discussion of the object and value of soil analyses is also included.

Carbon Dioxide from Fermentation. BY DR. P. FISCHER. *Pharm. Rev.*, 16, 214–220.—The author discusses the recovery of carbon dioxide from the fermentation of beer. 100 pounds of beer-wort will furnish about 4 pounds of carbon dioxide. Calculated on this basis the annual production in the United States, most of which, however, still escapes into the atmosphere, is more than 300,000,000 pounds. The process and plant used by the Pabst Brewing Company, of Milwaukee, is described. Here air-free gas is collected directly from the fermenting tubs, and is subsequently liquefied and purified. In 1897 this company sold over 1,500,000 pounds of the liquefied gas.

A Preliminary Report on the Soils of Florida. BY MILTON WHITNEY. *U. S. Dept. Agr., Div. of Soils, Bull.* 13.—This report contains a brief description of the principal types of soil and the characteristic vegetation, and a discussion of the chemical composition, of the physical texture, and of the water content, with reference to the differences in agricultural value.

An Electrical Method of Determining the Moisture Content of Arable Soils. BY MILTON WHITNEY, FRANK D. GARDNER, AND LYMAN J. BRIGGS. *U. S. Dept. Agr., Div. Soils, Bull.* 6.—The principles of physical chemistry involved in the method are discussed, and a description and illustrations of the apparatus are given.

Biological Chemistry. 115

An Electrical Method of Determining the Temperature of Soils. BY MILTON WHITNEY AND LYMAN J. BRIGGS. *U. S. Dept. Agr.*, *Div. Soils, Bull.* 7.

An Electrical Method of Determining the Soluble Salt Content of Soils. BY MILTON WHITNEY AND THOS. H. MEANS. *U. S. Dept. Agr.*, *Div. Soils, Bull.* 8.

Methods for the Examination of Milk. BY BYRON STANTON. *Ohio San. Bull.*, 2, 17–25.—This paper is a critical discussion and comparison of the various methods in use for detecting adulteration in milk. It does not contain anything new.

A Comparison of Utah Feeding Stuffs. BY LUTHER FOSTER and LEWIS A. MERRILL. *Utah Agr. Expt. Sta. Bull.*, 54, 119–140.

Digestion Experiments with Shredded Corn-fodder, Lucern, Timothy, and Wheat Bran. BY JOHN A. WIDTSOE. *Utah Agr. Expt. Sta. Bull.*, 54, 141–151.

Cotton and Corn Experiments. BY C. C. PITTUCK. *Tex. Agr. Expt. Sta. Bull.*, 45, 978-1008.

E. H. RICHARDS, REVIEWER.

Nutrition Investigations in Pittsburg. BY ISABEL BEVIER. *U. S. Dept. Agr. Bull.*, 52, 1–48.—In the course of these studies a special point was made of the composition and cost of bread made at home in comparison with that purchased from bakeries. There is also a discussion of the apparent loss of nutrients during the process of baking.

Nutrition Investigations at the University of Tennessee in 1896 and 1897. BY CHAS. E. WAIT. *U. S. Dept. Agr. Bull.*, 53, 1–46.—In addition to three dietaries this bulletin contains valuable data upon the composition of Tennessee beef, mutton, and chicken. There is also an account of twenty-one digestion experiments with men, from which it appears that, in the cases given, a diet of bread and beef yielded only 85.5 per cent. of the total energy in the food eaten ; one of bread, milk, and eggs, 88.6 ; an average of 11 mixed diets, 89.9 ; while bread and milk alone returned 92.3 per cent. of the calculated energy. Although the time of the experiments—two days—was far too short for any very conclusive results, the facts have a certain value in the dearth of similar studies under American conditions.

Nutrition Investigations in New Mexico in 1897. BY ARTHUR GOSS. *U. S. Dept. Agr. Bull.*, 54, 1-20.—In addition to a die-

tary of a poor Mexican family, the author gives the analyses of the different cuts of a side of New Mexico range beef, which show the remarkable absence of fat in this class of meat when killed in the spring.

A Method for the Differentiation of Organic Matter in Water. By A. G. WOODMAN. *J. Am. Chem. Soc.*, 20, 497–501.—The reagent used is potassium bichromate and the method is a modification of that given by Joseph Barnes, which is essentially the determination of the ratio between the amount of oxygen used up by potassium permanganate and by bichromate from the same water. The results given indicate that the method is worth further trial.

The Operation of a Slow Sand Filter. By CHAS. E. FOWLER. *J. N. E. Water Works Assoc., 1898,* 209–244.

The Origin of Free and Albuminoid Ammonias in Polluted Waters. By ELMER G. HORTON. *J. Am. Public Health Assoc.,* 23, 199–205.—The author discusses the decomposition of urea under the conditions obtaining in the distillation for ammonia.

The Public Water Supplies of the State. By A. W. SHAFTER. *N. C. Board of Health, Bull.* 13, 2, 1–23.—The report includes a bacteriological and a partial chemical examination of the supply of 14 towns together with an examination of the watershed.

GENERAL AND PHYSICAL CHEMISTRY.

A. A. NOYES, REVIEWER.

The Atomic Weight of Cadmium. By H. N. MORSE AND H. B. ARBUCKLE. *Am. Chem. J.*, 20, 536–542.—In order to determine any correction that may exist for the presence of retained gas in the cadmium oxide, the authors have repeated the atomic weight determinations of Morse and Jones (*Am. Chem. J.*, 14, 241) using essentially the same method and the same samples of cadmium. The gas evolved on dissolving the oxide was collected as in the authors' experiments with zinc oxide (*this Rev.*, 4, 55), and, as before, was found to be very appreciable in quantity, and to consist of oxygen and nitrogen. The uncorrected result, 112.084, is very close to that previously obtained, 112.071, while the corrected one, 112.377, now agrees well with those obtained by Bucher, 112.39 and 112.38, in his work on the chloride and bromide of cadmium.

A Table of Atomic Weights. By THEODORE WILLIAM RICHARDS. *Am. Chem. J.*, 20, 543–554.—The author has made a critical study of the reliability of the existing atomic weight determinations, and has compiled a new table of atomic weights. In the case of seven elements his values differ markedly from those of Clarke, as is shown below :

	Richards.	Clarke.
Antimony	120.0	120.43
Cadmium	112.3	111.95
Calcium	40.0	40.07
Magnesium	24.36	24.28
Platinum	195.2	194.89
Tungsten	184.4	184.83
Uranium	240.0	239.59

The Reliability of the Dissociation Values Determined by Electrical Conductivity Measurements. By ARTHUR A. NOYES. *J. Am. Chem. Soc.*, 20, 517–528.—The author endeavors to show that an error has been made by van Laar (*Ztschr. phys. Chem.*, 21, 79) in the derivation of his heat-of-solution formula, and that therefore his conclusion based thereon in regard to the unreliability of the dissociation values determined by electrical conductivity is entirely unjustifiable. The question is also discussed from several other points of view. Incidentally a new, rigidly exact and general expression is derived from the relation between the heat of solution of dissociated substances and the change in their solubility with the temperature.

Molecular Weights of Some Carbon Compounds; A Few Words More. By C. L. SPEYERS. *J. Am. Chem. Soc.*, 20, 546–547.—This note is a reply to the criticism by the reviewer of a previous article by the same author (*this Rev.*, 4, 55). In explanation of his presentation to his readers of several pages of numerical results without a word of discussion, the author states that he himself was at a loss to account for the peculiar results obtained. He rejects the suggestion of the reviewer as to the possibility of error from imperfect elasticity of the bulb.

The Atomic Mass and Derivatives of Selenium. By VICTOR LENHER. *J. Am. Chem. Soc.*, 20, 555–579.—A current of hydrochloric acid gas was passed over silver selenite, first in the cold, then at a general heat ; the silver chloride formed was weighed, and reduced with hydrogen; and the residual silver again weighed. Eleven determinations of the ratio $Ag_2SeO_3 : 2AgCl$ gave 79.329 as the atomic weight of selenium and eight determinations of the ratio $Ag_2SeO_3 : 2Ag$ gave an identical value. From ammonium bromoselenate the selenium was precipitated by hydroxylamine hydrochloride, the precipitate collected on a Gooch filter and weighed. Eight determinations of the ratio $(NH_4)_2SeBr_6 : Se$ lead to the atomic weight 79.285. In conclusion, a large number of organic and inorganic bromoselenates are described ; and evidence of the non-existence of a selenium monoxide is furnished by several experiments.

Osmotic Pressure. By C. L. SPEYERS. *J. Am. Chem. Soc.*, **20**, 579–585.—The author attempts to show with the help of hypotheses relating to the constitution of solutions, that the specific gravity value which enters in the thermodynamical relation between osmotic pressure and vapor pressure is that of the pure solvent and not that of the solution. This conclusion had, however, already been reached by other authors. The author, to be sure, considers the mechanical arrangement involved in the demonstration of it recently given by Noyes and Abbot to be "purely imaginary and not possible in fact," and "the results depending on their theoretical deductions to be valueless."

ANALYTICAL CHEMISTRY.

ULTIMATE ANALYSIS.

W. H. WALKER, REVIEWER.

On the Lindo–Gladding Method of Determining Potash. By A. L. WINTON AND H. J. WHEELER. *J. Am. Chem. Soc.*, **20**, 597–609.—With a view of meeting certain criticisms of the Lindo-Gladding method for determining potash, the authors have made an extensive study of the recent work bearing on the subject. Numerous data are presented which show that the objections made to the process are, so far as practical considerations go, without foundation.

On the Estimation of Manganese Separated as Carbonate. By MARTHA AUSTIN. *Am. J. Sci.*, **155**, 382–384.—The separation of manganous carbonate is shown to be both rapid and complete when made in a warm solution and in the presence of considerable ammonium chloride. The precipitate cannot, however, be weighed as carbonate, as carbon dioxide escapes before all the moisture has been expelled ; but, by the addition of sulphuric acid and subsequent weighing as anhydrous sulphate (*this Rev.*, **4**, 60) an accurate determination can be made.

A New Volumetric Method for the Determination of Copper. By RICHARD K. MEADE. *J. Am. Chem. Soc.*, **20**, 610–613.—A volumetric method for the determination of copper more accurate and also more widely applicable than either the cyanide or iodine method is found to be the following : The copper is precipitated from an acid solution by addition of ammonium or potassium thiocyanate, and the cuprous salt thus obtained converted to cuprous oxide by warming with potassium hydrate. To this is added a solution of ferric chloride or sulphate, and the ferrous salt formed by the oxidation of the cuprous oxide is then titrated with standard permanganate solution.

A Short Study of Methods for the Estimation of Sulphur in Coal. By G. L. HEATH. *J. Am. Chem. Soc.*, 20, 630–637.—The author makes a comparative study of the five more common methods for the determination of sulphur in coal, and notes such modifications and precautions as he finds necessary to obtaining the most accurate results.

On the Condition of Oxidation of Manganese Precipitated by the Chlorate Method. By F. A. GOOCH AND MARTHA AUSTIN. *Am. J. Sci.*, 155, 260–268.—Although the "chlorate process" is at present the method most widely used by practical chemists for the separation of manganese, yet the degree of oxidation of the precipitated oxide has never been definitely determined. The author finds the more soluble sodium chlorate a better precipitant than the potassium salt, but that the precipitate produced by neither of these corresponds exactly to the dioxide, MnO_2. A procedure is given by which the precipitate can be made to correspond to the formula, but it is not recommended by the authors as a rapid analytical method.

F. J. MOORE, REVIEWER.

Use of Hydrofluoric Acid in the Determination of Manganese in Iron and Ores. By ALLEN P. FORD AND I. M. BREGOWSKI. *J. Am. Chem. Soc.*, 20, 504–506.—An objection to the Williams method is the clogging of the filter by separated silicon. This may be completely dissolved and the filtration accelerated by the addition of a few drops of hydrofluoric acid just after the precipitation of the manganese peroxide. In the analysis of ores by this method, the presence of hydrofluoric acid has the peculiar effect of making the precipitation more complete, thus rendering a second precipitation unnecessary.

The Determination of Lead in Alloys. By W. E. GARRIGURS. *J. Am. Chem. Soc.*, 20, 508–510.—Lead may be separated from copper and zinc by precipitation as chromate in ammoniacal solution. The following separation of lead from tin is outlined: The alloy is decomposed by nitric acid, the metastannic acid is dissolved in concentrated sulphuric acid, and, after specified dilution, filtered off from the insoluble lead sulphate.

Inaccuracies in the Determination of Carbon and Hydrogen of Combustion. By CHARLES F. MABERY. *J. Am. Chem. Soc.*, 20, 510–513.—The author enumerates four important sources of error : inefficiency of the purifying train ; loss of gaseous hydrocarbons ; imperfect absorption of carbon dioxide ; and loss of moisture from the potash bulb. A statement of how the difficulties may be overcome is postponed till the completion of work now on hand.

Some Further Applications of Hydrogen Peroxide to Quantitative Analysis. By PERCY H. WALKER. *J. Am. Chem. Soc.*, **20**, 513–515.—Hydrogen peroxide prevents the precipitation of titanium by ammonia and of uranium by sodium hydroxide. This affords a means of separation of both metals from iron, and of uranium from zirconium.

The Error in Carbon Determinations Made with the Use of Weighed Potash Bulbs. By GEORGE AUCHY. *J. Am. Chem. Soc.*, **20**, 528–534.—In a series of combustions, sulphuric acid bulbs were used to determine the amount of moisture carried over from the prolong of the potash bulbs. Some anomalous results obtained were traced to the condensation of moisture upon the outside of the bulbs. It was found that the error from this source may, in moist weather, exceed that caused by neglecting to use the sulphuric acid bulbs at all. Furthermore, this error cannot be compensated for by the use of tared bulbs, as the condensation proceeds with utter irregularity under apparently similar conditions. This renders true blanks in damp weather impossible. The conclusions reached are supported by a large amount of experimental data.

Note on Drown's Method of Determining Silicon in Steel. By GEORGE AUCHY. *J. Am. Chem. Soc.*, **20**, 547–549.—The author modifies Drown's method by conducting the evaporation with aqua regia and sulphuric acid, instead of sulphuric and nitric acids. He claims a gain in accuracy as well as practical convenience.

Electrolytic Determination of Tin in Tin Ores. By E. D. CAMPBELL AND E. C. CHAMPION. *J. Am. Chem. Soc.*, **20**, 687–690.—Tin ores are decomposed by fusion with sodium carbonate and sulphur, the resulting sulphostannate transformed to the double oxalate of tin and ammonia, and the latter electrolyzed. The transformation to the double oxalate is accomplished as follows : The sulphostannate solution is neutralized with hydrochloric acid, and sodium peroxide is added. This oxidizes the tin to stannic chloride. The sulphur is now filtered out, and ammonia and ammonium oxalate are added to the solution. From the solution so obtained the tin is satisfactorily deposited by a current of 0.10 ampere and 4 volts.

Iodometric Estimation of Tellurium. BY JAMES F. NORRIS AND HENRY FAY. *Am. Chem. J.*, **20**, 278–283.—In the presence of a large excess of alkali, tellurous acid is oxidized quantitatively by potassium permanganate to telluric acid according to the equation : (1) $2KMnO_4 + 3TeO_2 = K_2O + 2MnO_2 + 3TeO_3$. If to such a solution potassium iodide and sulphuric acid are added, the following reaction takes place : (2) $2MnO_2 + 4KI + 4H_2SO_4 = 2MnSO_4 + 2K_2SO_4 + 4H_2O + 4I$. Any excess of permanganate will react as follows : (3) $2KMnO_4 + 10KI + 8H_2SO_4 = 2MnSO_4 + 6K_2SO_4 + 8H_2O + 10I$. Combining equations (1) and (2) the oxidation of tellurium dioxide may be expressed in this way : $2KMnO_4 + 3TeO_2 + 4KI + 5H_2SO_4 = 3TeO_3 + 3K_2SO_4 + 2MnSO_4 + 5H_2O + 4I$. There is therefore found to be a deficit of two atoms of iodine, or two molecules of sodium thiosulphate for each molecule of tellurium dioxide present. In estimating tellurium by this method a portion of the substance, about 0.150 gram, is treated with 20 cc. of a ten per cent. solution of sodium hydroxide, potassium permanganate is added until the meniscus of the brown solution shows a deep pink. The solution is then diluted to 400 cc. with ice-water, and there are added 10 cc. of potassium iodide solution containing 2 grams of the salt, and dilute sulphuric acid until the solution becomes clear. The liberated iodine is titrated with sodium thiosulphate. The value of the permanganate solution is found in the same way by liberating iodine from potassium iodide and titrating with thiosulphate. The deficit in sodium thiosulphate represents the tellurium dioxide present. The method was tested against carefully purified tellurium dioxide made from recrystallized basic nitrate, and against the double bromide of tellurium and potassium. The process is rapid, gives accurate results, and can be used in the presence of halogen acids.

PROXIMATE ANALYSIS.

The Commercial Analysis of Bauxite. BY WM. B. PHILLIPS AND DAVID HANCOCK. *J. Am. Chem. Soc.*, **20**, 209–225.—In the interests of those concerned in the alum manufacture from bauxite, it is essential that a plan be devised by which the amount of easily soluble alumina may be determined, as well as the total amount. A scheme is presented by which it is possible to distinguish between the alumina present as trihydrate, that is in a state of lower hydration, and that as clay. It is suggested that the alumina soluble in sulphuric acid of 50° B. at 100° C. in one hour be called "free alumina", that soluble only when the solution is evaporated until the acid fumes appear

" available alumina," while the difference would be designated as " combined alumina," this system to be adopted by producer, broker, and consumer alike.

A. H. GILL, REVIEWER.

A Comparison of Various Rapid Methods for Determining Carbon Dioxide and Monoxide. By L. M. DENNIS. *J. Am. Chem. Soc.*, 19, 859–870.—The results show the comparative accuracy of the various forms of apparatus in use and are summarized in the article.

The Econometer: a Gas Balance for Indicating Continuously the Proportion of Carbonic Acid Gas in the Flow of Furnace Gases. REPORT BY A COMMITTEE. *J. Franklin Inst.*, 145, 205. —Instead of absorbing the carbon dioxide, this apparatus weighs the dried and filtered gas continuously. The weight being known, the percentage of carbon dioxide is easily deduced. There are said to be over twelve hundred of these apparatus in use.

MINERALOGICAL AND GEOLOGICAL CHEMISTRY.

W. O. CROSBY, REVIEWER.

Orthoclase as Gangue Mineral in a Fissure Vein. By WALDEMAR LINDGREN. *Am. J. Sci.*, 155, 418–420.—After noting the rather sparing occurrence of the feldspars in true veins, the author describes a silver-gold vein near Silver City, Idaho, having a gangue of quartz and orthoclase. The orthoclase is of the variety *adularia*, and the evidence of its aqueous origin is followed by an analysis yielding: SiO_2, 66.28; Al_2O_3, 17.93; K_2O, 15.12; Na_2O, 0.25; undetermined, 0.42; total, 100.00.

Notes on Rocks and Minerals from California. By H. W. TURNER. *Am. J. Sci.*, 155, 421–428.—This paper describes: 1. A peculiar quartz-amphibole diorite, with very complete analyses of the diorite and its component amphibole; 2. A new amphibole-pyroxene rock from Mariposa County; 3. A quartz-alunite, with an analysis of the alunite, which occurs as an efflorescence; 4. Zircon from gravels; 5. Molybdenite from several localities; 6. Tellurium, selenium, and nickel in gold ores; 7. Carbonaceous material in quartz from gold veins east of the Morter Lode; 8. Berthierite from Tuolumne County.

Mineralogical Notes on Anthophyllite, Enstatite, and Beryl (Emerald) from North Carolina. By J. H. PRATT. *Am. J. Sci.*, 155, 429–432.—The anthophyllite and enstatite are from the great dunite dikes of western North Carolina, and two analyses of each are given. The emerald is from a vein of pegmatite in Mitchell County, and was not analyzed.

The Jerome Kansas Meteorite. By HENRY S. WASHINGTON. *Am. J. Sci.*, 155, 447–454.—This meteorite, about 65 pounds in weight, is a deeply oxidized mass made up of numerous chondrules of bronzite and olivine, with fragmental crystals of these minerals and pyroxene, and small angular masses of nickel-iron (4.3 per cent.). No troilite was recognized, but the analysis indicates that it was originally present to the extent of 5.2 per cent. An approximate chemical analysis and an analysis of the nickel-iron, which contains 10.01 per cent. of nickel, are followed by exhaustive analyses of both the soluble and insoluble portions ; and from these analytic data the mineralogical composition is calculated, the chief constituents, in order of abundance, being olivine, bronzite, limonite (secondary), oligoclase, troilite, pyroxene, nickel-iron, and orthoclase.

On the Origin of the Corundum Associated with the Peridotites in North Carolina. By J. H. PRATT. *Am. J. Sci.*, 156, 49–65.—The peridotite (dunite or olivine rock) is a basic, magnesian, plutonic rock forming lenticular dikes and bosses in gneiss, and the corundum is invariably found on the borders of these masses, between the dunite and gneiss. The author's conclusion, which appears to be well sustained by the facts, is that the corundum is not in any sense a secondary mineral, but dates from the original solidification of the dunite, having existed in the solution of the molten mass of the dunite at the time of its intrusion and separated out among the first minerals as the mass began to cool. The dunite magma holding in solution the chemical elements of the different minerals would be like a saturated liquid, and as it began to cool the minerals would crystallize out, not according to their infusibility but according to their solubility in the molten magma. The more basic portions, according to the general law of cooling and crystallizing magmas, being the most insoluble, would be the first to separate out. These would be the oxides containing no silica, such as chromite, spinel, and corundum. The important experiments of Morozewicz with molten basic glasses are cited as fully corroborating this view ; and it is noted that the crystallization of the corundum and other oxides would begin on the outer border of the mass where cooling was most rapid. Convection currents would then tend to bring new supplies of material carrying alumina into this outer zone, where it would be deposited as corundum. This is essentially Becker's theory of fractional crystallization ; and it is noted that the high fluidity of these very basic magmas is a very favorable condition.

Erionite, a New Zeolite. By ARTHUR S. EAKLE. *Am. J. Sci.*, 156, 66–68.—This mineral occurs in very fine, white, pearly and woolly threads, associated with opal in a rhyolite tuff from Durkee, Oregon.   Analysis gives : SiO_2, 57.16 ; Al_2O_3, 16.08 ; CaO, 3.50 ; MgO, 0.66 ; K_2O, 3.51 ; Na_2O, 2.47 ; H_2O, 17.30 ; total, 100.68.   Allowing one molecule of water as hydroxyl, as the dehydration experiments indicate, we obtain the formula $H_3Si_4Al_3CaK_2Na_2O_{11} + 5H_2O$.   This is analogous to the formula for stilbite with the calcium largely replaced by alkalies ; but in other respects the new zeolite has no resemblance to stilbite. The name refers to its woolly appearance.   An analysis of the associated milky opal gave : SiO_2, 95.56 ; H_2O, 4.14 ; and a trace of alumina.

Metamorphism of Rocks and Rock Flowage.   By C. R. VAN HISE.   *Am. J. Sci.*, 156, 75-91 ; *Bull. Geol. Soc. Am.*, 9, 269-328.—This important contribution to dynamical geology, which is condensed from a partly written treatise on metamorphism and the metamorphic rocks, is mainly a physical study ; but the important cooperation of chemical agencies is fully recognized in the paragraphs on chemical action and its relations to heat and pressure, the upper and lower physico-chemical zones, etc. Van't Hoff's law, that '' on the whole, the preponderating chemical reactions at lower temperatures are the combinings (associations) which take place with the development of heat, while the reactions preponderating at higher temperatures are the cleavings (dissociations) which take place with the absorption of heat, is made a basic principle of the discussion ; and the contrasts of the upper and lower zones of the earth's crust resulting from this law and the natural antagonism of heat and pressure are traced out in hydration and dehydration, the mutual replacements of oxygen and sulphur, carbon dioxide and silicon dioxide, and the tendency to develop in the upper zone minerals of lower specific gravity with consequent expansion of the rocks, and in the deeper-seated zone of minerals of higher specific gravity with consequent contraction of the rocks.   In both the physical and chemical categories, alike at lesser and greater depths, water is recognized as the one important and essential medium of alteration ; and an almost inappreciable proportion of water is regarded as sufficient for extensive and rapid metamorphism, in which it may act solely as agent, suffering neither gain or loss. In this connection, the author cites the experiments of Barus, according to which 180° C. is a critical temperature for the solution of glass in water, the action being very slow below this temperature and astonishingly rapid above it.   The solution of the glass and crystallization of its derived minerals are essentially contemporaneous and continuous processes, involving, in the absence of hydrous derivatives, no necessary diminution of the water, which may continue its work as a mineralizer indefinitely and so rapidly as to dissolve and deposit in crystalline form a volume of glass equal to that of the water in about half an hour from which the author calculates that, even if the rate for rocks be only one-thousandth that for glass, a rock formation could be dissolved and recrystallized 50,000 times by one per cent. of water in a mountain-making period of 150,000 years.

Mineralogical Notes. By C. H. WARREN. *Am. J. Sci.*, 156, 116–124.—This paper describes: 1. Melanotekite, a basic silicate of iron sesquioxide and lead, from Hillsboro, New Mexico, the analyses of exceptionally pure material indicating for this species and kentrolite, the corresponding basic silicate of manganese sesquioxide and lead, the formula $Fe_4(Mn_4)Pb_3Si_3O_{11}$, instead of $Fe_4(Mn_4)Pb_3Si_3O_9$, heretofore accepted. 2. Pseudomorphs after phenacite, from Greenwood, Maine, in which gigantic crystals up to twelve inches in diameter having the form of phenacite have been completely replaced by a mixture of quartz and cookeite, with not a trace of beryllium remaining. 3. Similar pseudomorphs after large crystals of topaz from the same locality. 4. Crystallized tapiolite (tantalate of iron and manganese) from Topsham, Maine, which is distinguished by its tetragonal form from its orthorhombic dimorph, tantalite, and by its composition from the corresponding dimorphous niobates, mossite and columbite. 5. Crystallized tantalite from Paris, Maine, which is shown by its very high specific gravity (7.26) not to be columbite, while the absence of manganese adds to its chemical interest. 6. Cobaltiferous smithsonite from Boleo, Lower California, which had been mistaken for the rare hydrated cobalt carbonate, remingtonite, but which is found by analysis to contain 39.02 per cent. of ZnO and only 10.25 per cent. of CoO.

Sölvsbergite and Tinguaite from Essex County, Mass. By HENRY S. WASHINGTON. *Am. J. Sci.*, 156, 176–187.—The sölvsbergite forms a dike four feet wide cutting granite, and is specially distinguished by the presence of glaucophane and riebeckite. One complete analysis is given and compared with four analyses from other regions; and from the analysis the mineral composition is computed, the chief constituents, in order of abundance, being albite, orthoclase, glaucophane, riebeckite, quartz, and titanite. In this connection an analysis is also given of the Quincy granite, in which T. G. White has reported a blue hornblende which he referred to glaucophane; analysis of four foreign granites are quoted for comparison; and the calculation of the mineral composition gives, in order of abundance, quartz, albite, orthoclase, riebeckite, and glaucophane, the riebeckite largely predominating over the glaucophane. The analyses are of special interest as pointing to the existence of a purely iron-alumina glaucophane. The tinguaite also occurs as a dike in the granite ; and its most notable characteristic is the occurrence in it, apparently as an original constituent, of a large proportion (37.4 per cent.) of analcite. As before, the ana lysis is compared with the similar rocks of other regions and the mineral composition is deduced therefrom.

Distribution and Quantitative Occurrence of Vanadium and Molybdenum in Rocks of the United States. By W. F. HILLE-BRAND. *Am. J. Sci.*, 156, 209–216.—The analytical data show the quantitative occurrence and distribution of vanadium in a large number (57) and variety of igneous rocks, in a few of the component minerals of these rocks, and in a few metamorphic and secondary rocks. Two of the samples in the last series were highly composite, one representing 253 sandstones and the other 498 limestones. The conclusions suggested by a comparison of these data are : that vanadium occurs in quite appreciable amounts in the more basic igneous and metamorphic rocks, up to 0.08 per cent. or more of V_2O_3, but seems to be absent or nearly so from the highly siliceous ones ; that the chief source of the vanadium is the heavy ferric-aluminous silicates—the biotites, pyroxenes, amphiboles ; that limestones and sandstones contain only very small amounts of vanadium ; that molybdenum is confined to the more siliceous rocks ; and so far has been found only in traces in these. .

An Occurrence of Dunite in Western Massachusetts. By G. C. MARTIN. · *Am. J. Sci.*, 156, 244–248.—The dunite or olivine rock, of which only two other occurrences are known in North America, forms an irregularly elliptical boss of distinctly igneous origin, about 1,000 by 2,000 feet in extent, in the town of Cheshire. The olivine is extensively serpentinized, and the original accessories include chromite, magnetite, and picotite. The olivine, purified by the Thoulet solution, gave on analysis : MgO, 51.41 ; SiO_2, 40.07 ; FeO, 4.84 ; Al_2O_3, 1.94 ; H_2O, 1.03; total, 99.29.

Anthracite Coal in Arizona.—By W. P. BLAKE. *Am. Geol.*, 21, 345–346.—This is a hard, graphitic anthracite forming heavy beds in the carboniferous strata of southern Arizona. It resembles the Rhode Island anthracite, but the percentage of ash is larger (13.20 to 30.00); it is hard to ignite, and its fuel value is practically *nil*. Five approximate analyses are given.

Studies on an Interesting Hornblende Occurring in a Hornblende Gabbro, from Pavone, near Ivrea, Piedmont, Italy. By FRANK R. VAN HORN. . *Am. Geol.*, 21, 370–374.—The approximate mineralogical composition of the gabbro is : plagioclase 33, hornblende 27, diallage and hypersthene 25, and magnetite and spinel 15 per cent. The hornblende, although one of the most basic constituents of the rock, is only occasionally approximately idiomorphic in the prismatic zone. The author suggests that this crystallographic peculiarity may be due to the high percentage of alkalies which it contains. A complete analysis is given and three others are quoted. The percentage of water is high (2.79), while the silica is very low (39.58). The discussion shows that the mineral is approximately an orthosilicate, and the author concludes that an orthosilicate molecule enters largely into the composition of the aluminous amphiboles.

Weathering of Diabase near Chatham, Virginia. By THOMAS L. WATSON. *Am. Geol.*, 22, 85–101.—The rock in question is a typical olivine diabase (plagioclase, augite, olivine, and magnetite) and forms great dikes intersecting both the crystalline schists and gneisses and the Triassic shales and sandstones. It weathers in concentric layers, yielding bowlder-like residual masses ; and the final product is a tough clay of a bright red color, while biotite, chlorite, and serpentine are intermediate derivatives. The analytic data include : 1. Bulk analyses of fresh, partially weathered, and completely decomposed diabase, the contrast of two and three being much more marked than of one and two, and the most essential changes being greatly increased hydration, partial loss of the alkalies (the soda, as usual, suffering more than the potash), almost complete loss of the alkaline earths (the magnesia yielding more readily and completely than the lime), great increase of iron oxide due both to loss of other constituents and peroxidation of FeO, and a notable diminution of the silica due to the elimination of the protoxide bases, the alumina alone remaining essentially unchanged. 2. Analyses of the augite and feldspar, as separated by the Thoulet solution, the feldspar proving to be a labradorite whose albite-anorthite ratio is Ab_4An_3. 3. Determination of the relative amounts of material in the fresh, altered and decomposed rock soluble in hydrochloric acid of different strengths, all other conditions remaining constant. The loss was greatest in the decomposed and least in the altered material, and was limited in the former almost wholly to the iron oxide and alumina, while in the latter and the fresh rock the lime and magnesia were also largely dissolved. 4. Analyses of the portions soluble in hydrochloric acid and sodium carbonate are compared with the bulk analyses to determine the percentage of each constituent lost and retained, the total loss on passing from the fresh to the decomposed rock amounting to 70.31 per cent., although previous investigations show that this loss rarely exceeds 50 per cent. of the total rock mass. 5. A mechanical analysis of the partially decomposed rock, showing the mineralogical character of the particles of different sizes.

Chemical and Mineral Relationships in Igneous Rocks. BY JOSEPH P. IDDINGS. */. Geol.*, 6, 219-237.—This is an attempt to correlate the mineral composition of igneous rocks with the chemical composition of their magmas; that is, of each rock as a whole. The chief●difficulties are : first, the variable composition of the rock-making minerals, quartz alone having an absolutely fixed composition and no element occurring only in one mineral ; second, the fact that no fixed association of minerals necessarily results from the crystallization of a magma, the result being largely controlled by the physical conditions. To avoid undue complexity, the author confines his attention to the more important rock-making minerals, including quartz, feldspathic minerals, micas, pyroxenes, amphiboles, olivine, and magnetite. The empirical and dualistic formulas are given for each species ; and the latter are classified in accordance with the ratios of the protoxide and sesquioxide bases to the silica. After quoting briefly some of the laws governing the relations of the mineral and chemical composition formulated in an earlier paper, the author discusses in greater detail and with the aid of diagrams, the relations particularly of quartz, and of leucite, nephelite and sodalite; thus making more evident the interdependence of the various minerals on one another and on the chemical composition of the magma.

A Study of Some Examples of Rock Variation. BY J. MORGAN CLEMENTS. */. Geol.*, 6, 372-392.—The rocks in question include diorites, gabbros, norites, and peridotites occurring in the Crystal Falls iron-bearing district of Michigan. Petrographic descriptions and chemical analyses of the several types are followed by a discussion of their chemical relations, the complete analyses, percentages of the chief oxides, and atomic proportions of the metals, being presented in tabular form ; and the author concludes that the rapid changes in mineralogical composition and texture in a single rock exposure, and the changes thus occasioned from one type into another through intermediate facies, show very clearly the intimate relationship of the rocks to one another, and warrants the assumption that they all belong to a geological unit.

Notes on Some Igneous, Metamorphic and Sedimentary Rocks of the Coast Ranges of California. BY H. W. TURNER. */. Geol.*, 6, 483-499.—The rocks considered in this paper include : 1. *Metabasalts and diabases,* formerly regarded as metamorphic sandstones, of which ten analyses are quoted without discussion. 2. *Serpentine,* which has also been regarded as, in part at least, altered sedimentary rocks, but which the author holds to be, in the main at least, of igneous origin. Nine analyses, representing five localities, are quoted, showing great uniformity of composition and indicating that olivine or rhombic pyroxene must have been a prominent constituent of all of the original rocks from which the serpentines were derived. 3. *The Franciscian or Golden Gate formation*. 4. *The San Pablo formation*, which contains layers of rhyolitic tuff or pumice, of which two partial analyses are given.

Syenite-porphyry Dikes in the Northern Adirondacks. By H. P. CUSHING. *Bull. Geol. Soc. Am.*, 9, 239–256.—These dikes, of which fourteen have been discovered, and which are shown by their field relations to be younger than the pre-Cambrian gneisses and anorthosites which constitute the mass of the Adirondacks and older than the Potsdam sandstone, consist of a sub-acid holocrystalline rock chiefly composed of acid feldspars (microperthite, albite, orthoclase, and microcline) and biotite, with less abundant quartz and hornblende, and accessory magnetite, hematite, apatite, and titanite, and various secondary species. Three original analyses are given, selected to represent the mean and extremes of composition ; and from these the percentages of the component minerals are deduced. In the discussion of the petrologic relationships of the dikes, numerous other analyses of related rocks are quoted.

Weathering of Alnoite in Manheim, New York. By C. H. SMYTH, JR. *Bull. Geol. Soc. Am.*, 9, 257–268.—The alnoite, which forms several small dikes in the calciferous '' sand rock'' on East Canada Creek, is an ultrabasic type consisting largely of biotite, and serpentine derived from original olivine, the olivine itself being extremely rare, and the minor constituents are magnetite, apatite, and perofskite, with secondary calcite. The investigation is largely based upon the methods established by Merrill. Both the fresh rock and its highly weathered facies were analyzed, the chief points of interest being, as usual, the increase of ferric oxide and water and diminution of ferrous oxide, alkalies, alkaline earths and silica in the weathered material. From the analyses the loss for the whole rock and the percentages of each constituent retained and lost are calculated. The titanic oxide is shown to be one of the most resistant constituents of the rock, its behavior being almost identical with that of alumina, so that the two are taken together as the basis for comparison. The iron oxides have proved almost equally insoluble, the apparent net increase being due, of course, to peroxidation of FeO. The large proportions of magnesia and lime in the weathered rock indicate that the process is far from complete ; and in harmony with this view 93.60 per cent. of the weathered rock was found to be soluble in hydrochloric acid and sodium hydrate solution. The contrast between the surface weathering and deep-seated alteration of rocks, the rate of decomposition of biotite, and the time of weathering, are also discussed. Biotite appears to weather rapidly in acid rocks and slowly in basic rocks simply because, while it is chemically one of the weakest constituents of the former, it is one of the most resistant constituents of the latter, the difference being relative only.

Clay Deposits and Clay Industry in North Carolina. By HEINRICH RIES. *N. C. Geol. Surv., Bull. 13,* 1-157.— Although regarded as a merely preliminary report, this is a fairly comprehensive, if not a detailed account, of the clays of a great state. But it is not of local interest only, for the admirable introductory sections, forming nearly half the work, and covering the chemical and physical properties, mining and preparation of clays in general, and more specifically of the kaolins or china clays, pottery clays, fire clays, and brick clays, must prove of general interest and value. Under the chemical properties of clays are discussed : the fluxing impurities, including the alkalies, compounds of iron, lime, and magnesia ; non-fluxing impurities, including silica, titanium, organic matter, and water ; analytical methods ; and the rational analysis of clays. The descriptive sections include nearly seventy original analyses of North Carolina clays by Prof. Chas. Baskerville, of the State University ; and these are repeated in tabular form at the end of the report. Each analysis gives the silica, alumina, ferric oxide, lime, magnesia, alkalies, moisture, and water ; and in certain cases the ferrous oxide, organic matter, sulphuric and titanic oxide were also determined. By way of rational analyses, on which the value of a clay chiefly depends, the clay substance (pure kaolin), free sand, and total fluxes are given in each case ; and for the china clays also the percentages of quartz and feldspar in the sand.

A. H. GILL, REVIEWER.

Preliminary Paper on the Composition of California Petroleum. By C. F. MABERY. *Am. Chem. J.,* 19, 796-804.—Oil from Ventura and Fresno counties was examined. The former is extremely heavy and dark ; it has a specific gravity of 0.888 at 20°, and contàins 0.84 per cent. of sulphur. On distillation it gave 9.7 per cent. below 150°, 29.1 between 150° and 300°, and 61.2 residue above 300°. It was more carefully distilled *in vacuo*. Small quantities of benzene, toluene, probably also xylene, hepta- and octo-naphthenes were obtained. These will be treated more at length in a later article.—Fresno County oil is of lighter color and lower gravity and of a greenish hue. It contains but 0.21 per cent. of sulphur. On distillation it gave 33 per cent. below 150°, 25 per cent. from 150°-200°, 21 per cent. between 200° and 250°, 12.4 per cent. from 250° to 300°, and but 9 per cent. above 300°. It is to be further investigated.

Notes upon the Chemical Composition of Natural Gas from Great Salt Lake. By F. C. PHILLIPS. *Proc. Eng. Soc. Western Pa.*, 13, 453.—An analysis by another observer has shown that the gas had apparently contained 16.6 per cent. of hydrogen, which would make it quite different from the gas in West Virginia. On passing the gas, however, for many hours through tubes containing pure and dry palladium chloride it was found that no hydrochloric acid was formed, which is proof that no hydrogen is contained in the gas.

ORGANIC CHEMISTRY.

J. F. NORRIS, REVIEWER.

The Action of Ethylic Oxalate on Camphor (III). By J. BISHOP TINGLE. *Am. Chem. J.*, 20, 318-342.— In previous papers (*J. Chem. Soc.*, 57, 652, and *this Rev.*, 3, 153) the preparation of camphoroxalic acid was described, and the formation by the acid of an additive compound with hydroxylamine was given as evidence that it is an unsaturated hydroxyl derivative of

$$\text{the formula } C_8H_{14}\left\langle \begin{array}{l} C:C.OH.COOH \\ \quad\mid \\ CO \end{array} \right.$$

In order to gain additional evidence for the structure of the compound, the work described below was undertaken. By the action of acetic anhydride on camphoroxalic acid three substances were obtained: One was probably the anhydride of the acid, as it was also obtained when benzoyl chloride was used instead of acetic anhydride; the second compound was a monoacetyl derivative of camphoroxalic acid; the third compound was not identified. The acetyl derivative readily combines with bromine vapor, giving an oily product, which quickly loses hydrogen bromide. When camphoroxalic acid is treated with bromine in chloroform solution, the evolution of hydrogen bromide begins immediately. The resulting oil was reduced with magnesium amalgam. The substance thus obtained closely resembled camphoroxalic acid in general properties, but an examination of the crystals showed that it was not identical with it. The compound will be studied further in order to establish its structure. When quickly distilled under ordinary pressure, camphoroxalic acid yields some camphor. Heating with barium hydroxide in a current of dry hydrogen causes almost complete conversion into camphor and oxalic acid. Attempts to convert camphoroxalic acid into an isomeric acid by prolonged heating at 150° were unsuccessful, whilst partial etherification with alcoholic hydrogen chloride failed to show any indication of non-homogeneity. When boiled with phenylhydrazine in anhydrous ether free from alcohol, the acid yields a crystalline salt. By the action of dilute sulphuric acid at 135° for three hours on camphoroxalic acid; a new acid of the formula $C_{11}H_{16}O_4$ was obtained. Ethyl camphoroxalate was obtained in pure condition by heating the acid with highly dilute absolute alcoholic hydrogen chloride. The ester readily combines with dry ammonia and with hydroxylamine. Attempts to obtain a benzoyl derivative of the ester were unsuccessful. Methyl camphoroxalate and phenyl hydrazine yield a phenylhydrazide, which is converted by glacial acetic acid into methyl camphylphenylpyrazolecarboxylate,

$$C_8H_{14}\left\langle\begin{array}{cc}C-C.CO.OCH_3\\ \| \quad \| \\ C \quad N\end{array}\right.$$
$$\underset{C.C_6H_5}{\vee}$$

. Isoamyl camphoroxalate was also prepared and converted into a phenyl hydrazide. Camphylphenylpyrazolecarboxylic acid, when distilled over barium hydroxide, yielded an oily substance which gave Knorr's pyrazoline reaction. It probably contained camphylphenylpyrazole. Sodium camphylpyrazolecarboxylate injected into the veins of dogs is without any marked physiological action. Sodium camphoroxalate, under similar circumstances, is distinctly and rapidly toxic.

Veratrine and Some of Its Derivatives. BY GEORGE B. FRANKFORTER. *Am. Chem. J.*, 20, 358–373.—The substance known in pharmacy as veratrine has been studied by a number of chemists, but concordant results have not been obtained, as different preparations of the alkaloid differ widely in composition and chemical, physical, and physiological properties. The author finds that crystallized veratrine is identical with the cevadine of Bossetti (*Arch. Pharm.*, *1883*, 82). The name veratrine, however, has been retained since cevadine is the common veratrum alkaloid used at present, and it seems advisable to retain the name which associates the alkaloid with the genus of plants from which it is obtained. The substance investigated was the so-called "Merck veratrine." The free alkaloid has the composition $C_{32}H_{49}NO_9.H_2O$ and was shown to contain a methoxy group. By triturating veratrine with a large excess of iodine, an iodide containing four atoms of iodine and three molecules of water of crystallization was obtained. When this salt was heated at 110° to constant weight, an iodide containing two atoms of the halogen resulted. With ammonia the tetraiodide gave a monoiodide. Chloral hydrate reacted readily with the alkaloid forming a compound of the composition $CCl_3CH(O.$ $C_{19}H_{21}NO_3)_2$. With methyl iodide an addition-product is formed. The iodine in this compound was removed by means of silver oxide and the resulting hydroxyl compound, $C_{20}H_{24}NO_3.$ CH_2OH, converted into a hydrochloride, which formed a double salt with gold chloride. The alkaloid also formed crystalline addition-products with ethyl bromide and allyl iodide.

On the Formation of Imido-1,2-Diazol Derivatives from Aromatic Azimides and Esters of Acetylenecarboxylic Acids. By A. MICHAEL, F. LUEHN, and H. H. HIGBEE. *Am. Chem. J.*, 20, 377-395.

—Phenylazimide readily unites with the esters of $\alpha\beta$-acids of the acetylene series. The resulting addition-products do not show the instability that characterizes the trinitrogen ring, but are stable towards heat and towards most reagents. Phenylazimide and acetylenedicarboxylic ester reacted according to the following equation :

$$C_6H_5-N\diagdown \begin{matrix} N \\ \| \\ N \end{matrix} + \begin{matrix} C-COOC_2H_5 \\ \||| \\ C-COOC_2H_5 \end{matrix} = C_6H_5-N\diagdown \begin{matrix} N=N \\ | \\ C=C \\ | \quad | \\ H_5C_2OOC \quad COOC_2H_5 \end{matrix}$$

The resulting *n*-phenylimido-1,2-diazoldicarboxylic ester was converted into the free acid, and a number of its salts were prepared. When the acid was heated *n*-phenyl-1,2-imidodiazol,

$$C_6H_5-N\diagdown \begin{matrix} N=N \\ | \\ CH=CH \end{matrix}$$

, a weakly basic substance which formed a platinum double salt, was obtained. In order to establish the above formula for the diazoldicarboxylic acid, it was transformed into the triazolcarboxylic acid, which was prepared by Bladin' (*Ber. d. chem. Ges.*, 26, 545) and which has the formula

$$HN\diagdown \begin{matrix} N-C.COOH \\ \| \\ N-C.COOH \end{matrix} \quad \text{or} \quad HN\diagdown \begin{matrix} N-C.COOH \\ | \quad \| \\ N-C.COOH \end{matrix} .$$

As direct oxidation did not remove the phenyl group from *n*-phenylimido-1,2-diazoldicarboxylic acid a nitro group was introduced and then reduced. The amino acid thus formed was readily oxidized to the above triazoldicarboxylic acid. The nitro derivative was prepared by condensing nitrophenylazimide with acetylenedicarboxylic ester, because a small yield of it was obtained when the diazoldicarboxylic acid was nitrated. *p*-Nitro-*n*-phenyl-1,2-imidodiazol was obtained by heating the corresponding dicarboxylic acid. Phenylazimide condenses with phenylpropiolic ester and forms *n*-phenylimidophenyl-1,2-diazolcarboxylic acid. The compounds mentioned and a number of their derivatives are described in detail. The conclusion can be drawn from the authors' work that the addition of substituted azimides to esters of the acetylene carboxylic acids to form imidodiazol derivatives is a general reaction, since the fatty monobasic acetylene acids add negative atoms more readily than phenyl propiolic acid.

On the Oxide of Dichlormethoxyquinonedibenzoylmethylacetal. By C. LORING JACKSON AND H. A. TORREY. *Am. Chem. J.*, **20**, 395-430.—The substance prepared by Jackson and Grindley (*Am. Chem. J.*, **17**, 644) by the action of sulphuric acid on dichlordimethoxyquinonedibenzoyldimethylacetal has been found to be the oxide of dichlormethoxyquinonedibenzoylmethylacetal. It is the first example of an orthoquinone and has the following structure: $C_6.OCOC_6H_6(1),Cl(2),OCH_3(3),-OCOC_6H_6(4),OCH_3(4),Cl(5),O(6,1)$. The fact that the compound is not a ketone was shown by the failure of hydroxylamine to react with it, and by the formation of the sodium salt of chloranilic acid with a hot solution of sodium hydroxide. The substance does not contain a hydroxyl group since it is inactive with cold sodium hydroxide. Isoamylamine converts the oxide into the isoamylamine salt of oxydichlorisoamylamidoquinone, and sodium methylate changes the oxide into the sodium salt of dichlordimethoxyquinonedimethylhemiacetal. It is shown that these facts are evidence against the oxide being a paraquinone. . The above reaction can be explained, however, if the substance is either an ortho- or a metaquinone, but the analogy between the reactions of ethylene oxide and those of the compound under discussion are so striking, that it is probable that the ortho structure exists in the oxide. Since a decision between the ortho and meta formulæ could not be founded on any study of the derivatives of the oxide, as both bodies give the same products when the oxygen bond was opened, the action of boiling sulphuric acid on pyrocatechin and resorcin, substances containing two hydroxyl groups in the ortho and meta positions, respectively, was studied. Resorcin gave resorcin ether and pyrocatechin gave colored products which were probably formed from a quinone. These experiments, therefore, did not furnish evidence for either structure. The compound formed by the action of isoamylamine on the oxide was shown to be the salt of the amine and a phenol of the composition $C_6Cl_2(C_5H_{11}NH)HOO_2$. For the sake of comparison with this body, dichlordiisoamylamidoquinone was made by the action of isoamylamine on chloranil. This compound, as well as the isoamylamine salt of chloranilic acid, differed entirely in properties from the product made from the oxide. Tetrabromguaiacol and tribromveratrol, which were made in connection with the work on the action of sulphuric acid on pyrocatechin derivatives, are described. The following acetyl acetals were prepared : $C_6Cl_2(OCH_3)_2(OCH_3)_2(OCOCH_3)_2$, and $C_6Cl_2(OC_2H_5)_2(OC_2H_5)_2(OCOCH_3)_2$. Some experiments on the action of sodium alcoholates on quinone and chloranil, which led to the formation of very unstable colored compounds, are described. It is shown that an alkaline solution of sodium chloranilate is converted into iodoform by treatment with iodine.

On the Colored Compounds Obtained from Sodic Alcoholates and Picryl Chloride. BY C. LORING JACKSON AND W. F. BOOS. *Am. Chem. J.*, 20, 444–454.—The authors have continued the work of Jackson and Ittner (*this Rev.*, 3, 107) on the colored products of the action of sodium alcoholates on certain aromatic nitro compounds. The substances formed from picryl chloride and the sodium derivatives of methyl, ethyl, propyl, isoamyl, and benzyl alcohol have been studied. They all have the formula $C_6H_2(NO_2)_3ORNaOR$, and are decomposed by acids giving picric ethers. According to the theory of Victor Meyer, these compounds are formed by the replacement of a hydrogen atom in the benzene ring by sodium, and the addition of a molecule of alcohol of crystallization. The fact, established by the authors, that the methyl compound can be heated to 130° without loss of weight, is evidence against the above theory. The more probable view is that the compounds are formed by the addition of the alcoholate to the picryl ether. Whether the alcoholate is added to the carbon of the benzene ring, or to a nitro group, is still an open question. By the decomposition of the corresponding colored compound with an acid, propyl, isoamyl, and benzyl ethers of picric acid were prepared. When a concentrated solution of picryl chloride in methyl alcohol was treated with an excess of an aqueous solution of barium hydroxide, a compound of the following formula was precipitated : $[C_6H_2(NO_2)_3OCH_3]_2Ba(OH)_2.10H_2O$.

On the Action of Orthodiazobenzenesulphonic Acid on Methyl and Ethyl Alcohol. BY E. C. FRANKLIN. *Am. Chem. J.*, 20, 455–466.—A detailed description of the preparation of *o*- and *m*-amidobenzenesulphonic acids from benzene is given, together with an account of the separation and purification of the acids. *o*-Diazobenzenesulphonic acid was decomposed with methyl alcohol at atmospheric pressure, at an increase of 850 mm., and at a decrease of 450 mm. of mercury : The methoxy product alone was obtained. With ethyl alcohol only the alkoxy reaction took place, although the yield was less than when methyl alcohol was used. Increased pressure favors the reaction. When *p*-methoxy or *p*-ethoxybenzenesulphonamide was treated with fuming nitric acid, *m*-dinitrobenzene and a nitro substitution-product of the corresponding sulphonic acid were formed. *o*-Methoxybenzenesulphonamide and fuming nitric acid gave, as one of the reaction-products, *m*-dinitrobenzene. With the meta compound, however, no dinitrobenzene was formed.

The Action of Nitric Acid on Tribromacetanilide. By WILLIAM B. BENTLEY. *Am. Chem. J.*, 20, 472–481.—On account of the conflict between the description of the nitro derivative of *s*-tribromaniline published by Körner (*Jsb. d. Chem., 1875,* 347) and that given by Remmers (*Ber. d. chem. Ges.*, 17, 266) the author undertook a study of the compound. According to Remmers, tribromnitraniline can be prepared by saponifying with ammonia the nitro compound formed by the action of nitric acid on tribromacetanilide. The author was unable to obtain a nitro derivative from tribromacetanilide. Fuming nitric acid either had no action or produced an oily mass, from which nothing crystalline could be obtained. With concentrated acid (sp. gr. 1.38) several products were formed—tetrabrombenzene, a volatile oil of irritating odor (probably dibromdinitromethane), bromanil, oxalic acid, and picric acid. If the nitric acid was dilute, or was used in glacial acetic acid solution, the products were the same as when concentrated acid was used. In attempting to establish the identity of the bromanil formed in the reaction, the action of sodium phenylate on it was studied. In absolute alcohol solution dibromdiethoxyquinone was formed, but if 95 per cent. alcohol was used dibromdiphenoxyquinone was obtained as described by Jackson and Grindley (*this Rev.*, 1, 427).

Researches on the Cycloamidines: Pyrimidine Derivatives. By H. L. WHEELER. *Am. Chem. J.*, 20, 481–490.—For convenience of reference the author calls cycloamidines those compounds which have the amidine formation $RC{\overset{\displaystyle NR''}{\underset{\displaystyle NXR'}{<}}}$ in which two of the radicals R, R', and R'' are replaced by a ring structure or a bivalent grouping. There are two types of cycloamidines, each type having a tautomeric form. The first contains one nitrogen atom in the ring; the second contains two. The object of the work, of which the present paper gives a preliminary account, was to compare the action of alkyl halides on cycloamidines and on simple amidines, in order to get further evidence of the correct structure of these compounds. According to Peckmann (*Ber. d. chem. Ges.*, 28, 2362 and 869) when a simple amidine reacts with alkyl iodides, the hydrogen X is directly replaced, and hence the reaction serves to determine the structure of the compound. It is the author's opinion, however, that the above reaction is not conclusive evidence, for the alkyl iodides may be added either to the amido or imido group, or, when the substituents are similar, to both, and also perhaps to the atoms joined by the double bond. As a confirmation of the above opinion the author finds that phenylmethylanilidopyrimidine acts with methyl and ethyl iodides forming stable addition-products, and that no substitution takes place. As these compounds yield alcohols, sodium iodide, and unaltered anilidopyrimidine quantitatively when treated with alkali, the alkyl iodide in all probability was joined to one of the tertiary nitrogen atoms. From analogy a similar structure might be expected in the case of the alkyl halogen addition-products of the simple amidines. If this is the case the structure of the latter compounds must be completely revised, and the tautomeric structure must be assigned to them. Phenylmethylpyrimidon, a cycloamidine of the second type, unites with one molecule of methyl iodide to form a compound, which is decomposed into hydrogen iodide and an alkylated pyrimidon when treated with alkali. The alkyl derivative differs from the product obtained by heating phenylmethylchlorpyrimidine with sodium methylate. The alkyl iodide, therefore, is added to one of the nitrogen atoms of the pyrimidon ring.

On Phenylglutaric Acid and Its Derivatives. By A. S. AVERY AND ROSA BOUTON. *Am. Chem. J.*, 20, 509–515.—By the condensation of benzalmalonic ester with sodium malonic ester an oil was formed which could not be purified, but which consisted principally of benzaldimalonic ester, as it yielded β-phenylglutaric acid when saponified with hydrobromic acid. The silver, copper, and barium salts and the anhydride of the acid were analyzed. From the anhydride β-phenylglutaranilic acid, β-phenylglutaranil, and β-phenylglutar-p-tolilic acid were prepared. By the action of fuming sulphuric acid on the anhydride an acid containing no sulphur was obtained. Its nature was not established.

On α-Methyl-β-phenylglutaric Acid. By A. S. AVERY AND MARY L. FOSSLER. *Am. Chem. J.*, 20, 516–518.—α-Methyl-β-phenylglutaric acid was prepared by the following method: The methyl ester of α-methylcinnamic acid was condensed with sodium malonic ester. When acidified the condensation-product yielded methylphenylpropanetricarboxylic ester, and, on saponification with potassium hydroxide, the corresponding acid. The impure acid, when distilled over a free flame, gave the anhydride of α-methyl-β-phenylglutaric acid. The latter was dissolved in ammonia, treated with copper sulphate, and the free acid obtained by decomposing the copper salt with hydrogen sulphide. The silver salt was analyzed. As the acid contains two asymmetric carbon atoms the authors propose to study it further.

Researches on the Cyclo Amides: α-Ketobenzmorpholine and α-Benzparaoxazine Derivatives. BY H. L. WHEELER AND BAYARD BARNES. *Am. Chem. J.*, 20, 555–568.—α-Keto-benzmorpholine has the structure of a cyclo amide and yields isomeric compounds derived from the following tautomeric formulæ :

$$(1)\quad C_6H_4 \Big\langle {}^{O \,-\, CH_2}_{NH-CO} \qquad\qquad (2)\quad C_6H_4 \Big\langle {}^{O \,-\, CH_2}_{N=COH}$$

From the sodium and silver salt of the cyclo amide and alkyl iodides, derivatives of formula (1) and formula (2) (oxybenz-paraoxazine), respectively, were obtained. The structures of the alkyl substitution-products were determined by the following facts : The product obtained by the action of methyl iodide on the sodium salt gave on prolonged heating with concentrated hydrochloric acid, at a high temperature, o-methylaminophenol, thus showing in the case of the sodium salts that the alkyl groups attach themselves to nitrogen. The products obtained from the silver salts immediately regenerated α-ketobenz-morpholine with cold dilute hydrochloric acid. This instability in the presence of acids, being a characteristic property of substituted imido esters, shows that in this case the alkyl groups are attached to oxygen. This analogy between the salts of α-keto-benzmorpholine and the salts of the anilides was shown further by the physical and chemical properties of the compounds. The cycloimido esters obtained from the silver salt have marked odors and decompose into ketobenzmorpholine on exposure to the air ; the isomeric compounds have no odor and are stable on exposure. The compounds from the silver salt react with bases giving amidines, which yield well-crystallized salts. The silver salt of the cycloamide reacts with acyl chlorides, like the silver salts of the anilides, giving characteristic derivatives which have the acyl group joined to nitrogen. It follows that α-keto-benzmorpholine behaves like formanilide, and, therefore, has the structure represented by formula (1). The authors are of the opinion that amides have the keto formula because their oxy compounds do not lose water and form lactones. If the amides had the enol structure they would be imido acids and their oxy derivatives would be expected to separate water like the oxy acids. The following substances are described : The sodium, methyl, and ethyl derivatives of α-ketobenzmorpholine ; the silver, methyl', ethyl, isopropyl, isobutyl, and isoamyl derivatives of oxybenzparaoxazine ; the acetyl, benzoyl, phenylamino, *m*-chlorphenylamino, β-naphthylamino, isobutylamino, and allylamino derivatives of benzparaoxazine ; and a number of the salts of the amino derivatives.

The Action of Amines on Acylimidoesters: Acyl Amidines. By H. L. WHEELER AND P. T. WALDEN. *Am. Chem. J.*, **20**, 568–576.—On account of the ease with which benzoylbenzamidine separates ammonia, giving dibenzamide, Pinner (*Die Imidoäther und ihre Derivate*) assigns to the compound the structure represented by formula (1) :

$$(1)\ C_4H_4C \Big\backslash{\substack{NHCOC_4H_4 \\ NH}} \ , \qquad (2)C_4H_4C \Big\backslash{\substack{NCOC_4H_4 \\ NH_2}} .$$

The authors find that benzoylbenzimido esters, $C_4H_4C \Big\backslash{\substack{NCOC_4H_4 \\ OR}}$,

react with ammonia, giving a compound which was shown to be identical with that obtained by Pinner, while an isomeric compound, formula (2), would be expected if the reaction takes place, as is usually assumed, by double decomposition, replacement, or substitution. It is probable that the action of amines on acylimidoesters, like the action of amines on the amidines themselves, does not take place by direct replacement, but involves the intermediate formation of an unstable addition-product, which in breaking down could form a compound of a type represented by either of the above formulas. That this view is correct and that this decomposition has no significance in regard to the structure of the acyl amidines is shown by the behavior of the analogously constituted acylimido esters, benzoylbenzimidoethyl ester, and benzoylphenylacetimidoethyl ester. These compounds easily decompose with water : The first gives dibenzamide and alcohol, and the second benzamide and ethylphenyl acetate. The structure of amidines cannot, therefore, be determined by their decomposition-products. A number of acyl amidines are described. They are stable towards cold alkali. Some of them dissolve in acids and can be precipitated unchanged by alkali or ammonia. They combine with one equivalent of hydrogen chloride to form unstable salts. In aqueous solutions these salts decompose, forming a diacylamide.

Preparation of Sodium Benzenesulphonate. By H. W. HOCHSTETTER. *J. Am. Chem. Soc.*, **20**, 549.—If, in the prep-

aration of sodium benzenesulphonate by the method of Gatter-
mann (*Die Praxis des organischen Chemikers*) the sulphonic acid
is added to a sodium chloride solution having a specific gravity
of 1.151 at 18°, the resulting salt is obtained in an almost pure
condition.   When a saturated salt solution is used the sulpho-
nate is contaminated with about thirty-seven per cent. of sodium
chloride.

The Action of Organic Acids upon Nitrils. By JOHN ALEX-
ANDER MATHEWS. *J. Am. Chem. Soc.*, **20**, 648–668.—The
author summarizes his work briefly as follows: (1) Cyanacetic
acid under certain conditions appears to rearrange to yield its
isomer malonimide.   (2) Benzoic acid and ethylene cyanide
give benzonitrile and succinimide.   (3) Phenylacetic acid and
ethylene cyanide give phenylacetonitrile and succinimide.   As
a secondary product phenylacetic acid and phenylacetonitrile
give diphenyldiacetamide.   (4) In (2) and (3) it seems probable that β-cyanpropionic acid is an intermediate product and re-
arranges to give succinimide, and that in general, when an
imide is produced by this reaction, it may be considered as resulting from an intermediate cyan acid.   (5) The substituted
monobasic acids, salicylic and anthranilic, give no similar results on account of the decomposition they undergo by heating.
(6) Phthalimide results from phthalic acid and propionitrile,
and phthalimide and succinimide result from phthalic acid and
ethylene cyanide.   (7) Phthalic anhydride and acetonitrile do
not react under the conditions presented. (8) Terephthalic
acid and propionitrile do not react under the conditions presented.
(9) Homophthalimide is not readily formed from homophthalic
acid and a nitrile.   It does not result by a rearrangement of
cyan-*o*-toluic acid. (10) Diphenimide results in nearly theoretical amounts from diphenic acid and acetonitrile.   (11)
o-Sulphobenzoic acid and acetonitrile yield a compound isomeric
with saccharin which may be unsymmetrical *o*-sulphobenz-
imide. (12) By varying the conditions three of the four possible imides of mellitic acid were produced by heating this acid
with acetonitrile.   The *p*-euchronic acid is a new compound;
o-euchronic acid and paramid were already known.   It has been
shown that aluminum amalgam can be used to give the euchron
test.

**The Action of Metallic Thiocyanates upon Aliphatic Chlorhy-
drins.** BY WILBER DWIGHT ENGLE. *J. Am. Chem. Soc.*, **20**,
668–678.—The author's results may be briefly summarized as
follows: Monochlorhydrin, α,γ-dichlorhydrin, and acetodi-
chlorhydrin form corresponding thiocyanates, which are very
unstable and immediately change to complex secondary compounds. α,β-dibromhydrin and its acetic ester form dithiocya-

nates, which can be separated and purified. Treated with tin and hydrochloric acid they give double chlorides of tin and a compound of the probable structure

$$\begin{array}{c} CH_2-S \\ | \qquad\qquad \diagdown \\ CH-S \diagup C : NHHCl. \\ | \\ CH_2OH \end{array}$$

Epichlorhydrin readily forms epithiocyanhydrin, which gives epihydrin sulphide with hydrogen sulphide, and epihydrindimethylsulphine iodide with ethyl iodide.

Acetonechloroform. BY FRANK K. CAMERON AND H. A. HOLLY. */. phys. Chem.*, 2, 322–335.—From the products of the reaction between acetone and chloroform in the presence of potassium hydroxide Willgerodt (*Ber. d. chem. Ges.*, 14, 2456) isolated two substances which had the same composition and chemical properties. These were called solid and liquid acetonechloroform and were assigned the following formulæ respectively: $(CH_3)_2C.CCl_3.OH$ and $(CH_3)_2C.CCl_2H.OCl$. The authors have studied these compounds and have come to the conclusion that the liquid variety consists of the solid contaminated by a small amount of acetone and water. When the pure solid compound was moistened and distilled, an oil was formed which had all the properties of the liquid described by Willgerodt. Molecular weight determinations of the pure solid made by the freezing-point and boiling-point methods, using benzene and acetone as solvents, gave results agreeing with a simple molecular formula. From a study of the changes produced in the freezing-point of acetonechloroform by the addition of varying quantities of water, the conclusion is drawn that the hydrate previously described does not exist. Since the added water cannot be removed mechanically the mixture is perhaps a case of a solid solution.

The Menthol Group. BY EDWARD KREMERS. *Pharm. Arch.*, 1, 107–121.—A critical review of the methods of preparation and properties of menthene and its nitrosochloride is given. Molecular weight determinations of the nitrosochloride were made, using ether, benzene, and chloroform as solvents. The results varied from 278 to 982, and were found to be dependent on concentration and temperature. Menthene nitrosate was shown to have twice the molecular weight represented by its formula.

Decomposition of Iodoform by Light. BY EDWARD KREMERS AND E. C. W. KOSKE. *Pharm. Arch.*, 1, 194-200.—According to Fleury (*J. de Pharm. et de Chem.* [6], 6, 97), when light acts upon a solution of iodoform, iodine is set free until the brown color in the solution prevents further decomposition. In one experiment when silver was added to unite with the liberated iodine the decomposition was complete. The authors show that iodoform is decomposed slowly by light which has passed through a solution of iodine in potassium iodide, and that nearly all of the iodine in iodoform is removed by silver in the dark after standing seven days.

ANALYTICAL CHEMISTRY.

PROXIMATE ANALYSIS.

A. H. GILL, REVIEWER.

On the Occurrence of Hydrogen Sulphide in the Natural Gas of Point Abino, Canada ; and a Method for the Determination of Sulphur in Gas Mixtures. By F. C. PHILLIPS. *J. Am. Chem. Soc.*, 20, 696–705.—The gas mentioned in the title contained about 0.8 per cent. by volume of sulphuretted hydrogen. To determine sulphur in gas mixtures the gas is burned with oxygen from a burner resembling a Bunsen blowpipe, surrounded by a cylinder arranged so that the products of combustion can be collected in sodium hypobromite solution, the sulphur being finally estimated as sulphate. Where small samples of gas are collected these are displaced by the use of carbonic oxide. The results obtained by the precipitation of lead sulphide agreed very well with those obtained by combustion.

The Chemical Composition and Technical Analysis of Water Gas. By E. H. EARNSHAW. *Am. Gas. Light J.*, 69, 488–490 ; 528–529.—Benzene is absorbed by alcohol saturated with gas over mercury, the alcohol vapor being afterwards absorbed by water ; carbon dioxide, by potassium hydrate ; illuminants, by a saturated solution of bromine water ; oxygen, by phosphorus ; and carbonic oxide, by cuprous chloride, using finally a fresh pipette. The determination of hydrogen and the remaining hydrocarbons is effected by exploding a part of the residue over mercury, and by burning another part with air by passing over palladium black, the object of this being to determine the amount of CO left unabsorbed by the cuprous chloride, amounting sometimes to 0.4 per cent. The method of calculation of the results of explosion giving the average composition of the illuminants and the higher marsh gas hydrocarbons, as well as the calculation of the heating value, are also detailed.

The Determination of Methane, Carbon Monoxide, and Hydrogen by Explosion in Technical Gas Analysis. By W. A. Noyes and J. W. Shepherd. *J. Am. Chem. Soc.*, 20, 343.—

The apparatus used was that of Orsat, a fourth pipette being added for explosions; water acidulated with sulphuric acid was used as the confining liquid, it lessening the absorption of carbon dioxide. The results obtained are fairly satisfactory for technical work, those for methane varying from 0.5 per cent. too high to 0.2 too low, for carbonic oxide from 1.9 to 0.3 too low, for hydrogen from 0.3 too high to 0.5 too low.

The Gas Composimeter. *Elec. Eng.*, 25, 311.—

The article which is illustrated describes an instrument for determining and recording the per cent. of carbon dioxide in chimney gases. It depends for its action upon the flow of gases through small apertures. Gas is sucked constantly out of the chimney, passed through potassium hydrate solution, and the difference in the rate of flow of the residue and of the original gas is noted.

A. G. Woodman, Reviewer.

Standards for White and Black Mustard Seed. By John Uri Lloyd. *Pharm. Rev.*, 16, 328–333.—

This work was undertaken with a view to establish a standard for starch in powdered black and white mustard seed. The author finds for black mustard seed that the presence of even 0.1 per cent. of starch is shown by iodine in the presence of potassium iodide. He recommends that this test be employed after the sample has been mixed with a definite quantity of seed previously ascertained to be free from starch. For white mustard seed it was found that by the above test as little as 0.05 per cent. of starch can be detected with certainty.

Report on an Investigation of Analytical Methods for Distinguishing between the Nitrogen of Proteids and that of the Simpler Amides or Amido Acids. By J. W. Mallet. *U. S. Dept. Agr., Div. Chem., Bull.* 54, 1–25.—

This report embodies the results of an investigation undertaken by the author at the suggestion of the Office of Experiment Stations. A study was made of a number of typical substances representing proteids, gelatinoids, and the simpler amides and allied substances. An attempt to separate the water-soluble proteids from the amides by Graham's method of dialysis gave fairly clean separations of leucin, aspartic acid, and kreatin from solutions containing albumen, but the process was found inconveniently slow. The reaction with nitrous acid and evolution of elementary nitrogen gave results which varied greatly and no differences were found upon which an analytical procedure could be based. Experiments on the interaction with sodium hypobromite, and with potassium permanganate in the presence of free acid or alkali, also gave no indication of distinctions sufficient for use in analysis. The only reagent that was found to give satisfactory results was a solution of phosphoduodeci-tungstic acid in dilute hydrochloric acid. By the use of this reagent as a precipitant, followed by thorough washing of the precipitate with hot water, it was found possible to effect a separation of the simpler amidic substances from all the proteids and proteid-like bodies except the peptones. This last group can be precipitated, however, by tannic acid. Full details of the proposed method are given. In order to calculate from the nitrogen found in each group after separation the amount of the proximate nitrogenous constituent present, the following factors are suggested : for proteids and allied substances, 6.25 ; for flesh bases and simpler amids of animal origin, 3.05 ; for simpler amids and amido acids of vegetable origin, 5.15 ; and for mixed amidic constituents of unabsorbed residua in digestion experiments, 9.45.

Separation of Proteid Bodies from the Flesh Bases by Means of Chlorin and Bromin. By H. W. WILEY. *U. S. Dept. Agr., Div. Chem., Bull.* 54, 27–30.—This article is a description of the method of separation proposed by Rideal and Stewart and used in the laboratory of the Agricultural Department. After the finely ground material has been extracted with ether to remove fat, it is thoroughly exhausted with cold or lukewarm water, and then with water nearly boiling. The water-soluble constituents of the nitrogenous constituents, which have been thus extracted, are then precipitated by agitation with bromine water. The sum of the nitrogen in the part insoluble in water and in the part precipitated by bromine is subtracted from the total nitrogen determined in the original sample, and the difference is the nitrogen in the flesh bases. The proteids are calculated by the factor N × 6.25, and the flesh bases by the factor N × 3.12.

Commercial Fertilizers. *Md. Agr. Coll. Quart.*, 1, 1–41 ; *Ky. Agr. Expt. Sta. Bull.*, 76, 97–105.

ASSAYING.

H. O. HOFMAN, REVIEWER.

The Taylor Improved Assay Furnace. By R. P. ROTHWELL. *Eng. Min. J.*, 65, 583.—This is a portable sheet-iron muffle-furnace weighing 250 pounds. It is 12 inches square, is lined with 2-inch fire-brick, takes a 6x12x4-inch muffle, and can be used as a melting furnace for a No. 16 graphite crucible after removing the muffle.

An Improved Method for Preparing Proof Gold and Silver.
By J. W. PACK. *Eng. Min. J.*, 66, 36.—The improvement on
the ordinary methods consists in precipitating gold and silver on
aluminum foil. From a dilute solution of auric chloride, poured
into a beaker containing aluminum foil, the gold is at once precipitated quantitatively ; argentic chloride is completely decomposed by aluminum. Any residual aluminum is removed by
warming the precipitated silver with hydrochloric acid.

APPARATUS.

A. H. GILL, REVIEWER.

**A Convenient Gas Generator and Device for Dissolving
Solids.** By T. W. RICHARDS. *Am. Chem. J.*, 20, 189-195.—
Six different forms of apparatus are figured and explained.
They are simply constructed, give a constant evolution of gas,
and the acid employed is completely used up ; furthermore, the
waste acid can be withdrawn without interrupting the action of
the generator.

An Efficient Gas Pressure Regulator. By PAUL MURRILL.
J. Am. Chem. Soc., 20, 801.—The apparatus is of the usual
floating bell type, and would seem to be very efficient and cheap.

Electric Furnaces for the 110-Volt Circuit. By N. M. HOP-
KINS. *J. Am. Chem. Soc.*, 20, 769-773.

Volumetric Apparatus. By G. E. BARTON. *J. Am. Chem.
Soc.*, 20, 731-739.—The article compares the capacities of liter
flasks standardized at various temperatures and advocates 22°
C. as a standard as being more nearly that of the laboratory.
The error of graduation of ten liter flasks from domestic and
foreign makers was found to vary from zero to three hundredths
of one per cent ; the probable error in checking the calibration
of a liter flask was ascertained to be not more than six-thousandths
of one per cent. It was determined further that a difference in
temperature of 0.1°, or a change of 15 mm. in the barometer,
produced a variation of 0.02 cc.

Lubricants for Glass Stop-cocks. By F. C. PHILLIPS. *J.
Am. Chem. Soc.*, 20, 678-681.—Two mixtures, one of 70 parts
pure rubber, 25 parts spermaceti, and 5 parts vaseline, the other
of 70 parts rubber and 30 parts unbleached beeswax, were found
to yield the best results. The rubber should be new and pure
and be melted at a low temperature in a covered vessel and the
wax added and the mixture thoroughly stirred. The lubricant
is unsaponifiable, but may be removed from the apparatus by
nitric acid.

On a New Form of Water Blast. By B. B. BOLTWOOD. *Am. Chem. J.*, 20, 577-580.—The apparatus cannot be well understood without the diagram ; the orifice offered for the entrance of air and water is of such size that any appreciable friction is avoided and the energy of the jet is transmitted to a much greater volume of water than in the Richards or Muencke apparatus. These latter forms furnish only 1.1 times the quantity of air for water used ; this apparatus gives three times the volume of air for water employed.

TECHNICAL CHEMISTRY.

G. W. ROLFE, REVIEWER.

The Clarification of Cane Juice with Lime, Sulphur, and Heat. By R. E. BLOUIN. *La. Planter and Sugar Mfr.*, 20, 375-379.—An important paper on clarification of cane juice. The author's conclusions which favor preliminary sulphuring of the cold juice, liming to neutrality, and subsequent heating as the most effective means of clarification, are borne out by the elaborate analytical data presented.

The Use of Lime and Sulphur in Sugar House Work. By E. W. DEMING. *La. Planter and Sugar Mfr.*, 20, 379-381.— This paper contains many useful hints and notes on the application of these clarifying agents in factory practice.

Double Sulphuration and Filtration. By A. L. BARTHOLEMY. *La. Planter and Sugar Mfr.*, 21, 28-29.—The author discusses the merits of the system.

Centrifugal Defecation and Carbonatation. *Sugar Beet*, 19, 79-80.   This article describes the Hignette method of cold defecation of beet-juice with lime and separation of the precipitate from carbonatation by centrifugals.   The advantages claimed are decrease of bulk of scums and consequent saving of sugar, amounting to 20 cents a ton.

BIOLOGICAL CHEMISTRY.

A. G. WOODMAN, REVIEWER.

Analysis of the Rhizome of Aralia Californica. By WILLIAM R. MONROE. *Am. J. Pharm.*, 70, 489-492.—The author has made a study of the rhizome in a fresh condition and has separated a volatile oil having a strong aromatic odor. Besides minute quantities of sugars and proteid matter he found considerable amounts of calcium oxalate.

The Chemical Life History of Lucern. BY JOHN A. WIDTSOE AND JOHN STEWART. *Utah Agr. Expt. Sta. Bull.*, **58**, 1–90.— This bulletin contains the results of a study of the chemical composition of lucern at different periods of its growth. The method of analysis has been mainly that of Dragendorff, the various extracts obtained having been subjected to a detailed examination. Experiments have also been made on the digestibility of lucern by methods of natural digestion.

Condensing Milk by Cold Process. BY BYRON F. MCINTYRE. *Sci. Am. Supp.*, **56**, 18976–18977.—In condensing milk by cold process it is found necessary to have milk free from animal odor, because any odor is intensified by the condensation. The flavor is therefore corrected if necessary by heating the milk to 80° F. in a vacuum pan. It is then run through a separator set for "heavy cream" (assaying not less than 50 per cent. fat). The fat-free milk is cooled and then placed in freezing closets for twenty-four hours, the ice being removed at intervals by the centrifugal machine. 100 gallons of milk are condensed to 13 gallons, then 5½ gallons of the heavy cream are added, so that if the final product is diluted with three times its volume of water it will contain an average of 3.6 per cent. of fat. The condensed milk has a sp. gr. of about 1.16; it has a smooth consistency, and a pronounced sweet taste. In keeping qualities it is superior to whole milk.

The Amount of Formaldehyde Gas Yielded by Different Lamps and Generators. BY E. A. DE SCHWEINITZ. *J. Am. Pub. Health Assoc.*, **23**, 118–120.—The author finds that the various generators which he has examined do not furnish the theoretical or guaranteed amount of gas, the amount obtained varying from 2½ per cent. to 20 per cent. of the guaranteed yield. The method found most satisfactory for the determination of the formaldehyde was the one depending upon the decomposition of hydroxylamine hydrochloride by formaldehyde gas. The formaldehyde acts upon the hydroxylamine hydrochloride so that hydrochloric acid is quantitatively replaced, according to the following formula: $NH_2O + HCl + HCHO = CH_2{>}N{-}OH + HCl + H_2O$. The hydrochloric acid set free was determined by titration with tenth-normal alkali.

Report of the Committee on Disinfectants. BY FRANKLIN C. ROBINSON. *J. Am. Pub. Health Assoc.*, **23**, 101–109.—This preliminary report contains a description of the modes of preparation, and of the chemical properties of formaldehyde, together with suggestions as to its use for disinfection.

On the Determination of Undigested Fat and Casein in Infant Feces. By HERMAN POOLE. *J. Am. Chem. Soc.*, 20, 765–769.—This paper is a continuation of one already published (*this Rev.*, 4, 115). Results are given of a study of three cases in which the children were fed on a milk prepared from the average milk of a large number of cows, care being taken to have the milk, as nearly as possible, of the same composition.

The Fertilizing Value of Street Sweepings. By ERVIN E. EWELL. *U. S. Dept. Agr., Div. Chem., Bull.* 55, 1–19.—This bulletin gives data concerning the disposition of street sweepings and a table of analyses.

The Chemistry of the Corn Kernel. By C. G. HOPKINS. *Univ. Ill. Agr. Expt. Sta. Bull.*, 53, 180a–180d.—This bulletin is an abstract of the work done at the Station in analyzing the constituents of the corn kernel in order to determine their exact composition.

Foods and Food Adulterants. By H. W. WILEY AND OTHERS. *U. S. Dept. Agr., Div. Chem., Bull.* 13, 1169–1374.—This portion of the bulletin is devoted to cereals and cereal products. It contains brief descriptions of the processes of milling and tables of analyses showing the composition of a great variety of cereals and cereal products, as bread, rolls, cake, etc. In most cases the calories of combustion have been calculated for the cereals, while in a few cases the heat of combustion has been determined by direct experiment.

G. W. ROLFE, REVIEWER.

Sugar Beets. By M. B. HARDIN. *S. C. Agr. Expt. Sta. Bull.*, 34, 3–9.

Sugar Beets in Idaho. By CHAS. W. McCURDY. *Univ. of Idaho Agr. Expt. Sta. Bull.*, 12, 37, 73.

New Mexico Sugar Beets, 1897. By ARTHUR GOSS. *New Mexico Coll. Agr. Expt. Sta. Bull.*, 26, 71–113.

Sugar Beets in Colorado in 1897. By W. W. COOKE AND W. P. HEADDEN. *State Agr. Coll. Expt. Sta. Bull.*, 42, 3–64.

Sugar Beet Investigations in 1897. By A. D. SELBY AND L. M. BLOOMFIELD. *Ohio Agr. Expt. Sta. Bull.*, 90, 123–162.

Wyoming Sugar Beets. By E. E. SLOSSON. *Univ. Wy. Agr. Coll. Bull.*, 36, 189–205.—These bulletins are typical of the class which are being issued by the Agricultural Experiment Stations throughout the country. They contain many interesting analytical data in various lines of research besides being popular manuals of beet culture.

A Soil Study. Part I. The Crop Grown : Sugar-Beets. BY W. P. HEADDEN. *Col. State Agr. Coll. Bull.*, 46, 3–63.—This treats of an investigation of the effect of certain mineral salts, particularly alkalies, on the growth of sugar-beets. A great number of valuable data are presented. The conclusions arrived at are too many to be detailed here. In general, they show that in amounts even as small as a few hundredths of a per cent. in the soil, alkalies have a strong toxic action on the growing plant. Even extremely small quantities of alkali increase the ash of the beet notably. Much light is thrown, by these experiments, on the distribution of mineral matter through the tissues of the plant and the conditions of its assimilation.

Laboratory Notes. BY F. H. STORER. *Bussey Inst. Harvard Univ. Bull.*, 2, 409–421.—In the main these notes describe a continuation of researches on the carbohydrates of tree trunks, dealing principally with the pentosan group. (A) With the idea that hemicellulose (xylan) is difficultly digestible and generally more stable than the other carbohydrates the author has examined decaying birchwood, but finds much less gum than in sound timber. (B) The details are given of an elaborate investigation of the carbohydrates and derived products. (C) The solvent action of alkaline solutions on the xylan of coniferous wood under various conditions, and its bearing on the accuracy of the results of analyses are treated at length. (D) The amount of wood gum in the strawberry is shown to be small. (E) The possible presence of xylan in the membrane of the starch grain in the light of experimental evidence collected is discussed. (F) An analysis is given of the ashes of Javan sugar-baskets, disposed of at the refineries by burning, which shows that they are inferior to wood ashes for fertilizer.

Notes on Taka-Diastase. BY W. E. STONE AND H. E. WRIGHT. *J. Am. Chem. Soc.*, 21, 639–647.—A comparative investigation of the hydrolyzing action of malt- and taka-diastase infusions on potato starch paste. Taka-diastase evidently acts more energetically than malt-diastase up to a certain point, but does not, under the hydrolyzing conditions used by the author, carry the conversion nearly as far. In starch determinations, the use of taka-diastase was found objectionable, owing to incompleteness of its solvent action. The reviewer suggests that much might be learned by an investigation of the carbohydrates formed under varying conditions of hydrolysis by methods similar to those used by Brown and Morris on malt-diastase conversions. It would certainly be interesting to compare the curve of copper reduction with that of malt-diastase hydrolyzed products.

METALLURGICAL CHEMISTRY.

H. O. HOFMAN, REVIEWER.

Improvements in Mining and Metallurgical Apparatus during the Last Decade. BY E. G. SPILSBURY. *Trans. Am. Inst. Min. Eng.*, 27, 452–465.—A presidential address.

Silver-Lead Blast Furnace Construction. BY H. V. CROLL. *Eng. Min. J.*, 65, 639–640.—The paper is an illustrated description of some of the leading improvements that have been made during the last ten years in the constructive details of lead blast furnaces, especially in Colorado and Utah.

Recent Advances in Silver-Lead Smelting. BY R. H. TERHUNE. *Min. Sci. Press*, 77, 132–133.—The paper represents an address delivered by the author last July, at Salt Lake City, before the International Mining Congress, and is therefore rather general in character. The work of the Ropp Straight-Line Mechanical Roasting Furnace, which the author introduced at the Hanauer Works in 1896, is given in more detail than any other subject discussed.

The Condition in which Zinc Exists in Lead. BY M. W. ILES. *School Mines Quart.*, 19, 197–200.—Zinc can be present in a lead blast furnace slag as oxide, silicate, and sulphide. Of these the oxide is readily soluble in dilute sulphuric acid, the silicate less so, the sulphide not at all. The author treated finely pulverized slag with sulphuric acid of different degrees of concentration, for 25 hours, and found that part of the zinc only went into solution, and then very slowly. He concludes from this that zinc is present chiefly in the form of silicate. He proved that some zinc is present as sulphide by blowing compressed air through molten slag and analyzing the fumes which had been caught by sucking the gases through muslin and woollen filters. The analyses showed the presence of lead, zinc, silver, traces of gold and copper, and of sulphur and sulphur trioxide.

The Distribution of Precious Metals and Impurities in Copper and Suggestions for a Rational Method of Sampling. BY E. KELLER. *Trans. Am. Inst. Min. Eng.*, 27, 106–123.—This paper has been previously published (*J. Am. Chem. Soc.*, 19, 243–258) and reviewed (*this Rev.*, 3, 98–99).

The Jeffrey Double-Strand Scraper Conveyor. By R. P. ROTHWELL. *Eng. Min. J.*, 66, 341.—A short illustrated description of a conveyor intended for the charging of ore into leaching vats. With the large capacities of the vats used in cyaniding low-grade gold ores, the introduction of mechanical devices for charging is becoming of increasing importance.

The Potsdam Gold Ores of the Black Hills. By F. C. SMITH. *Trans. Am. Inst. Min. Eng.*, 27, 404-428.—The main part of this paper discusses the occurrence and the general character of the siliceous silver-bearing gold ores near Deadwood, S.D. It gives also a brief outline of the three processes by means of which they are treated ; *viz.*, barrel chlorination (Golden Reward Co., 39,000 tons; Kildonan Milling Co., 36,000 tons), cyaniding (Black Hills Gold and Extraction Co., 6,500 tons), and smelting with pyrite and copper ore for matte (Deadwood and Delaware Smelting Co., 65,700 tons). There are shipped outside 4,800 tons, making a total annual product of 152,000 tons.

Note on the Influence of Temperature in Gold Amalgamation. By F. F. SHARPLESS. *Eng. Min. J.*, 66, 183.—The author, contributing to the discussion given in the title, found that with increase of temperature of battery-water, the amalgam became harder until with the water at 83° F., quicksilver ran off from the lower plates into the trap, which forced him to hang up the stamps. He was using old plates, and from the upper parts the silver had all been worn off. These worked well with warm water, while the lower parts, which were still well silvered, would not retain even the quicksilver that rolled off from upper parts.

Testing Oxidized and Pyritic Gold Ores. By P. H. VAN DIEST. *Min. Sci. Press*, 76, 568.—The paper embodies practical hints for the amalgamator and contains a few remarks about chlorinating and cyaniding.

The Chlorination Mill at Colorado City. By H. V. CROLL. *Eng. Min. J.*, 66, 425-426.—The paper is an illustrated description of a barrel-chlorination plant intended to treat Cripple Creek gold ores. Special attention appears to have been given to the cheap handling of ores. There is a large ore-bedding floor, having a capacity of 7,000 tons. The ore is dried and roasted in mechanical furnaces. The mill has 10 barrels, 6' in diameter and 12' long, receiving each a charge of 9 tons of ore, which is chlorinated in from 1 to 3 hours. The barrels have coarse sand filters inclosed in finely perforated sheet lead. These are to retain the larger part of the pulp ; the fine slimes passing through them are removed by filter boxes placed in a separate building.

Cyanide and Chlorination in Colorado.   By R. B. Turner. *Rep. State Bureau Mines, 1897,* 127–132.—The paper discusses the economic rather than the chemical features of the two processes as carried out to-day in Colorado.   Thus the cost of cyaniding in a 50-ton plant is given as $2.70 per ton of raw ore, to which $0.75 may have to be added for roasting and $0.45 for sampling.

The Cyanide Process.   By S. B. Christy.   *Trans. Am. Inst. Min. Eng.,* 27, 821–846.—A discussion by E. B. Wilson, A. James, and G. A. Packard.

Analyses of Cyanide Mill Solutions.   By W. J. Sharwood. *Eng. Min. J.,* 66, 216.—The changes mill solutions undergo are shown by nine complete analyses made by the author.   The table gives also the tons of solutions from which the samples were taken, the time they were in use, the general character, size, and tons of ore treated by the solutions.   The accompanying text discusses in detail the character of the ores, the manner of working in the mills, and the methods followed in analyzing.

The Precipitation of Gold by Zinc Thread from Dilute and Foul Cyanide Solutions.   By A. James.   *Trans. Am. Inst. Min. Eng.,* 27, 278–283.—The precipitation of gold from very dilute solutions is found to be defective in some cases; in others the gold from solutions with less than 0.05 per cent. free cyanide is reduced to below one grain per ton.   The author found that the imperfect precipitation is sometimes caused by the shavings being too coarse and not packed tightly enough in the precipitation bases, which often are too shallow.   Their depth ought to be greater than either the width or the length.   Further, by giving the solutions a longer contact with the zinc and by making them slightly alkaline with caustic soda, they are easily freed from their gold ; thus, a flow at the rate of 1 ton solution per 24 hours for each cubic foot of zinc shavings and an addition of one-half pound of caustic soda per ton of solution are usually sufficient.   With very foul solutions, containing no free cyanide, a flow of one-half ton solution per cubic foot of shavings in 24 hours, will remove the gold.   Solutions containing 0.2 per cent. copper were effectively treated as long as the flow was slow enough and the acidity had been neutralized with caustic soda.

The Influence of Lead on Rolled and Drawn Brass.   By E. S. Sperry.   *Trans. Am. Inst. Min. Eng.,* 27, 485–505.—The author examined systematically the influences varying percentages of lead have on a brass composed of sixty per cent. copper and forty per cent. zinc, keeping the copper constant and replacing parts of the zinc by lead. Cuttings from such a brass free from lead are long and tenacious, so that a very low speed must be employed in cutting; when, however, two per cent. lead are added (leaded brass), the chips are very short and a high speed can be employed in cutting. Clock-brass and screw, or drill-rod brass, are common varieties of leaded brass.

The Influence of Lead on Rolled and Drawn Brass. By F. FIRMSTONE. *Trans. Am. Inst. Min. Eng.*, 27, 977–978.—The author discusses the subject treated in the article abstracted in the preceding review, and brings out the fact that as early as 1818 Berthier had called attention to the effect of lead on brass.

A Combination Retort and Reverberatory Furnace. By COURTNEY DeKALB. *Trans. Am. Inst. Min. Eng.*, 27, 430–436.—The paper is an illustrated description of a reverberatory furnace with a single zinc-retort, $9\frac{1}{2}$x36 inches outside dimensions. It is the one used in the metallurgical laboratory of the Rolla School of Mines for the reduction of roasted zinc ores.

Zinc Smelting in the Joplin District. By E. HEDBURG. *Mines and Minerals*, 19, 103–104.—The main value of this paper lies in the figures of cost that it gives of the smelting of blende in Southwestern Missouri. It enumerates the leading zinc works in the Central Western States, gives an outline of roasting in a 2-hearth hand reverberatory furnace, and of reducing in a double direct-fired Belgian furnace, with 56 retorts, (7' long, and 8" in diameter) on a side. The cost is given in full for a plant having six double retort-furnaces, or 682 retorts in all. For details the reader is referred to the original.

Notes on Six Months' Working of Dover Furnace, Canal Dover, Ohio. By A. K. REESE. *Trans. Am. Inst. Min. Eng.*, 27, 477–485.—The iron blast furnace, the work of which is described in the paper, was built in 1895. It is 75' high, the crucible 10'6" in diameter, and 5'6" deep, the bosh 16'6" and the stock line 12'6" in diameter. It has a Y-shaped downcomer, taking the gases from two openings in the furnace. For details the reader is referred to the original.

The Recovery of Iron from Cupola Cinder. By W. J. KEEP. *Iron Age*, 61, 6.—In iron foundries there is a considerable loss of metal in the form of small shots in cinders and skulls. Thus, every ton of dirt that goes from the common cinder mill to the dump contains about 300 pounds of shot, and the skulls from 100 ladles about 75 pounds iron. In order to recover this waste, the author uses a horizontal wet-crushing revolving cylinder,

30" in diameter and 40" long on the inside, having a longitudinal, star-shaped iron roller which breaks up the cinder while the barrel revolves.   The finely-crushed cinder, coke, dirt, etc., are carried out automatically, leaving behind the clean shots of metal.

Sulphur in Embreville Pig Iron. By G. R. JOHNSON. *Trans. Am. Inst. Min. Eng.*, 27, 243–249.—The author investigated the distribution of sulphur in Embreville (Tenn.) pig iron. He obtained fractions from 43 pigs and took 9 samples from each fracture and analyzed them separately for sulphur.   In addition, he determined from the average of each set of nine samples the sulphur, silicon, and phosphorus.   The results show that the top of a pig is always richer in sulphur than the bottom.   The author attributes this fact to the pig's giving off, during solidification, included gases, particularly sulphur dioxide, which cannot escape completely.   He believes also that the rough surfaces especially of pig irons low in silicon are due to the escape of sulphur dioxide rather than to a small percentage of silicon.

The Micro Structure of Steel and the Current Theories of Hardening. By A. SAUVEUR.   *Trans. Am. Inst. Min. Eng.*, 27, 846–944.—A discussion by A. Ledebur, R. A. Hadfield, H. C. Jenkins, J. O. Arnold, H. D. Hibbard, P. H. Dudley, E. D. Campbell, F. Osmond, H. M. Howe, and the author.

The Calorific Value of Certain Coals as Determined by the Mahler Bomb. By N. W. LORD AND F. HAAS.   *Trans. Am. Inst. Min. Eng.*, 27, 259–271.—The principal aim of the paper is to determine the calorific powers of a number of coals in general use in Ohio and to find out whether analytical data can be safely used for their values.   For this purpose forty samples of coal, each weighing 50 pounds, were obtained ;   viz., 11 samples from Upper Freeport coal (O. and Pa.), 7 from Pittsburg coal, 11 from Middle Cittanning (7 Darlington coal, Pa., and 4 Hocking Valley coal, O.), 4 from Thacker coal (W. Va.), 5 from Pocahontas coal (Va.), and 1 from Mahoning coal (O.). Ultimate and proximate analyses were made of each.   The calorific powers were determined with the Mahler bomb and calculated from the Dulong formula including the values for sulphur.   The tables of results give also : The differences between the determinations and the calculations of the calorific powers ; the average analyses and heating powers of the single seams ; and the heating power of each sample calculated from the average heating power of the seam to which it belongs.   The results show that for the coals examined the values calculated from the ultimate analyses agree within 2 per cent. with those obtained with the Mahler bomb. They also prove that the method suggested by Kent (*The Mineral Industry*, 1, 97) of calculating the calorific power from the fixed carbon obtained in the proximate analysis does not give satisfactory results. Finally, by comparing the calorific powers of the different samples from a single seam with an average of all the calorimetric tests, ash and moisture being excluded, results are obtained which show that the actual coal of a given seam, at least over considerable areas, has a uniform heating value.

The Calorific Value of Certain Coals as Determined by the Mahler Bomb. By W. Kent. *Trans. Am. Inst. Min. Eng.*, 27, 946–961.—This is a discussion of the preceding paper.

Review of American Chemical Research.

INDEX OF SUBJECTS.

INDEX OF AUTHORS.

ND - #0008 - 210323 - C0 - 229/152/28 [30] - CB - 9780332140780 - Gloss Lamination